AF885919

Climate Change Adaptation in the Water Sector

Climate Change Adaptation in the Water Sector

Edited by Fulco Ludwig, Pavel Kabat,
Henk van Schaik and Michael van der Valk

earthscan
publishing for a sustainable future

London • Sterling, VA

First published by Earthscan in the UK and USA in 2009

Copyright © Fulco Ludwig, Pavel Kabat, Henk van Schaik and Michael van der Valk, 2009

All rights reserved

ISBN: 978-1-84407-652-9

Typeset by FiSH Books, Enfield, Middx.
Cover design by Ruth Bateson

For a full list of publications please contact:

Earthscan
Dunstan House
14a St Cross St
London EC1N 8XA, UK
Tel: +44 (0)20 7841 1930
Fax: +44 (0)20 7242 1474
Email: earthinfo@earthscan.co.uk
Web: **www.earthscan.co.uk**

22883 Quicksilver Drive, Sterling, VA 20166-2012, USA

Earthscan publishes in association with the International Institute for Environment and Development

A catalogue record for this book is available from the British Library

Library of Congress Cataloging-in-Publication Data

Climate change adaptation in the water sector/edited by Fulco Ludwig ... [et al.].
p. cm.
Includes bibliographical references and index.
ISBN 978-1-84407-652-9 (hardback)
1. Climatic changes. 2. Water-supply—Management. I. Ludwig, Fulco, 1972–
QC981.8.C5 C51134555
333.91—dc22

2008042257

At Earthscan we strive to minimize our environmental impacts and carbon footprint through reducing waste, recycling and offsetting our CO_2 emissions, including those created through publication of this book. For more details of our environmental policy, see www.earthscan.co.uk.

This book was printed in the UK by MPG Books, an ISO 14001 accredited company. The paper used is FSC certified and the inks are vegetable based.

FSC Mixed Sources
Product group from well-managed forests and other controlled sources
www.fsc.org Cert no. SA-COC-1565
© 1996 Forest Stewardship Council

Contents

PART II: CASE STUDIES

Edited by Peter Droogers

List of Figures, Tables, Boxes and Plates

Figures

Tables

Boxes

Plates

List of Contributors

Editors

Fulco Ludwig is a research scientist with the climate change group of Wageningen University and Research Centre (WUR). Previously he has worked for the Co-operative Programme on Water and Climate (CPWC) and at the Commonwealth Scientific and Industrial Research Organisation (CSIRO), where he worked for the Water for a Healthy Country flagship programme. His research examines climate change impacts and adaptation on water resources, agriculture and nature with a focus on developing countries.

Pavel Kabat is Full Professor and Chair Holder of the Earth System Science and Climate Change Group at WUR, Co-Chair of the International Scientific Steering Committee of IGBP/ILEAPS, Science Director of CPWC and Science Director of the Dutch National Research Programme on Climate Change and Spatial Planning. His scientific focus is on land–atmosphere interactions, climate hydrology and the water cycle, climate system, and climate change.

Henk van Schaik is Programme Co-ordinator of CPWC. His activities include stimulating attention and research to the impacts of climate change upon water systems at national, regional and basin level, the assessment of vulnerabilities at local levels, the preparation of coping measures, and the encouragement of initiatives in water sector policies and organizations. In this capacity he has been (co-)author of several publications and he was the (co-)organizer of several international conferences on water and climate.

Michael R. van der Valk is hydrologist and coordinator of the Communication and Information portfolio of CPWC. Since its initiation in 1993 he has been final editor of *Stromingen*, the professional magazine of the Netherlands Hydrological Society, where he is also board member for international relations. Besides his work as Scientific Secretary of the Netherlands National Committee IHP-HWRP (UNESCO and WMO; formerly at the Royal Netherlands Meteorological Institute), he is board member of the Netherlands' chapter of the International Association for Hydrogeologists (IAH) and director of CrossVision Communications.

Authors

Jeroen Aerts
Institute of Environmental Studies
Vrije Universiteit, Amsterdam, The Netherlands

Marloes Bakker
Co-operative Programme on Water and Climate (CPWC)
PO Box 3015, 2601 DA Delft, The Netherlands

Bryson C. Bates
CSIRO Marine and Atmospheric Research
Underwood Avenue, Floreat, WA 6014, Australia

Eelco van Beek
Deltares/University of Twente
Rotterdamseweg 185, 2600 MH Delft, The Netherlands

Casey Brown
Department of Civil and Environmental Engineering
12B Marston Hall, University of Massachusetts, 130 Natural Resources Road, Amherst, MA 01003-9293, USA

Joost J. Buntsma
Ministry of Transport, Public Works and Water Management
Postbus 20904, 2500 EX Den Haag, The Netherlands

John M. Callaway
UNEP–RISØ Centre
Roskilde, Denmark

Esther Conrad
International Research Institute for Climate and Society
Earth Institute, Columbia University, 138 Monell Building, 61 Route 9W, PO Box 1000, Palisades, NY 10964, USA

Peter Droogers
FutureWater
Costerweg 1G, 6702 AA Wageningen, The Netherlands

Dulce Elazegui
Institute for Strategic Planning and Policy Studies
College of Public Affairs, University of the Philippines Los Baños, Laguna, Philippines

Tira Foran
Unit for Social and Environmental Research
Faculty of Social Sciences, Chiang Mai University, Chiang Mai 50000, Thailand

Po Garden
Unit for Social and Environmental Research
Faculty of Social Sciences, Chiang Mai University, Chiang Mai 50000, Thailand

Steve W. Gillham
Umgeni Water
PO Box 9, Pietermaritzburg 3200, KwaZulu-Natal, South Africa

Jac A. M. van der Gun
International Groundwater Resources Assessment Centre (IGRAC)
Princetonlaan 6, 3584 CB Utrecht, The Netherlands

Molly Hellmuth
International Research Institute for Climate and Society
Earth Institute, Columbia University, 138 Monell Building, 61 Route 9W,
PO Box 1000, Palisades, NY 10964, USA

Michelle J. A. Hendriks
Ministry of Transport, Public Works and Water Management
Postbus 20904, 2500 EX Den Haag, The Netherlands

Graeme Hughes
Water Corporation of Western Australia
Perth, WA, Australia

Bart van den Hurk
Royal Netherlands Meteorological Institute
Wilhelminalaan 10, 3732 GK De Bilt, The Netherlands

Daniela Jacob
Max Planck Institute for Meteorology
Bundesstraße 53, 20146 Hamburg, Germany

Pavel Kabat
Earth System Science and Climate Change Group
Wageningen University and Research Centre, Droevendaalsesteeg 4,
6708 PB Wageningen, The Netherlands

Britta Kastens
Institute of Environmental Systems Research
University of Osnabrück, Barbarastrße 12, D-49076 Osnabrück, Germany

Louis Lebel
Unit for Social and Environmental Research
Faculty of Social Sciences, Chiang Mai University, Chiang Mai 50000, Thailand

Daniël B. Louw
University of the Free State
Bloemfontein, South Africa

Fulco Ludwig
Earth System Science and Climate Change Group
Wageningen University and Research Centre, Droevendaalsesteeg 4,
6708 PB Wageningen, The Netherlands

Jesse B. Manuta
School of Arts and Sciences
Ateneo de Davao University, The Philippines

Marcus Moench
Institute for Social and Environmental Transition (ISET)
948 North Street, Suite 7, Boulder, CO 80304, USA

Sabine Möllenkamp
Institute of Environmental Systems Research
University of Osnabrück, Barbarastrße 12, D-49076 Osnabrück, Germany

A. Sankarasubramanian
International Research Institute for Climate and Society
Earth Institute, Columbia University, 138 Monell Building, 61 Route 9W,
PO Box 1000, Palisades, NY 10964, USA

Henk van Schaik
Co-operative Programme on Water and Climate (CPWC)
PO Box 3015, 2601 DA Delft, The Netherlands

Shiv Someshwar
International Research Institute for Climate and Society
Earth Institute, Columbia University, 138 Monell Building, 61 Route 9W,
PO Box 1000, Palisades, NY 10964, USA

Mark J. Summerton
Umgeni Water
PO Box 9, Pietermaritzburg 3200, KwaZulu-Natal, South Africa

Michael R. van der Valk
Co-operative Programme on Water and Climate (CPWC)
PO Box 3015, 2601 DA Delft, The Netherlands

Jeroen Veraart
Climate Change Spatial Planning Programme
c/o Wageningen University and Research Centre, Earth Systems Science and Climate Change Group, PO Box 47, 6700 AA Wageningen, The Netherlands

Acknowledgements

This book is a major result of the Co-operative Programme on Water and Climate (CPWC), a Dutch-funded international programme that aims to stimulate activities in the water sector that contribute to managing the effects of climate variability and change, particularly for the most vulnerable countries. After the successful publication of *Climate Changes the Water Rules* (2003), the goal was to produce a coping compendium of climate change adaptation options for the water sector. This coping compendium evolved into this book: *Climate Change Adaptation in the Water Sector*.

CPWC is the main sponsor of this book. The Partners for Water programme, a joint initiative of six departments of the Government of the Netherlands, is the principal financer of CPWC's activities. Several European Union (EU) projects have contributed to the publication as well. The NeWater project contributed through one of the case studies while NeWater results are presented in several chapters. The EU FP6 project WATCH has contributed to the publication of this book by co-funding the work of Fulco Ludwig and Pavel Kabat.

The editors of this book would like to thank all of the authors for their contributions. Many thanks also go to Peter Droogers, who compiled and edited an excellent and well-distributed set of case studies in Part II. Marloes Bakker has been a great help in reviewing and editing several chapters. Many thanks also go to Penelope Keenan, Janine Treves and Hamish Ironside, who proofread and corrected most of the book to avoid any abundance of 'Dunglish'.

List of Acronyms and Abbreviations

AMRIS	Angat–Maasim River Irrigation System
ARGE Elbe	Arbeitsgemeinschaft für die Reinhaltung der Elbe (working group for the protection of the Elbe)
ARK	Nationaal Programma Adaptatie Ruimte en Klimaat (national programme for spatial adaptation to climate change)
BERG	Berg Dam
BERGSUP	Berg Supplemental Site
BMU	Bundesministerium für Umwelt, Naturschutz und Reaktorsicherheit (German Federal Ministry for the Environment, Nature Conservation and Nuclear Safety)
BPIMO	Bulacan Provincial Irrigation Management Office
BRDSEM	Berg River Dynamic Spatial Equilibrium Model
BSIK	Besluit Subsidies Investeringen Kennisinfrastructuur
CH_4	methane
cm	centimetre
CO_2	carbon dioxide
COAG	Council of Australian Governments
COP	Conference of the Parties (of the UNFCCC)
CPC	Climate Prediction Center
CPWC	Co-operative Programme on Water and Climate
CPWF	Challenge Program on Water and Food
CSAG	Climate Systems Analysis Group
DDPM	Department of Disaster Prevention and Mitigation (Thailand)
DF	distant future
DJF	December, January and February
DNLP	dynamic multi-regional, non-linear programming
ECMWF	European Centre for Medium-Range Weather Forecasts
EEA	European Environment Agency
ENSO	El Niño Southern Oscillation
EOF	empirical orthogonal function
EPS	Ensemble Prediction System (of the ECMWF)

ESM	Earth system model
EU	European Union
EuroSIP	European Multi-model Seasonal-to-Interannual Prediction system
FGG Elbe	Flussgebietsgemeinschaft Elbe (river basin community Elbe)
FP6	6th Framework Programme for Research and Technological Development (of the European Commission)
GAMS	General Algebraic Modelling System
GCM	general circulation model
GCM	global climate modelling/model
GDP	gross domestic product
GEF	Global Environment Facilty
GHG	greenhouse gas
GIS	geographic information system
GLOF	glacial lake outburst flood
GOCC	government-owned and controlled corporation
GWP	Global Water Partnership
HAD	High Aswan Dam
ICPE	International Commission for the Protection of the Elbe
ICPR	International Commission for the Protection of the Rhine
IOCI	Indian Ocean Climate Initiative
IOD	Indian Ocean Dipole
IPCC	Intergovernmental Panel on Climate Change (of the United Nations)
IPO	Interprovinciaal Overleg (interprovincial discussion platform in The Netherlands)
IRI	International Research Institute for Climate and Society
ISDR	International Strategy for Disaster Reduction (of the United Nations)
IWCM	integrated water cycle management
IWRM	integrated water resources management
IWSS	Integrated Water Supply Scheme
K	Kelvin
km	kilometre
KNMI	Royal Netherlands Meteorological Institute
LDC	least developed country
m	metre

m^3/s	cubic metres per second
MAP	mean annual precipitation
MDG	Millennium Development Goal
mg	milligram
mm	millimetre
MW	megawatt
MWSS	Metropolitan Waterworks and Sewerage System (Manila)
N_2O	nitrous oxide
NABU	Naturschutzbund Deutschland (nature and biodiversity conservation union, Germany)
NAO	North Atlantic Oscillation
NAPA	National Adaptation Programme of Action
NBW	Nationaal Bestuursakkoord Water (national governmental agreement on water, The Netherlands)
NCEP	National Center for Environmental Predictions
NCR	National Capital Region (of Manila)
NF	near future
NGO	non-governmental organization
NIA	National Irrigation Administration (the Philippines)
NPC	National Power Corporation (the Philippines)
NWRB	National Water Resources Board (the Philippines)
NWS	National Weather Service (of the National Oceanographic and Atmospheric Administration, US)
OAGCM	ocean–atmosphere general circulation model
PAGASA	Philippines National Meteorological Service
PDCO	Bulacan Provincial Development Coordinating Office (the Philippines)
PDF	probability distribution function
PDO	Pacific Decadal Oscillation
PDS	partial duration series
PEMC	Philippine Electricity Market Corporation
POT	peaks-over-threshold approach
RCM	regional climate model
RID	Royal Irrigation Department (Thailand)
ROC	relative operating characteristics
SES	socio-ecological system

SOI	Southern Oscillation Index
SPKD	Spatial Planning Key Decision process
SST	sea surface temperature
START	global change SysTem for Analysis, Research and Training
SWIM	Soil and Water Integrated Model
TWAT	Theewaterskloof Dam
UBA	Umweltbundesamt (Federal Environment Agency, Germany)
UK	United Kingdom
UN	United Nations
UNESCO	United Nations Educational, Scientific and Cultural Organization
UNFCCC	United Nations Framework Convention on Climate Change
US	United States (of America)
UvW	Unie van Waterschappen (union of water boards in The Netherlands)
VNG	Vereniging van Nederlandse Gemeenten (association of Dutch municipalities)
VROM	Ministry of Environment and Spatial Planning (The Netherlands)
WB21	Waterbeheer 21^{e} Eeuw (water management in the 21st century)
WFD	Water Framework Directive
WHO	World Health Organization
WMO	World Meteorological Organization (of the UN)
WMRS	Wemmershoek Dam
WSA	water service authority
WSP	water service provider

1

Introduction

Fulco Ludwig, Peter Droogers, Michael van der Valk, Henk van Schaik and Pavel Kabat

Climate variability results in significant impacts on water availability and safety. Every year, millions of people are affected by droughts and floods. In the future, climate change is likely to increase both the number and magnitude of hydrological extremes.

The importance of climatic variability and change, resulting in long-term, far-reaching and widespread impacts on livelihoods, is clearly acknowledged by most scientists and policy-makers. Not as widely recognized, however, are changes in attitude towards water management that are required to successfully adapt to the impacts and challenges associated with climate change. Many people within the water sector are aware that climate is affecting water resources management, but do not know how to integrate climate change information within water management. Planners and developers find it hard to use climate scenarios and projections because of their inherent levels of uncertainty.

The main purpose of this book is to inform water managers and decision-makers about climate change, its impacts and how to adapt to these changes. It offers water professionals a comprehensive introduction to climate science, climate projection methodologies, their relevance and limitations for water management. It offers guidance and examples on how water management can and should reduce its vulnerability to future changes in the climate system. During the last decade, the availability of information and tools in relation to managing climate variability and change has rapidly expanded. After reading this book, water professionals and advanced students should feel much more comfortable in using climate information in decision support and in managing water resources. Readers will also become more familiar with the institutional challenges that are involved in climate change adaptation.

In the past, water managers have generally been conservative with regards to climate change. Examples in this book show how water managers struggle with using state-of-the-art information from new developments such as seasonal climate forecasting or climate change scenarios. Traditionally, the design of water management systems has been based on historical climate and hydrological data, assuming stationarity of weather and water system behaviour. However, the forecasted changes in climate no longer allow for such assumptions, and historical data are no longer adequate to meaningfully plan for variability and extremes. The impact of climate change on hydrological systems is expected to be such that new approaches are necessary to better ensure that investments will not be lost. This book provides initial guidelines and examples of how water management could be altered in order to reduce its vulnerability to climatic changes. In these chapters, the design of infrastructure is discussed alongside how institutions are adapting to new approaches that use climate change information for decision-making on investments and resource management. This book informs water managers on how to move from using only historical data to a decision-making system that includes information on climate variability and change. The information presented here could also be used to train the next generation of water managers in becoming familiar with these new approaches.

Climate Change Adaptation in the Water Sector is divided into two parts. Part I describes theoretical and methodological aspects of the climate system, and what options are available for the water sector to adapt to climate change and to cope with climate variability. In Part II, case studies on adaptation to climate change from all over the world focus on a variety of issues.

The book starts with an introduction on the climate system. Recent changes in climate are described and the science of predicting climate variability at a seasonal timescale is discussed. The last part of Chapter 2 focuses on climate projections at the decadal timescale. Chapter 3 focuses on the use of climate change scenarios. It describes issues such as regional climate change scenarios, and how tailor-made climate scenarios can be developed and used in different sectors. Chapter 4 provides a brief and general description of the impacts of recent and future climate change on water resources management. It discusses the impacts of droughts, floods and water quality and some possible institutional impacts.

Before discussing how water management can be adapted to cope with climate change, we take a look, in Chapter 5, at how the water sector has managed climate variability in the past. Current practices of using historical climate data for the design of water infrastructure are discussed and the concept of integrated water resources management and its relation to climate are introduced. Large seasonal variation in rainfall is a major challenge for water managers. At the start of the season, it is often unclear how much water will be available for different users. Seasonal forecasts can be used to partly reduce uncertainties so that water management can be improved on a seasonal basis. Chapter 6 discusses how to use seasonal forecasts and includes several practical examples.

Adaptation to climate change is discussed in Chapter 7. The major focus is on risk management, as well as issues such as dealing with uncertainty and adaptive management. The final chapter in Part I, Chapter 8, introduces the concept of climate-proofing, which has gained significant support during recent years. The idea is not to eliminate all climate risks but to use a combination of hard and soft measures to minimize and spread risks to acceptable levels. It is argued that climate change is not only a threat, but can also be seen as an opportunity.

Climate change adaptation is a relatively new challenge and most projects are still in their infancy: the effectiveness of practices, whether precautionary or proactive, is yet to be assessed. As a result, only a few well-documented cases of adaptation can be found. For this book we have collected a set of eight informative cases from different countries describing how people are adapting to climate change.

The first case study in Part II, Chapter 9, describes the management of floods and disasters in Thailand with a special emphasis on social justice. Starting with an evaluation of historical policies and practices, it draws inferences about the key challenges posed by altered flood regimes resulting from climate change and adaptation policies. These underline the importance of a policy of adaptation that emerges from contested and changing perceptions and experiences of risks. The main conclusion from this case study in Thailand is that persistent social injustices could be made worse by both inaction and misguided climate change adaptation policies. The chapter ends with a strong message that we should not wait for more catastrophic confirmations of climate change: there are many actions today that would benefit disadvantaged and vulnerable groups which do not need climate change as their justification.

The second case study from The Netherlands (Chapter 10), focuses on flooding as well. However, the socio-economic contexts of Thailand and The Netherlands are so disparate that adaptations in both cases are quite different. In the past, water management in The Netherlands was dominated by controlling fluctuations in water levels in order to protect the 50 per cent of the country located below sea level. Climate change, however, requires another approach towards water, and a policy shift from 'fighting against water' to 'living with water' has been advocated in the case study. The main issues required for this policy shift are discussed in Chapter 10. The most important message is that water management and spatial planning should be considered in a far more integrated way. A more collective approach is also required where various governmental bodies agree to act jointly to adapt to climate change in three focal areas: urban and rural development; areas close to rivers; and coastal areas. Besides this so-called practical approach where direct actions are taken, a forward-looking approach has been initiated. The latter includes two main activities: first, to initiate several large-scale integrated research programmes funded by the Dutch government; and, second, to develop joint strategies on 'adaptation spatial planning and climate' (ARK) between various ministries, provinces, municipalities, water boards, the research community and the private sector. The chapter concludes that this transition in management cannot be completed in the short term; it is a gradual and iterative process in which all parties must reset their visions on dealing with problems and solutions.

The Yemen case study in Chapter 11 addresses water shortages, with special emphasis on groundwater resources and the impact of climate change. The chapter starts by stating that groundwater systems are comparatively resilient to short-term and seasonal shortage of rainfall, but are very vulnerable to longer-term changes. Groundwater is often, especially in arid regions, the most reliable source of water – if not the only one – for domestic water supply and irrigation of crops. The case study explores to what extent groundwater in alluvial aquifers in arid regions may be affected by climate change during the 21st century. One of the main conclusions is that coping with the consequences of degenerating groundwater resources in Yemen's alluvial aquifers is difficult. Technical measures, such as artificial recharge and improved water-use efficiencies, will not be sufficient to overcome the negative impacts of climate change. Improved rigorous water resource planning and management is necessary. The study concludes that unconventional and innovative measures need to be developed – including control of demographic pressure and transition to a less water-dependent economy. The challenges as described in the chapter are not specific to Yemen; the overall conclusions may be extrapolated to alluvial aquifers elsewhere in arid zones. The mechanisms are similar: ever-increasing human pressure on scarce and dwindling groundwater resources with its related set of complex problems, escalated by climate change.

Chapter 12 focuses specifically on drinking water and the impacts of, and adaptation to, climate change. Umgeni Water in South Africa serves about 5 million people in Durban, Pietermaritzburg and their surroundings with a total of 340 million cubic metres of potable water annually. The utility's water resources assessment techniques consist of two, quite distinct, assessments. The first one, described as the current situation, is based on short timeframe analyses where current water demands are balanced against current supply availability, leading to possible changes to the system operation rules. The second one, the future situation, is based on long timeframe analyses where future water demands are balanced against future supply availability. Climate change would impact most significantly upon the latter type of assessment. It is interesting that climate change was never high on the agenda of the utility. A workshop on climate change, involving the top management of the utility, proved to be a milestone: consideration of climate change impacts were elevated to a higher level based on a better understanding of the topic. Currently, they rank climate change as the third highest risk associated with the management of the natural environment. Lessons learned from the case study are that Umgeni Water has developed a process to assess the hydrological impacts of climate change, and they are currently at the early stages of implementing this process. However, completing the process and tabling the current results is not considered to be the final answer to the problem as the process is dynamic and further analyses will be required as driving factors change.

Chapter 13 also addresses drinking water as the main topic, but now in the Australian context. The case study describes the adaptation measures taken by the Metropolitan Water Supply for Perth in Western Australia. Restrictions on urban water use have been imposed frequently in the past and Water Corporation Perth is beginning to incorporate climate change within its planning processes by a combination of increasing the supply and simultaneously trying to decrease the demand. However, the study claims that underpinning decision-making with information obtained from the latest developments in climate science is still in its infancy. Despite this lack of knowledge, Australia's urban water industry is responding to the stressors of climate change by changing its operating environment, developing or at least considering additional and alternative sources of water (e.g. water reuse and desalination), and being sensitive to the views and concerns of its customers. The described adaptive responses of water planners in Perth may set a pragmatic precedent for water planners elsewhere. The nature of their adaptive response will be shaped by the physical, hydrological, socio-economic and political settings that they confront and the financial resources available. Perth has also benefitted from a coastal location (making seawater desalination a feasible option), ready access to shallow as well as deep groundwater supplies, and an extensive array of dams and pipelines facilitating inter-basin transfer of water supplies.

Chapter 14 includes water economics as a means of assessing and adapting to the impact of climate change. The focus of this case study is the Berg River Basin in the Western Cape Region of South Africa. The basin is an economically important water supply system, providing the bulk of the water for household, commercial and industrial use in Cape Town. It also provides irrigation water to the lower part of the basin to cultivate roughly 15,000ha of high-value crops. The Berg River Dam, with its 130 million cubic metres of storage capacity, is expected to be operational some time during the period of 2008 to 2010. It is, however, unclear to what extent reservoir operation is consistent with expected climate change. Based on a combined water–climate–economic policy-planning model, a set of scenarios is analysed, resulting in alternative uses of the water from the Berg River Dam. The study describes the technical details of the model used to evaluate the impact of climate change. Moreover, the model has been used to assess the most optimal water resources allocation and reservoir operational rules to maximize economic returns of water under various climate change scenarios. The most relevant conclusions from the model evaluations are that climate change will reduce total water availability by 11 per cent in the near future and by 17 per cent in the distant future, and that climate change will reduce basin-wide welfare by between 6.3 per cent and 8.4 per cent in the near future, and by between 11.5 per cent and 15.6 per cent in the distant future, depending upon the water allocation option that will be implemented. This case study can serve as a typical example of how such a policy-planning tool could be used in other cases where water allocation issues should be assessed in the context of climate change.

The seventh case study, in Chapter 15, emphasizes the institutional adaptation to climate change for the Elbe Basin in Germany. The study investigates whether the current river basin management institutions in the Elbe Basin allow for adaptation to climate change impacts. It considers institutions as a broad set of rules, decision-making procedures and programmes. The current institutional adaptive capacity to climate change impacts in the Elbe Basin was based on the perceptions of 11 interviewed experts. In-depth interviews were conducted in spring 2007 with representatives from different organizations at international, national and sub-national levels. The focus was on a so-called analytical framework of seven elements as criteria for adaptation to climate change. These seven elements were essential for the entire study and comprise: (i) availability and communication of information; (ii) polycentric governance; (iii) participation; (iv) sectoral integration; (v) openness for experimentation; (vi) flexibility; and (vii) planning horizons, political support and economic resource. The study provides a general overview of the current situation of institutional adaptation in the German Elbe Basin. The main conclusion is that adaptation is still at an early stage, while a relatively high awareness of the issue already exists. Remarkably, the information on, and discussion about, adaptation is not as prominent as the current discussions concerning climate change mitigation. However, at the same time, adaptation strategies already exist at lower organizational levels of water management. One of the main conclusions from the study is that effective adaptation to climate change requires leadership and support by political decision-makers.

The last case study described in this book, in Chapter 16, originates from the Philippines where the use of seasonal climate forecasts to manage a reservoir system is explored. The study concentrates on the Angat Reservoir, which provides the primary source of water for Metropolitan Manila. The Philippines has an extremely variable climate largely due to the El Niño Southern Oscillation (ENSO), which can serve at the same time as the source of predictability and seasonal climate forecasts of inflows to the Angat Reservoir. The study shows that seasonal climate forecasts can be used to dynamically change the reservoir operational rules such that it reflects the probability of dry conditions in a given year instead of the long-term probability. Based on the conditions of the ocean and atmosphere, it can be determined that the probability of dry conditions is greater than or less than the long-term average. In years when the probability of dry conditions is less than average, more water could be released. However, the actual implementation of these seasonal climate forecasts in reservoir operation depends not only upon the potential benefits, but also upon the institutional context. The study concludes that the potential to apply seasonal climate forecasts to water management appears straightforward. However, the probabilistic nature of the forecasts, the uncertainty associated with any new innovation, and the institutional context within which water is managed all complicate the potential application of these techniques.

The overall conclusion of these eight case studies is that technical opportunities should be combined with institutional changes to adapt to climate change. The case studies also show that – despite climate change being a global problem – the solutions required for adaptation are mostly local.

Part I

Climate Change and Water

2

The Art of Predicting Climate Variability and Change

Bart van den Hurk and Daniela Jacob

The global climate is variable. Climate change always exists at many scales, from the global mean to the regional and local scales, and for temperature and many other variables. However, the notion that humankind 'very likely' has an influence on the global mean climate that is discernible from natural variability at seasonal and decadal timescales (IPCC, 2007) has raised concern about our vulnerability to various aspects of this climate variability. Various assessments conducted by the Intergovernmental Panel on Climate Change (IPCC) and many summaries and interpretations of climate change and climate variability have been published. An increasingly detailed picture arises of variations in temperature, precipitation, atmospheric humidity and soil moisture, as well as numerous other variables at many spatial and temporal scales.

The nature, amplitude and predictability of this variability strongly depend upon the spatial or temporal scales considered. The *global mean climate* changes in response to variations in solar forcing; the amplification of these responses within the climate system due to feedbacks (such as the snow/ice albedo feedback at high latitudes); internal oscillations of large-scale phenomena (such as the El Niño Southern Oscillation, or ENSO); the composition of the atmosphere (which varies according to volcanic activity and levels of greenhouse gas emissions); and the biophysical state of the land surface and oceans. On the *regional scale* (defined here as areas the size of subcontinents or major river basins), climate variability is further enhanced by variations in the atmospheric circulation and local land–atmosphere feedbacks.

The adaptation of humankind to climate change and variability is probably as old as humankind itself. However, due to increasing awareness of our influence on the global climate, the perspective of this adaptation process is slowly changing. While warnings of imminent rainstorms have, for some time, caused water managers to take advance measures, adaptation to changes at longer timescales is beginning to receive increasing attention. The most relevant horizons of climate variability for the water sector are:

- the *synoptic timescale*, where individual weather systems may result in extreme hydrological events;

- the *seasonal timescale*, where persistent anomalies in precipitation may accumulate in enhanced risks for droughts or flooding;
- the *decadal timescale*, where an outlook of global and regional trends during the coming decades are relevant for planning and implementing water resource management measures; and
- the *century timescale,* where changes in the mean (e.g. 30-year average) climatology of meteorological variables may affect the design of hydrological infrastructure for safety, traffic or water resources.

One may view climate variability as a combined result of different processes acting on different timescales. To put it in a simpler way: synoptic weather events are affected by atmospheric circulations and local feedbacks. Seasonal anomalous weather is related to large-scale variations in sea surface temperature (SST) or stored soil moisture and snow. Changes at even longer timescales are related to the slow variations in the ocean heat content and to large-scale changes in atmospheric composition. However, there is a clear link between short and the longer time horizons (Palmer et al, 2008). The response of local weather to a major SST anomaly such as an El Niño event is not equally strong in all places of the world. A good prediction of this local weather response requires an accurate representation of short-term variability and local processes. Likewise, projections of the change in global mean temperature in the next few decades are dependent upon assumptions of the amount of heat stored in the ocean, which, in turn, depends strongly upon atmosphere–ocean interactions at much shorter timescales. Since the major El Niño event of 1997 and 1998 (with significant effects on seasonal precipitation and temperature across the entire world), awareness has grown that adequately addressing climate variability at the seasonal timescale may help to anticipate climate change at longer timescales (see, for example, Hartmann et al, 2002, and Chapter 6 of this book).

In this chapter we will further explore climate variability and change, and its predictability at the seasonal and decadal timescales. We argue that an adequate awareness of tools and knowledge concerning the seasonal timescale may increase our ability to deal with climate variability at longer timescales. This hypothesis is further elaborated in Chapter 6. The following section will address the observed recent changes in some relevant climate variables that are compared to natural variability, concluding with the notion that on multi-year timescales, climate change is ongoing and detectable. 'Predictability of climate variability at the seasonal timescale' is devoted to the subject of predictability, in general, and seasonal predictability, in particular. This section describes and evaluates different techniques that allow for seasonal projections, and provides an overview of operationally available products. Finally, 'Climate projections on the decadal timescale' describes a system that can make projections for the next decennia. Global and regional climate change scenarios on an even longer timescale (half a century or more) are described in Chapter 3, including some examples of scenario products tailored for local (impact assessment) applications.

Climate change in the recent past

The publication of the *Fourth Assessment Report* of the IPCC (IPCC, 2007) leaves little doubt about the fact that global mean surface temperatures have been increasing since the mid 1970s, and that it is very likely that humans are contributing to this change. The implications of this global temperature increase for the water sector are widespread and are documented in many reports, papers and books. Chapter 4 further discusses these implications. Here we focus on the evidence of increases in recent decades and point to some references to observations and studies that may be relevant for water resource managers.

Major findings of the IPCC *Fourth Assessment Report*

Global mean concentrations of greenhouse gases such as carbon dioxide (CO_2), methane (CH_4) and nitrous oxide (N_2O) have increased due to human activities. This increase is very likely the cause of the worldwide increase in land and ocean temperatures (approximately 0.7°C/100 years), melting of snow and ice-caps, and sea-level rise (17cm/100 years) (see Plate 1 in the centre pages). Many long-term changes in the climate system have been observed, including changes in the temperature and sea-ice extent in the Arctic region, in large-scale precipitation patterns, in ocean salinity, in wind patterns and also in aspects of extreme weather. The simple paradigm that mean precipitation increases in the wet high latitudes and the tropics, and decreases in drier subtropical areas is confirmed by the observed increases in Northern Europe (5 to 10 per cent) and in the US, and in reductions in Northern Africa. However, the observed decreases in precipitation in Western and Central Africa, and the increase in Southern America and North-Western Australia do not fit into this simple picture. No changes were found in, for instance, the diurnal cycle of surface temperature, the sea-ice extent near the Antarctic continent, and small-scale phenomena such as tornadoes and lightning.

Detection of trends in a fluctuating signal

Although it may seem easy to define a trend as a signal that is increasing or decreasing over time, the detection of trends may be far from simple. Without exception, observed trends must be detected in time series that strongly fluctuate at daily, monthly, seasonal and inter-annual timescales. Whether a trend in a fluctuating signal is significant depends upon a few factors. The *length of the time record* (relative to the timescale of the phenomenon) plays a role because it is easier to detect a trend when the time series is longer. No meaningful statements on trends can be made from individual events or short episodes. The fact that the warmest year since the late 19th century is 1998 (and not, say, 2007) does not preclude the existence of a significant trend over the 20th century. The *variability of the phenomenon* is important as well: high variability makes trend detection more difficult. This makes trends in temperature (relatively strong trends compared to inter-annual variability and highly linked to the radiative effects of greenhouse gases) easier to detect than trends in mean precipitation or wind (which have a much noisier signature and a weaker link to these radiative effects). However, the erratic nature of precipitation makes trends in extreme precipitation less complicated to detect than trends in the mean (Groisman et al, 2005).

Detection of trends relies on homogeneous and undisturbed observational records. This is not trivial since many routine meteorological observation stations have changed position, sensor type, calibration or surrounding environment, often undocumented. An example of a carefully homogenized climate data set is the European Climate Assessment (Klein Tank et al, 2002; see http://eca.knmi.nl, accessed 1 July 2008). This assessment has been designed for analysis of trends in many climate indices, including extreme precipitation or heat waves (see Plate 2 in the centre pages for an example).

Observed recent trends

The time series of global mean temperatures are becoming long enough to enable the detection of clear trends and to compare them to the projections published by the IPCC *Fourth Assessment Report*. Rahmstorf et al (2007) conclude that the projections are conservative with respect to the observed trends in temperature and sea level since 1990 (see Plate 1 in the centre pages). The observed trends are found high in the range of the IPCC projections. Although the overlap between the observations and projections is too short to exclude the influence of natural variability, the results underpin the concern of climate actually changing.

Furthermore, the temperature record in Europe has shown a strong departure from the global mean temperature increase: depending upon the location, the regional temperature between 1950 and 2007 has increased up to 2.5 times the global mean temperature increase, particularly during spring and summer (Van Oldenborgh et al, in press). In future projections, land masses and high latitudes are shown to increase faster than ocean and tropical areas, and the observations seem to confirm this feature. In addition, (extreme) precipitation trends become significantly detectable. Zolina et al (2008) show an increase of heavy and extreme precipitation in Germany since 1950 in winter, spring and autumn. During the summer, heavy and extreme precipitation has decreased (see also Plate 2, centre pages).

Predictability of climate variability at the seasonal timescale

Predictions (predictability), projections and scenarios are different terms, although they are often interchanged. A *prediction* is a forecast of what will happen in future. This can be a deterministic forecast ('tomorrow it will be raining') or a probabilistic forecast ('there will be a more than average chance that tomorrow it will rain'). The predictability of a phenomenon can be defined as the degree to which its evolution can be deduced from the known initial conditions and the known evolution of factors that affect the phenomenon. It thus depends significantly upon the spatial and temporal scales of the phenomenon. *Projections* with, for instance, climate models can be made; but they cannot be considered as a prediction. A range of projections is often interpreted as a probabilistic forecast; but this is difficult as long as the quality of the projection (and therefore its likelihood to occur) cannot be firmly determined. A *scenario* is a projection following on from a set of basic 'what if' assumptions. For instance, for an assumed time evolution of greenhouse gas concentrations, the global mean temperature rise is deduced from an ensemble of climate model projections. However, within a given (concentration) scenario, the future climate can still evolve in multiple directions and, strictly speaking, cannot be predicted.

Global and regional climate *predictability* and the information that gives rise to predictability vary with the timescale and region considered. Predictability arises from at least two sources: initial conditions and changing external forcing. Predicting *synoptic weather* requires a good-quality initial condition of the atmosphere and land, and a decent meteorological model to describe the evolving dynamic weather features. Scientific and computational developments leading to improved initial conditions have extended the time range of sufficiently accurate weather predictions by approximately one day per decade since the late 1970s, up to approximately seven days at present. However, operational forecasting applications in the water sector usually rely on probabilities that extreme hydrological events occur, and the mean forecast quality (often denoted by the term 'predictive skill') is of less importance. Probabilistic weather forecasts have been used since the mid 1990s to assess the risks of, for instance, extreme river discharge, heavy precipitation events, hurricane tracks or other weather phenomena that have an impact upon society (see Figure 2.1). Applications focusing on this synoptic timescale are widely used and well known, and are not the subject of this book.

On longer timescales (such as the *seasonal timescale*), a likewise good initial condition of the slower components in the climate system is required: the temperature of the upper layers of the ocean and the sea surface temperature, ice cover extent, slowly varying signals in the stratosphere, and soil moisture and snow conditions on land. In addition, predictability at the seasonal timescale varies largely with seasons and across the globe since the chaotic nature of atmospheric motion destroys correlations as time proceeds. Seasonal predictions are routinely produced by a number of major weather services across the world. The El Niño Southern Oscillation (see the sub-section on 'Sources of predictability at the seasonal timescale') is an important source of predictability at seasonal timescales.

Seasonal forecasting tools

Seasonal forecasting tools have rapidly emerged in the past decade. Such tools are particularly powerful in areas and seasons where strong connections to slowly varying SST and other climate variables exist, and where the seasonal variability of the weather is substantial. In areas with small seasonal and year-to-year variability of mean seasonal

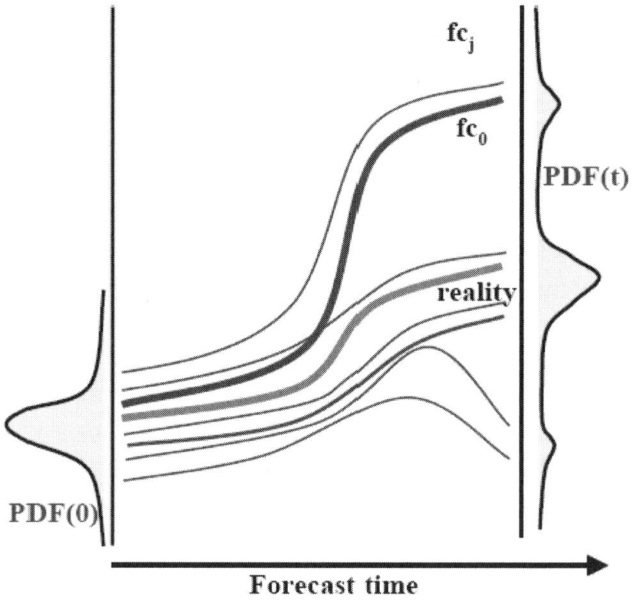

Figure 2.1 *Graphical display of probabilistic forecasting*

Note: As forecast time proceeds, the probability distribution function (PDF) of a given event evolves and may occasionally break up in different regimes. The 'reality' line represents a retrospective check of the probabilistic forecast.

Source: Taylor and Buizza (2004)

precipitation, such as the mid-latitudes or desert regions, less opportunities to predict anomalous climate conditions are present than in areas with strong variability (such as monsoon climates or land areas in the (sub)tropical regions). Seasonal forecasting tools do not aim to forecast a specific event at a given day, but rather the probability that the seasonal mean precipitation or temperature is higher or lower than the climatological mean. The existing tools can be roughly divided into two classes: statistical and numerical methods (Palmer and Anderson, 1994). In some applications, a mixture of the two is used.

Statistical methods use observed correlations between SST and regional weather patterns to make forecasts for the future. El Niño variations are an important source of predictability (see the following sub-section). Apart from giving a probability of anomalously high or low precipitation, they are often used to choose historical analogue years that serve as input to hydrological or agricultural applications (see, for example, Stone et al, 1996; Hamlet and Lettenmaier, 2000). However, they often suffer from limited observational record length needed for the calibration of the tools. And they are not able to cope with changes in statistical correlations induced by changes in the external forcings. Although future climate projections do not show strong shifts in El Niño frequency or structure (Van Oldenborgh et al, 2005a), nor in the structure of the teleconnections (Van Oldenborgh and Burgers, 2005; Sterl et al, 2007), the statistical relations found today may be different for tomorrow's climate conditions.

Numerical methods use an ensemble of projections with coupled ocean–atmosphere general circulation models (OAGCMs) initialized with an 'observed' state of the ocean, land and ice conditions. This approach copes with the inherent uncertainty introduced by the chaotic nature of the climate system. However, the quality of the initial states is fairly poor owing to the lack of routine observations in the ocean and on land. A well-known operational system for seasonal forecasting is the multi-model EUROSIP system (EUROpean multi-model Seasonal to Inter-annual Prediction) system (see www.ecmwf.int/products/forecasts/seasonal/forecast/forecast_charts/eurosip_doc.htm, accessed 1 July 2008), where seasonal predictions from three European meteorological services are combined into a single application database. By combining multiple modelling systems, the model's uncertainty can be assessed.

Both the statistical and numerical tools for seasonal prediction rely on existing sources of predictability at the seasonal timescale. The major source is the oceanic surface temperature (of which El Niño is the strongest expression); but other sources are being investigated as well. The major sources are briefly discussed in the following sub-section.

Sources of predictability at the seasonal timescale

El Niño Southern Oscillation (ENSO)

Irregular but persistent sea surface temperature variations in the equatorial Pacific Ocean are associated with the El Niño Southern Oscillation phenomenon, which returns on average every three to seven years. During an El Niño, SSTs are warmer than normal around the Equator in the eastern half of the Pacific Basin, usually starting early in the year and peaking during November to January. This results from an interaction between the ocean and the atmosphere, where changes in the ocean surface temperatures affect tropical rainfall patterns and atmospheric winds over the Pacific Ocean, which in turn affect ocean temperatures and currents. Details about the mechanisms and the degree to which they are reproduced adequately in present-day climate models are given by Neelin et al (1998). Popular documentation is given on many websites, including the National Center for Environmental Predictions (NCEP)/Climate Prediction Center (CPC) website (see www.cpc.ncep.noaa.gov/products/precip/CWlink/MJO/enso.shtml, accessed 1 July 2008).

The strength of an ENSO event is usually expressed as a SST anomaly in a particular Pacific section at the Equator. Different regions are used, leading to different (related) indices. For instance, one index (NINO1.2) concentrates on SST anomalies near the coast of Ecuador and Peru and is indicative of coastal precipitation variability. NINO3.4 is located in the centre of the Pacific and is related to weather phenomena around the world.

The ENSO phenomenon has a clear impact on (hydro-)climate in many regions of the world (see Plates 3 and 4, centre pages). The strongest relationships are found in the Pacific equatorial zone and coastal areas bordering the Pacific Ocean. Apart from a weak positive impact of ENSO upon precipitation in South-Western Europe during spring (Van Oldenborgh et al, 2000) a teleconnection between ENSO and European climate variability is not detectable.

Other variability modes

Apart from ENSO, a number of other modes of atmospheric and oceanic variability exist that bear some seasonal predictability in some regions of the world: the Pacific Decadal Oscillation (PDO), the North Atlantic Oscillation (NAO) and the Indian Ocean Dipole (IOD). For instance, rainfall in Eastern Africa is correlated to SST anomalies in the Western Indian Ocean. The NAO index (usually expressed as an anomaly in the pressure difference between Iceland and Lisbon) is positively correlated to precipitation in the northern part of Europe and reversed in Southern Europe (mainly during the winter season; see Plate 5, centre pages). Although these signals are weaker than ENSO teleconnections, they are used in some statistical seasonal forecasting tools.

Land–atmosphere interactions

An active field of research is the possible predictive skill (or forecast quality) present in the slowly varying terrestrial components of the climate system, such as snow and soil moisture. Statistical analyses have demonstrated a detectable positive correlation between springtime snow amounts and temperature up to one month later in North-Western Europe (Shongwe et al, 2007). Extreme hydrological events in Europe and the US (like the European 2003 summer heat wave) have incited a number of studies demonstrating increased likelihood of anomalous heat-wave intensities during summer when the winter/spring soil moisture content is relatively low (see, for example, Ferranti and Viterbo, 2006). These studies justify investments in widespread observation and data assimilation of these quantities in order to increase the quality of the terrestrial initial conditions of numerical seasonal prediction tools.

Regional differences in the predictability of the climate

The persistent ENSO feature is a powerful source of climate predictability at the seasonal timescale. If the initial condition related to the anomalous ENSO state is captured well, the relatively high correlations to weather phenomena elsewhere in the world enhance the quality of the forecast owing to the large persistence of the phenomenon.

Nevertheless, regional differences in the predictability of the climate exist. For instance, due to the high heat capacity of ocean water, the predictability of the temperature is higher over oceans than over land. Systematic changes in air circulation related to ENSO are also a source of high predictability. For example, the predictability of precipitation in high-rainfall regions in the tropics is strong due to a clear effect of ENSO.

Van Oldenborgh et al (2005b) compared the skill of seasonal predictions from a statistical forecast model and a number of European Centre for Medium-Range Weather Forecasts (ECMWF) dynamic coupled modelling systems performing a three-month forecast (Plates 6 and 7, centre pages). For December, January and February (DJF), positive skill in terms of 2m temperature is seen in areas where strong teleconnections with ENSO are present. The comparison shows that the dynamic model is better in areas where other factors than ENSO play a role, such as the Indian Ocean, or over many land areas outside the tropics. In general, forecasting precipitation is much more complex than temperature. Consequently, the precipitation forecast skill is generally lower (not shown).

Availability and formats of seasonal forecasts

Many (climate) institutes around the globe offer seasonal forecasts or outlooks for different periods and different parts of the globe. It is beyond the scope of this book to give an exhaustive list of products, but a few examples are provided here.

An example of statistical seasonal forecast products is provided by the Australian Bureau of Meteorology, which issues probability maps of above or below median rainfall for a three-month period (see www.bom.gov.au/climate/ahead/rain_ahead.shtml, accessed 1 July 2008). In addition, guidance is given on the results and the interpretation. The Canadian Weather Office (see http://text.www.weatheroffice.gc.ca/saisons/index_e.html, accessed 1 July 2008) provides similar maps for lead times longer than three months, but uses numerical forecasts for shorter lead times. The US Climate Prediction Center (see www.cpc.ncep.noaa.gov/products/predictions, accessed 1 July 2008) issues tercile maps (probabilities above, at or below normal) based on a mix of statistical and numerical methods. The statistical methods use various observed correlations, including ENSO and a soil moisture index. The skill of each method is assessed separately. Outlooks are given for the US only.

Products from comprehensive multi-model systems are given by the International Research Institute for Climate and Society (IRI; see http://portal.iri.columbia.edu, accessed 1 July 2008) and the EuroSIP consortium (see www.ecmwf.int/products/forecasts/seasonal/forecast/forecast_charts/eurosip_doc.htm, accessed 1 July 2008). The EuroSIP continues an earlier successful seasonal prediction project DEMETER (Palmer et al, 2004), which has generated a widely used data set of seasonal forecasts for evaluating model quality, climate variability at the seasonal timescales, maximum predictability across the world, and optimal interfaces between the meteorological products and end users in hydrology, agriculture, safety management and other applications. The most common products emerging from EuroSIP include an ENSO forecast plume and a range of probability maps for anomalously high or low precipitation, SST or temperature (see Plates 8 and 9, centre pages).

All seasonal forecast products are issued with a skill assessment. However, the skill parameters and criteria vary among the groups. A simple quality (or skill) measure is the correlation between forecasted and observed seasonal mean temperature, precipitation or surface pressure. More advanced quantities include the relative operating characteristics (ROC) score, which expresses whether a forecast is actually successful in predicting a certain event to happen. A useful data portal to compare skill and seasonal forecasts from a range of forecast centres is the Climate Explorer (see http://climexp.knmi.nl, accessed 1 July 2008). Here the seasonal forecasts are collected near real time and can be processed or verified using a common metric and verification database.

Climate projections on the decadal timescale

Predicting *decadal timescales* is even more difficult because of noise and other non-linear interactions that introduce uncertainty as time proceeds. Predictability at these timescales is considered to be successful depending upon a mix of an adequate initial condition of the long-term climate variables, such as the temperature distribution in the oceans, on the one hand – knowing the projected changes in the external forcings (in particular, greenhouse gas emissions and land-use changes) – and, on the other, adequately assessing the response of the climate system to these forcings. A first attempt to predict the global mean temperature for the next decade is presented by Smith et al (2007) and Keenlyside et al (2008). These are largely based on improved estimates of the ocean heat content and are discussed in more detail in the following sub-sections. It is likely that projected changes in radiative forcing are an important source of climate predictability, whose importance may grow with the timescale considered (Hurrell et al, 2007).

Use of climate models

Future projections of the global and regional climate are necessarily carried out with computer models representing our complex climate system. This climate system contains many mechanisms and interactions that can amplify, delay, dampen or transform disturbances in the so-called external forcings (e.g. solar radiation and atmospheric composition), resulting in smaller or larger responses. For instance, a relatively small change in the global and annual mean radiation received from the sun has led to large fluctuations of the surface temperature, giving rise to glacial and interglacial episodes in climate history. For realistic future projections, climate models must be able to adequately reproduce these feedbacks and responses. A continuous effort of testing, comparing, calibrating, revising and extending has resulted in a gradual improvement of the ability to reproduce the global mean and regional climate, similar to the gradual improvement of the quality of the weather forecasting tools. Confidence in the climate response to enhanced greenhouse gas concentrations is increasing, although (inherently) remaining model uncertainty still necessitates considering a range of possible future conditions (IPCC, 2007).

Projections for the near future

In their *Fourth Assessment Report*, the IPCC (2007) concludes that in the next few decades, global mean temperature is likely to rise by approximately 0.2°C per decade for a range of emission scenarios (see also Plate 1, centre pages). Up to the middle of the 21st century, uncertainty in greenhouse gas emissions is of less importance than uncertainty in the initial condition (in ocean, land and ice masses), in internal climate variability and in the predictive capacity of models. Projections of expected global mean precipitation change are less straightforward. Models and observations agree that increases are likely for the tropics and mid- and high-latitude zones, and reductions are probable in the subtropics. But spatial detail, internal climate variability and model disagreement are all stronger than for temperature.

Projections for the near future (at the so-called decadal timescale) are currently being carried out and explored by a number of global climate modelling (GCM) groups. A pioneering application was presented by Smith et al (2007), who designed a procedure to estimate the ocean heat content, which was used to launch GCM projections ten years into the future. The study identifies both the initial condition (i.e. the ocean heat content) and ongoing global warming as the most important sources of predictability at this timescale. Plate 10 in the centre pages is a cautious prediction of the global mean temperature up to 2014, where ongoing warming is compensated for by the cooling effect of internal climate variability during the years up to 2008. Although the work is a significant step towards designing an operational decadal forecasting system (see also Cox and Stephenson, 2007), the skill at the regional scale needs improvement. Our understanding of the (limitations of the) predictability at this timescale, as expressed by the quality of the models and initialization procedures, remains limited. Additional improvements are required before a truly useful forecasting system can be applied.

Projections for the 21st century summarized by the IPCC

At longer timescales (from, say, the second half of the 21st century onwards), the uncertainty of greenhouse gas emission rates gains importance in the range of projected global mean temperature: different emission scenarios and atmospheric greenhouse gas concentrations lead to discernible differences in the global mean temperature projections (IPCC, 2007). Even when concentrations will not increase from the level reached in 2000, the global mean temperature will continue to rise by approximately 0.1°C per decade during the 21st century due to the delayed response of the slow components in the climate system. Increases of emissions at the current rate or faster will lead to further temperature rise and changes in other climate variables, likely to be larger than observed during the 20th century. Global warming and sea-level rise will continue long after the 21st century, even with a stabilization of greenhouse gas emissions.

These very general conclusions are reported in the IPCC *Fourth Assessment Report*, published in 2007. Considerable additional detail is available from the large suite of GCM projections, observation analyses and regional downscaling tools carried out before and after the IPCC report. Insight into climate variability and change is continuously increasing. New components are added to climate models, which are turning into even more comprehensive Earth system models (ESMs), including additional important feedbacks from ecosystems, human influence on land use and atmospheric composition, and geophysical components such as deep water reservoirs, wetlands, oceanic slow currents and others. Climate projections will likewise continue to evolve; climate scenarios are a moving target and will change and be refined as scientific development progresses. The next chapter discusses the global and regional climate scenarios in more detail, paying attention to the relevance of the topic to the water sector.

References

Cox, P. and D. Stephenson (2007) 'A changing climate for prediction', *Science*, vol 317, pp207–208

Ferranti, L. and P. Viterbo (2006) 'The European summer of 2003: Sensitivity to soil water initial conditions', *Journal of Climate*, vol 19, pp3659–3680

Groisman, P. Y., R. W. Knight, D. R. Easterling, T. R. Karl, G. C. Hegerl and V. N. Razuvaev (2005) 'Trends in intense precipitation in the climate record', *Journal of Climate*, vol 18, pp1326–1350

Hamlet, A. F. and Lettenmaier, D. P. (2000) 'Long-range climate forecasting and its use for water management in the Pacific Northwest region of North America', *Journal of Hydroinformatics*, vol 02.3, pp163–182

Hartmann, H. C., T. C. Pagano, S. Sorooshian, and R. Bales (2002) 'Confidence builders: Evaluating seasonal climate forecasts for user perspectives', *Bulletin of the American Meteorological Society*, vol 83, pp683–698

Hurrell, J., D. Bader, T. Delworth, B. Kirtman, J. Meehl, H.-L. Pan and B. Wielicki (2007) *White Paper on Seamless Prediction*, www.cgd.ucar.edu/cas/jhurrell/Docs/SeamlessModellingDraft03302007.pdf

IPCC (Intergovernmental Panel on Climate Change) (2007) *Climate Change 2007: Synthesis Report, Fourth Assessment Report*, Cambridge University Press, Cambridge

Keenlyside, N. S., M. Latif, J. Jungclaus, L. Kornblueh, and E. Roeckner (2008) 'Advancing decadal-scale climate prediction in the North Atlantic sector', *Nature*, vol 453, pp84–88

Klein Tank, A. M. G. et al (2002) 'Daily dataset of 20th-century surface air temperature and precipitation series for the European Climate Assessment', *International Journal of Climatology*, vol 22, pp1441–1453

Neelin, J. D., D. S. Battisti, A. C. Hirst, F.-F. Jin, Y. Wakata, T. Yamagata, and S. E. Zebiak (1998) 'ENSO theory', *Journal of Geophysical Research*, vol 103, pp14,261–14,290

Palmer T. N. and D. L. T. Anderson (1994) 'The prospect for seasonal forecasting – a review paper', *Quaternary Journal of the Royal Meteorological Society*, vol 120, pp755–793

Palmer T. N. et al (2004) 'Development of a European multi-model ensemble system for seasonal to inter-annual prediction', *Bulletin of the American Meteorological Society*, vol 85, pp853–872

Palmer, T. N., F. J. Doblas-Reyes, A. Weisheimer and M. Rodwell (2008) 'Towards seamless prediction: Calibration of climate-change projections using seasonal forecasts', *Bulletin of the American Meteorological Society*, vol 89, pp459–470

Rahmstorf, S., A. Cazenave, J. A. Church, J. E. Hansen, R. F. Keeling, D. E. Parker and R. C. J. Somerville (2007) 'Recent climate observations compared to projections', *Science*, vol 316, pp709–709

Shongwe, M. E., C. A. T. Ferro, C. A. S. Coelho and G. J. van Oldenborgh (2007) 'Predictability of cold spring seasons in Europe', *Monthly Weather Review*, vol 135

Smith, D. M., S. Cusack, A. W. Colman, C. K. Folland, G. R. Harris and J. M. Murphy (2007) 'Improved surface temperature prediction for the coming decade from a global climate model', *Science*, vol 317, pp796–799

Sterl, A., G. J. van Oldenborgh, W. Hazeleger and G. Burgers (2007) 'On the robustness of ENSO teleconnections', *Climate Dynamics,* vol 29, pp469–485

Stone, R. C., G. L. Hammer and T. Marcussen (1996) 'Prediction of global rainfall probabilities using phases of the Southern Oscillation Index', *Nature*, vol 384, pp252–255

Taylor, J. and R. Buizza (2004) 'A comparison of temperature density forecasts from GARCH and atmospheric models', *Journal of Forecasting*, vol 23, pp337–355

Van Oldenborgh, G. J. and G. Burgers (2005) 'Searching for decadal variations in ENSO precipitation teleconnections', *Geophysical Research Letters*, vol 32, p15

Van Oldenborgh, G. J., G. Burgers and A. M. G. Klein Tank (2000) 'On the El Niño teleconnection to spring precipitation in Europe', *International Journal of Climatology*, vol 20, pp565–574

Van Oldenborgh, G. J., S. Y. Philip and M. Collins (2005a) 'El Niño in a changing climate: A multi-model study', *Ocean Science*, vol 1, pp81–95

Van Oldenborgh, G. J., M. A. Balmaseda, L. Ferranti, T. N. Stockdale and D. L. T. Anderson (2005b) 'Evaluation of atmospheric fields from the ECMWF seasonal forecasts over a 15 year period', *Journal of Climate*, vol 18, pp3250–3269

Van Oldenborgh, G. J., S. Drijfhout, A. van Ulden, R. Haarsma, A. Sterl, C. Severijns, W. Hazeleger and H. Dijkstra (in press) 'Western Europe is warming faster than climate models predict', *Climate of the Past*

Zolina, O., C. Simmer, A. Kapala, S. Bachner, S. Gulev and H. Maechel (2008) 'Seasonally dependent changes of precipitation extremes over Germany since 1950 from a very dense observational network', *Journal of Geophysical Research*, vol 113

3

Climate Change Scenarios at the Global and Local Scales

Daniela Jacob and Bart van den Hurk

Changes in climatological and hydrological conditions, as well as changes in political, economic, social and legal contexts, have posed new challenges to water management. Since the pressure resulting from these changes is very specific to local conditions, it is of eminent importance to use and further develop concepts and methodologies for water management within individual catchments. Considering changes and adaptation in water management systems and water use, possible future adaptation strategies need to balance water availability and use, as well as protection and risks.

The climate information chain

Meteorological and hydrological observations demonstrate that during the last decade the climate has changed. As reported by the Intergovernmental Panel on Climate Change (IPCC, 2001, 2007), a mean increase of near surface air temperature by 0.09K per decade was observed globally from 1951 to 1989. Until now (2008), this trend has continued. Europe experienced an extraordinary heat wave in the summer of 2003, with daily mean temperatures locally approximately 10 degrees warmer than the long-term mean. The water level in the Rhine River in The Netherlands reached critically low levels for cooling power plants. This event cannot be directly related to climatic changes, but it can be seen as a good example of what might happen in the future. As a result, the phenomenon increased awareness of the consequences of climate change. The increase of temperature varies with region and season. If the temperature of the atmosphere increases, it should be assumed that the water cycle is also intensified. However, there is still an ongoing debate about the extent to which global warming will increase precipitation (Lambert et al, 2008).

Global climate models (GCMs) have been developed to study the Earth's past and future climate system, driven by assumptions on the evolution of drivers of climate change. The drivers are, for example, the amount and distribution of aerosols and greenhouse gases (GHGs) in the atmosphere, which depend directly upon natural and man-made emissions. Emission scenarios are developed using so-called *story lines*, describing possible developments of the socio-economic system (Nakicenovic et al, 2000). These emission scenarios can be considered to express a first source of *uncertainty* in assessing future climate conditions. The emissions are translated into greenhouse gas concentrations, used by GCMs to make projections of future climate. The IPCC guided a large set of GCM projections, emphasizing that the limited skill of GCMs forms an important second source of uncertainty. Both scenario and GCM uncertainty result in an increasing spread of the projection of global mean temperature changes up to a range of between 1.5°C and 5.5°C in 2100 (IPCC, 2007).

GCMs are mathematical representations of the Earth's system, in which physical and biogeochemical processes are described numerically to simulate the climate system as realistically as possible. A considerable amount of research has been devoted to comparing GCM results to independent observations and to the subsequent improvement of models. An ensemble of past episodes is simulated and the results are compared against measurements before the models are used for climate change studies. Different ensemble members show differences that reflect a third source of uncertainty: the internal variability in the climate system that is inherently present and cannot be avoided even with perfect models. The individual ensemble members are usually in good agreement with atmospheric analyses (e.g. ERA40 re-analyses, which have been reconstructed using a large set of observations in a global modelling system; this data set is as close to reality as possible using state-of-the-art tools), but sometimes show a systematic bias (e.g. being approximately 0.5 degrees warmer than the reconstructed observations for members of the global coupled climate modelling system ECHAM5/MPIOM, not shown here). The observed increase during the last decades is clearly visible in the GCM simulations (see also IPCC, 2007).

Global mean temperatures are quite unrepresentative for the local-to-regional scale (where 'regional' refers to a domain of typically 500km to 1000km in this chapter). At this smaller spatial scale, variability is even larger than for the global mean climate, which introduces a fourth source of uncertainty. Even today, global climate models provide information only at a relatively coarse spatial resolution, often not suitable for regional climate change assessments.

To overcome this deficiency, two different principles are used to bring the information from the global model to the region of interest. *Statistical downscaling* techniques use an observed relation between large-scale phenomena (often fairly well represented in coarse-scale GCMs) and local quantities (such as daily precipitation or daytime temperature). This relation is subsequently applied to GCM output to obtain local and regional climate change signals. A major disadvantage of this approach is the implicit assumption that the calibrated relationships for present-day climate conditions are also applicable to future climate conditions. This is debatable when climate change leads to significantly different climate regimes.

Alternatively, *dynamical downscaling* represents the use of high-resolution regional climate models (RCMs), which are nested within GCMs (Jacob, 2009). Large-scale phenomena are inherited from the host GCM; but additional detail is provided concerning the land use, coast lines, topographical structures and better-resolved spatial gradients in physical fields (see Plate 11, centre pages). This additional information can substantially alter the regional flow pattern and give more credit to local feedback processes such as snow albedo/temperature or soil moisture/temperature feedback. RCMs therefore generally improve on the higher-order statistics of the meteorological variables. A drawback of RCMs is their large demand on computer resources and the complexity of their operation, which requires trained staff. However, dynamical downscaling methods are increasingly being used, and will be discussed in more detail below.

As for GCMs, the model quality of RCMs needs to be analysed before addressing climatic changes. For this purpose, RCMs are nested within re-analysis data, which is as close to reality as possible. The results of the RCM simulations of past decades are compared against independent observations, and means as well as extremes are considered.

As an example, simulated precipitation climatologies calculated with the regional climate model REMO of the Max Planck Institute for Meteorology (Jacob, 2001) with two different horizontal grid sizes are compared against observations that were compiled from observational records at the Swiss Federal Institute of Technology (ETH) (Schwarb et al, 2001). The 0.44 degree grid (~ 50km) is much finer than standard GCM grids (about 150km to 250km); but regional details on sub-catchment scale are still not visible. The total amount of precipitation and the horizontal pattern are much better resolved using the very high horizontal resolution of about 10km (see Plate 11, centre pages). Regional maxima, such as the ones in the Black Forest, and minima, as in the centre of the Alps, are detectable. However, the resolution is still too coarse to describe features in individual alpine valleys.

Regional climate simulations in Europe

A well-known ensemble of RCM simulations for Europe is collected in the context of the European Union project PRUDENCE (see http://prudence.dmi.dk; Christensen and Christensen, 2007). An ensemble approach is used to assess the magnitude of the uncertainty when downscaling global projections to the regional scale (Déqué et al, 2007; Jacob et al, 2006). Most simulations were carried out using GCMs driven by the A2 climate change scenario, projecting a relatively strong future increase of greenhouse gases up to the year 2100 and a subsequent global mean temperature increase of about 3.5 degrees (IPCC, 2001). Note that this approach does not allow assessments of the uncertainty associated with different greenhouse gas emission scenarios.

An analysis of the hydro-meteorological conditions for different river catchments shows significant differences between the projected changes for Northern and Central Europe for the time period of 2070 to 2100 compared to the current climate (1961 to 1990) (Hagemann and Jacob, 2007). For the Baltic Sea catchment, an ensemble mean precipitation increase of about +10 per cent for the annual mean is projected, with the largest increase of up to +40 per cent in winter, while a slight reduction of precipitation is calculated for the late summer. Evapotranspiration will increase during the entire year with a maximum relative increase in winter, even though the absolute changes are about twice as large between April and July. These increases in wintertime precipitation and evapotranspiration lead to an increase of river discharge into the Baltic Sea of more than 20 per cent in winter and early spring. Here, the seasonal distribution of discharge is largely influenced by the onset of spring snowmelt.

For the catchments of the Rhine, Elbe and Danube, a different change in the water balance components is computed. While the annual mean precipitation remains almost unchanged in these projections (except for the Danube, where it is even projected to decrease by about 5 per cent), it increases in late winter (January to March) and decreases significantly in summer. The evapotranspiration increases during the entire year, except in the summer, with a maximum relative increase in winter. These changes lead to a large reduction of 10 to 20 per cent in the annual mean discharge (see Plate 12, centre pages). Especially for the Danube, the projected summer drying has a strong impact upon the discharge, which is reduced by up to 20 per cent throughout the year except for late winter (February to March) when the increased winter precipitation causes a discharge increase of about 10 per cent. These projected changes in the mean discharge would have significant impacts upon water availability and usability in the affected regions. Note that although the projected quantitative changes are attached with some uncertainty, the ten models generally agree on the direction of the changes.

Of primary interest is the possible change in precipitation intensities (i.e. the amount of precipitation within a certain time period). The simulation of precipitation intensities or extreme precipitation events depends upon the applied model resolution. For example, the influence of the topography of the Alps on the formation of precipitation over the Rhine catchment is of interest. Simulation of intense showers requires a considerably higher resolution than the RCM results presented above.

Tailored climate scenarios

The chain of information on possible future climate developments – starting with a range of greenhouse gas scenarios, running an ensemble of GCMs and downscaling via regional climate models or statistical post-processing – is characterized by a continuous increase in the volume of numbers and possible pathways. For users in professional sectors with an interest in future climate evolutions, further guidance, data reduction or data transformation are needed. It is the rule rather than the exception that general climate change scenarios, even at the regional scale, need additional 'tailoring' to meet the user's needs.

This process of tailoring encompasses a wide range of procedures. For example, it can be the outcome of a discussion on the choice for the most relevant scenario for a given sector from an available plume. It can be a quantitative translation of a meteorological variable (such as the change of the daily mean temperature at the average hottest day in the year) into a quantity that is better related to the concerned sector (e.g. the likelihood of having temperatures in excess of 30°C that may harm the development of insects present in an ecological food chain; see Figure 3.1). It can be a detailed time series of daily precipitation at a given location consistent with assumptions about the future climate developments, such as to test sewerage design. Or it can be the change of the likelihood of extreme storm surges with return periods much longer than the observational record, to be derived from general scenario data by means of statistical extrapolation of extreme events.

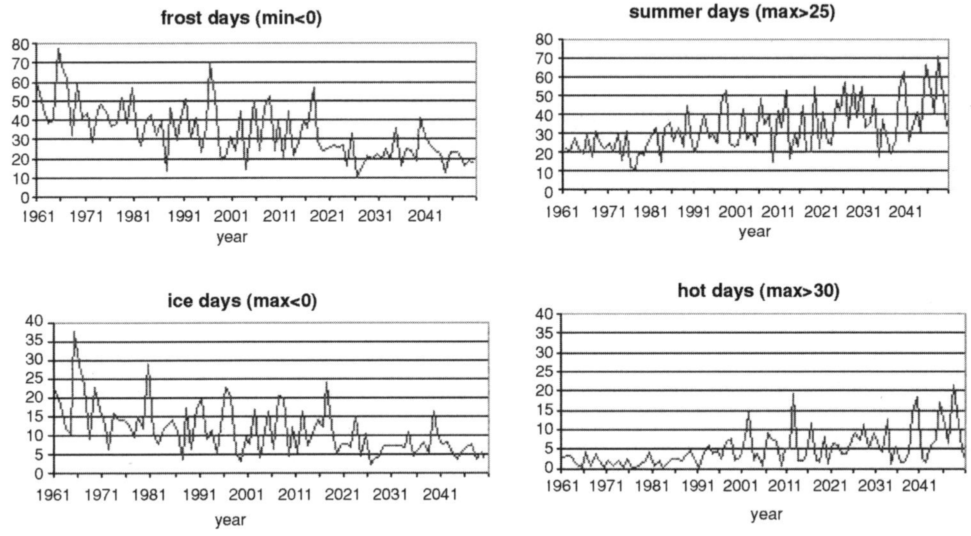

Figure 3.1 *REMO B2 scenario for the Rhine catchment: Frost days (upper left), ice days (lower left), summer days (upper right) and hot days (lower right)*

Source: Jacob, 2009

As an example, a time series of simulation results for the Rhine Basin is presented for a B2 scenario until 2050. Between 1960 and 2050, the near surface temperature might rise by about 3°C and the number of summer days and hot days will increase (see Figure 3.1). In addition, the number of consecutive periods with summer days with a daily maximum temperature above 25°C will be higher in future decades (not shown). Winter temperature also increases, leading to a decrease in frost and ice days.

This section provides a brief description of a number of examples of tailored climate scenarios. It is not intended to be a complete overview, but serves as an illustration of a necessary step in order to bring relevant climate change information to the professional end user.

Definition of a standard year

The calculation of groundwater dynamics in The Netherlands is operationally applied using a detailed hydrological model chain (Vermulst et al, 1998). The atmospheric input consists of a precipitation and evaporation time series obtained from routine meteorological observation stations. To assess the impact of climate change on groundwater dynamics, a new synthetic input database needs to be generated that is consistent with future climate conditions. Owing to the substantial computer resources required for this application, numerous calculations with multiple scenarios and multiple years are not feasible. The user first requested defining a 'standard' meteorological year that is representative of present-day climate conditions, and then modifying the meteorology of this standard year in a way that is consistent with one of the recently published KNMI'06 climate change scenarios (Van den Hurk et al, 2006). Since the spatial and temporal variability of precipitation (and evaporation) is considerable, the definition of a 'standard' year is not a trivial task. The implicit assumption in this request was that average meteorological conditions (especially rainfall) would also lead to average highest and lowest groundwater levels. This is not automatically true. Depending upon the quantity addressed (mean groundwater table, highest/lowest groundwater table during the growing season and cumulative surface evaporation during the growing season), the optimal procedure to generate this standard year varies (Van der Scheur et al, 2006).

The next step, the construction of a synthetic meteorological forcing for the target year 2050, comes down to a modification of the chosen standard year so that it is consistent with a chosen scenario for 2050. The KNMI'06 scenarios give changes in seasonal mean and extreme temperature and precipitation, as well as changes in the number of wet days per season, averaged for the entire country without regional detail. The tailoring procedure consists of modifying the time series of precipitation and temperature in each set of sub-domains consistent with the so-called W+ scenario. Although the W+ scenario does not specify regional differences in changes, the spatial patterns of many climate variables (e.g. the intensity of the wettest day or the number of days with a temperature greater than 25°C) in the newly constructed time series look different from the control data set. Plate 13 (centre pages) provides an example of the resulting change in the highest (springtime) and lowest (end of growing season) groundwater table in The Netherlands. In this example, the signature of the W+ scenario is clearly present: more winter precipitation leads to higher mean groundwater tables at the start of the growing season, whereas higher summer temperature and reduced summer precipitation causes the lowest groundwater table to be even lower.

A time series transformation tool, developed for this application, is provided on a public website (Bakker and Bessembinder, 2007; see http://climexp.knmi.nl/Scenarios_monthly, accessed 2 July 2008), in which users can choose archived data or upload their own data, and transform these according to one of the KNMI'06 scenarios.

Precipitation scenarios for the Rhine

Multiple studies have addressed the impact of climate change on the mean and extreme discharge of the Rhine River. For instance, in co-operation with the Federal Environment Agency (Umweltbundesamt, or UBA), REMO was used for a control simulation from 1950 to 2000 and three transient runs for the IPCC SRES scenarios A2, A1B and B1. The calculations with REMO show an increase of high precipitation events over the Alpine part of the Rhine catchment, especially in summer, assuming an A1B emission scenario (Jacob et al, 2008). In the high-resolution simulations applied here (10km), the regional pattern of temperature change displays a stronger warming in the south and south-east of the domain covering Germany, the Alps and Switzerland for the time period of 2071 to 2100 compared to 1961 to 1990 (Plate 14, centre pages), associated with a decrease of precipitation during the summer. An increase of precipitation in south and south-west regions during the winter was simulated. The winter precipitation is mostly rain and less precipitation falls as snow.

Climate change is one of the many changing variables that affect the discharge behaviour of this major European river system: safety infrastructure, demography, economic developments, water use and many more factors need to be integrated in an assessment of future developments in the area. A complicating factor of the Rhine Basin is the international dimension: it has five riparian countries, which all have different management structures, operation practices and resource interests. The difference in the historical way of defining climate change scenarios plays a role. For instance, unlike the German REMO scenario calculations described above, the Dutch approach does not directly use regional climate model output. Instead, a broad assessment is given of changes in a number of key climate variables under a range of varying assumptions not directly consistent with the different greenhouse gas emission scenarios.

For a project aiming to evaluate the effectiveness of a range of flood protection measures for a given future climate scenario (ACER; see http://ivm5.ivm.vu.nl/adaptation/project/acer, accessed 14 July 2008), a detailed hydrological model is used that requires meteorological input at high spatial and temporal resolution. The required data, provided by regional climate models, were cast into a long synthetic sequence of meteorological years, and the hydrological model was run for a number of multi-day episodes with either a high or a low cumulative precipitation amount. The long sequence is generated by re-sampling from an archive of bias-corrected regional climate model simulations.

For this study, the tailored climate scenario to be provided consisted of three components. First, a regional climate model output time series needed to be selected that matched the selected KNMI'06 scenario for a number of key variables. The match is required to link the results of this assessment to other studies addressing the consequences of climate change for the water studies that rely on a consistent coupling with the Rhine discharge projections, such as assessments for the drinking water resources, shipping traffic or power plant cooling. Second, the model output needed to be corrected for spatially and temporally varying systematic errors. And, third, the bias-corrected model sequences for present-day and future conditions were re-sampled in a long time series, while preserving the basic important statistical coherence of the meteorological forcing.

It is beyond the scope of this book to fully describe all steps in detail. As an example, results are shown of a comparison of RCM model output to the KNMI'06 scenarios in Plate 15 (centre pages). Obviously, the RCM run provides considerably more spatial detail than the single number in the KNMI'06 scenario, pointing at a spatial gradient in the changes. Averaged over the Rhine Basin, the summertime comparison yields the best resemblance to the so-called G and G+ scenarios (giving small positive and negative changes, respectively), whereas for the winter months a better correspondence with the more extreme W and W+ scenarios is shown. For other variables extracted from the model run (wet day frequency, extreme precipitation, temperature), this will result in other KNMI'06 scenarios to provide the best match. Thus, the comparability to the KNMI'06 scenarios depends upon the season and relevant meteorological quantities. For extreme discharge conditions, the selected RCM matches well with the KNMI'06 scenario with the strongest signal (dominated by autumn/winter ten-day accumulated precipitation) and is thus useful for evaluating the effectiveness of the infrastructure. This comparison is exemplary for the different climate change scenario approaches in the various Rhine riparian countries.

Regional climate effects atlas

An integrated means of communicating climate change and its effects to a wider professional user population is by creating geographic information system (GIS)-layered atlas maps with a range of variables. For the Dutch Climate Changes Spatial Planning Programme (see www.klimaatvoorruimte.nl, accessed 2 July 2008), a so-called *climate effects atlas* is constructed by a group of mixed professionals, ranging from climate scenario experts, climate change effect specialists and spatial planners. The main aim of this atlas is to provide a solid base for the many political and economic discussions regarding the adaptation to climate change. The atlas is designed to provide information on primary climate effects (changes in meteorological variables) and secondary effects (changes in agricultural damage, groundwater levels and river discharge) for a wide range of sectors at the regional (province) level.

The climate effects atlas consists of a number of stacks with geographically explicit information, and an extensive description of the interpretation and background of these maps. The first stack of maps is a set of climate change scenarios expressed by a range of meteorological variables for a range of KNMI'06 climate scenarios. For each scenario and variable, an observed time series at distributed meteorological stations is transformed consistent with the climate scenario and is spatially interpolated. A second stack of maps consists of effects of the anticipated climate change variables on groundwater, water quality, biodiversity, etc. Finally, a third stack contains regional planning projections, including developments in infrastructure, housing, agriculture, etc.

It is the combination of high-resolution maps that makes it possible to identify areas where spatial planning, climate change and its effects on ecology, water and infrastructure are facing a possible future conflict.

Plate 16 (centre pages) shows an example of a map of primary effects for Noord-Brabant (The Netherlands). Based on the KNMI'06 climate change scenarios, time series of temperature and precipitation were transformed in order to match the W+ scenario, and the resulting change in the summer precipitation is shown. Likewise, changes in the number of tropical days, wet days, extreme precipitation classes and related variables are contained in the atlas.

Plate 17 (centre pages) is an example of a secondary effect at this regional level: the expected increase in annual drought damage in the agricultural sector according to the G scenario. Changes in the temperature and precipitation are evaluated in terms of changes in agricultural yield using a crop model and are expressed as damage due to reduced yields.

This first version of the atlas is a preparation for a second edition, in which adaptation options for agriculture, nature, water management, tourism, traffic and other sectors are further detailed.

Conclusions and perspectives

The translation of climate change information to impact assessment involves many complex processes and is often non-linear, which means that a 10 per cent increase in precipitation is not automatically, for example, a 10 per cent increase in water levels. Therefore, the construction of regional climate change scenarios and the interpretation of their results need special attention. Currently, substantial effort is given to creating regional climate change information. Besides the careful analyses of embedded uncertainties, which arise from the choice of the modelling chain (GCM-RCM), the simulation domain and the horizontal resolution, aggregate calculations are performed under well-constrained conditions (see the ENSEMBLES website at www.ensembles-eu.org). These simulations will help to understand the robustness of the spatial and temporal pattern in the climate change signals, and provide the necessary amount of data for analysing the changes in a probabilistic sense.

Comparing the simulation results with observations is important in order to be able to judge the quality of the model results and will lead to further model development, an ongoing process to better simulate the Earth's climate. More complex climate models will include more physical and biogeochemical processes; but this will automatically have more degrees of freedom, which again contributes to uncertainty. Currently, large efforts are under way to understand and define the relevant degrees of uncertainty.

The largest uncertainty, however, lies in the possible future development of greenhouse gas and aerosol emissions in the atmosphere. Natural contributions from major volcanic eruptions, as well as anthropogenic emission rates, cannot be forecasted for the following centuries. Therefore, one must consider possible future climatic changes under different emission scenarios (e.g. IPCC SRES scenarios) for water management issues. Weak and strong emission scenarios will lead to small or substantial warming (associated with precipitation changes), referred to as so-called *climate change corridors*, describing the range in which possible climatic changes might happen. This means that, in any case, a small and a large change should be considered, and this might be sufficient for several questions related to water.

Tailoring is not just applying an advanced model chain; it is also about giving guidance on the interpretation of scenarios. Often this comes down to advising the evaluation of multiple scenarios within the range of sensitivity of the specific sector or region, and not only relying on just a single calculation (even when this calculation is expensive in terms of resources). It is dangerous to stick with a single scenario: it gives rise to selective warning, dependent upon the policy-maker's interests or background (e.g. a contrast between a strong doom scenario versus a 'nothing wrong' attitude).

Ideally, simulation results from a well-developed ensemble of regional climate change scenarios (taking into account different emission scenarios and several GCM-RCM chains) should directly be introduced into hydrological models. The associated ensemble of hydrological simulations is used to translate the changes in hydrometeorological quantities into changes in mean river discharge, but also in extremes such as flood frequency or low flow periods. Changes in probability density distributions can be analysed.

Another very important issue is the communication with stakeholders in individual case studies and sectors in order to gain experience in distributing climate change information in an understandable way, in enhancing the information exchange, and in bridging cultural and language gaps. This can be achieved through close contact between scenario developers, stakeholders and interested individuals, which should be included in the process of generating climate change information at a very early stage. As a first step, it is important to analyse how the specific action/sector might be affected by climate change (e.g. which hydro-meteorological quantities influence the operational management and investments). Then, the ideal set of climate change information will be identified and tailored to the needs of the client. Finally, the impact assessment will be carried out, including changes in mean quantities as well as in extremes; levels of robustness will be analysed and uncertainties determined.

References

Bakker, A. and J. Bessembinder (2007) 'Neerslagreeksen voon-de KNMI'06', *H_2O*, vol 22, pp45–47

Christensen, J. H. and O. B. Christensen (2007) 'A summary of the PRUDENCE model projections of changes in European climate by the end of this century', *Climatic Change*, PRUDENCE special issue, vol 81, supplement 1, pp7–30

Déqué, M., D. P. Rowell, D. Lüthi, F. Giorgi, J. H. Christensen, B. Rockel, D. Jacob, E. Kjellström, M. de Castro and B. van den Hurk (2007) 'An intercomparison of regional climate simulations for Europe: Assessing uncertainties in model projections', *Climatic Change*, vol 81

Hagemann, S. and D. Jacob (2007) 'Gradient in the climate change signal of European discharge predicted by a multi-model ensemble', *Climatic Change*, PRUDENCE special issue, vol 81, supplement 1, pp309–327

IPCC (2001) *Climate Change 2001: Contribution of Working Group I to the Third Assessment Report of the IPCC*, Cambridge University Press, Cambridge, UK

IPCC (2007) *Climate Change 2007: Synthesis Report, Fourth Assessment Report*, Cambridge University Press, Cambridge, UK

Jacob, D. (2001) 'A note to the simulation of the annual and inter-annual variability of the water budget over the Baltic Sea drainage basin', *Meteorology and Atmospheric Physics*, vol 77, pp61–73

Jacob, D. (2009) 'Regional climate models: Linking global climate change to local impacts', in *Springer Encyclopedia of Complexity and System Science*, in press

Jacob, D., L. Bärring, O. B. Christensen, J. H. Christensen, S. Hagemann, M. Hirschi, E. Kjellström, G. Lenderink, B. Rockel, C. Schär, S. I. Seneviratne, S. Somot, A. van Ulden and B. van den Hurk (2006) 'An inter-comparison of regional climate models for Europe: Design of the experiments and model performance', *Climatic Change*, PRUDENCE special issue, vol 81, supplement 1, pp31–52

Jacob D., H. Göttel, S. Kotlarski, P. Lorenz and K, Sieck (2008) *Klimaauswirkungen und Anpassung in Deutschland – Phase 1: Erstellung regionaler Klimaszenarien für Deutschland*, Abschlussbericht zum UFOPLAN-Vorhaben 204 41 13

Lambert, F. H., A. R. Stine, N. Y. Krakauer and J. C. H. Chiang (2008) 'How much will precipitation increase with global warming?', *EOS Newsletter*, vol 89, no 21

Nakicenovic, N., J. Alcamo, G. Davis, B. de Vries, J. Fenhann, S. Gaffin, K. Gregory, A. Grübler, T. Y. Jung, T. Kram, E. L. La Rovere, L. Michaelis, S. Mori, T. Morita, W. Pepper, H. Pitcher, L. Price, K. Raihi, A. Roehrl, H.-H. Rogner, A. Sankovski, M. Schlesinger, P. Shukla, S. Smith, R. Swart, S. van Rooijen, N, Victor and Z. Dadi (2000) *IPCC Special Report on Emissions Scenarios*, Cambridge University Press, Cambridge, UK, and New York, US

Schwarb, M., C. Daly, C. Frei and C. Schär (2001), 'Mean annual and seasonal precipitation in the European Alps 1971–1990', in *The Hydrological Atlas of Switzerland*, http://hydrant.unibe.ch/hades/hadeshome.htm

Van den Hurk, B., A. Klein Tank, G. Lenderink, A. van Ulden, G. J. van Oldenborgh, C. Katsman, H. van den Brink, F. Keller, J. Bessembinder, G. Burgers, G. Komen, W. Hazeleger and S. Drijfhout (2006) *KNMI Climate Change Scenarios 2006 for the Netherlands*, KNMI Scientific Report WR 2006-01, KNMI, De Bilt, The Netherlands

Van der Scheur, W., F. Keller, A. Bakker and T. Kroon (2006) 'Op zoek naar een klimaat representatief standaardjaar', *Toetsing van landelijke en regionale hydrologische kenmerken*, RWS-RIZA/KNMI, November

Vermulst, J. A. P. H., T. Kroon and W. J. de Lange (1998) 'Modelling the hydrology of the Netherlands on a nation wide scale', in H. Wheater and C. Kirby (eds) *Hydrology in a Changing Environment*, John Wiley & Sons Ltd, UK

4

The Impacts of Climate Change on Water

Fulco Ludwig and Marcus Moench

A large proportion of solar energy is used to drive the hydrological cycle. This energy is mainly used for evaporation and subsequent precipitation. Due to higher greenhouse gas concentrations, more energy is available at the Earth's surface, which intensifies the hydrological cycle (Kabat and van Schaik, 2003; IPCC, 2007). As discussed in the previous chapters, climate variability and change have a large impact upon precipitation patterns and changes in rainfall are expected for the future. Some regions will receive more rainfall, while subtropical regions, in particular, are likely to see less rain. This chapter reviews recent changes in water resources availability and extreme events; the most important impacts of climate change upon the water sector are then discussed.

Recent changes in the water cycle

Over the last few decades, water availability in rivers, lakes and groundwater has changed significantly. Some of these changes are due to a different climate; but other factors have also had a major impact upon water availability. Water demand and withdrawals have increased rapidly over the last decades due to population growth and economic development. As a result of increased water use, lake levels have dropped, with Lake Aral in Central Asia being the most dramatic example (Kabat and van Schaik, 2003). In some cases, changes in climate have contributed to dropping lake levels. For example, in Western Africa, the water level of Lake Chad declined due to both human activities and reduced rainfall.

In general, it is difficult to detect if changes in water availability are caused by climate change or whether they are due to other impacts. Historical surface water levels, preserved through stream-flow gauge records, show large decadal and multidecadal variations; it is therefore often difficult to detect the impact of climate change in these signals. However, some changes in surface water discharge can be clearly linked to climate change. For example, the timing of river flows in regions with winter snowfall has significantly changed (Barnet et al, 2005). Due to higher temperatures, snow melts earlier in the season, and during the winter more precipitation falls as rain instead of snow. The combination of earlier melt and higher winter precipitation leads to higher river discharge during (early) spring and less stream flow during the summer. Often, peak water demand occurs during the summer, so this change in timing of river flows can have large impacts upon water resource management, causing water shortages during the summer. Higher temperatures have also reduced snow cover and glacier shrinkage has been observed around the globe (Oerlemans, 2005). In Peru, for example, the area covered by glaciers has been reduced by 25 per cent in the last three decades (Barnet et al, 2005). In the Andes, the disappearance of glaciers can have serious consequences for water resources because most people living west of the Andes rely on glacier-supplied river water for their water resources (Mark and Seltzer, 2003). Both shorter snowfall seasons and shallower snow packs have been observed here during the last decades. Higher temperatures have also increased runoff in the Himalayas from melting glaciers, which has caused pro-glacial lakes to fill. When moraine walls fail, outbursts of glacial lakes and mudflows may occur. Reduction of permafrost leads to less soil stability, also enhancing mudflows, rock fall and avalanches.

In some cases, reduced rainfall has caused problematic reductions in stream flows. This is especially the case in (semi-)arid regions where small changes in rainfall can cause substantial changes in runoff. One of the best-documented cases is Western Australia, where lower rainfall since the 1970s has caused large reductions in stream flow, which has reduced water availability for the Perth metropolitan area.

As a result of climate change, extreme rainfall events are predicted to increase (IPCC, 2007). However, since extreme events are, by definition, rare – and in many regions there is large natural variability in the occurrence of extreme rainfall events – it is often difficult to find clear trends. On a global scale, however, it is clear that the frequency of extreme rainfall events has increased over the last decades and the number of intense hurricanes (categories 4 and 5) also seems to have increased over the last decades (Webster et al, 2005). Examples of regions where more extreme events have been observed are Southern Africa (Usman and Reason, 2004) and Northern Australia.

The higher number of extreme rainfall events has probably played a role in the recent increase of flood frequency. The total number of floods and economic losses related to floods has sharply increased during the last decades (Bates et al, 2008). However, it is still unclear what the role of climate change has been on the higher number of floods. The authors of the Intergovernmental Panel on Climate Change's (IPCC's) *Fourth Assessment Report* concluded that there is no categorical evidence that the trend is related to climate (IPCC, 2007). Increase in flood damage is also driven by socio-economic factors, such as concentrations of people and economic activities in vulnerable areas. Nowadays, many more people than ever live in large cities that are located along the coast or near major river systems, and similar floods used to be much less disastrous than they are today. In addition, land-use changes have contributed to the increased number of floods. Forest and bushland clearing increases runoff, which makes the chance of floods more likely in the case of intense rainfall events. With the removal of mangrove forests, natural protection against coastal floods has disappeared.

Globally, the intensity and duration of droughts have increased since the 1970s (IPCC, 2007). This is particularly the case in the tropic and subtropics. The increase in droughts is caused by a combination of diminishing rainfall and higher temperatures. The Sahel region, in particular, has suffered from more intense and longer droughts during the last 30 years. There are, however, some indications that rainfall has recovered in the Sahel since 1998 (Nicholson, 2005). Southern and Eastern Australia has become drier over the last decades and, since 2003, Eastern Australia has suffered from the worst drought on record (Smith, 2004). This drought has severely affected both dryland and irrigated agriculture. Many farmers have gone bankrupt and water available for agriculture has dropped dramatically. The drought has also affected industrial and domestic water supply. Almost all of the major cities in Australia have restrictions on domestic water supply and water companies are actively looking for new sources of water (see Chapter 13). Semi-arid regions in North America, such as the south-western US and parts of southern Canada, have seen an increase in the number of droughts over the last decades due a drop in rainfall. As a result of population and economic growth in many areas of North America, water demands have increased, which has made these regions much more vulnerable to droughts (IPCC, 2007).

Impacts of climate change and variability on water resource management

River flows

Perhaps the most important effect of climate change is the impact that it has on river discharge. Climate change affects total annual stream flow, as well as seasonal dynamics (e.g. due to changes in snowmelt period). In general, the impacts are relatively simple: higher rainfall will result in higher stream flow and reduced rainfall will decrease the stream flow. However the correlation between changes in stream flow and rainfall are very different in different climates. Especially in semi-arid regions, river flows are sensitive to changes in rainfall. Generally, in semi-arid regions only a small portion of the rainfall results in runoff and most rainfall will evaporate or infiltrate into the soil. Due to the very small difference between rainfall and evaporation in these dry regions, a small reduction in rainfall can cause rivers to dry up. For example, in Africa, in regions with an annual rainfall of less than 500mm, a 10 per cent reduction in precipitation causes a 50 per cent lower runoff (de Wit and Stankiewicz, 2006). Similarly, small increases in rainfall can already cause new areas to become floodplains. This is especially the case if increased rainfall results from more days with heavy rainfall, a likely future scenario for several regions around the world. The stream flow of many rivers in semi-arid regions is already very variable both within seasons and between different years. Climate change is likely to increase rainfall variability, which will result in higher stream-flow variability.

Since a large segment of society and industry depends upon rivers, the societal effects of changes in discharge can be enormous. Existing conflicts about water appropriation – such as dams, irrigation and wetland conservation – are likely to expand when water stress increases.

In colder climates, a major proportion of annual stream flow comes from snowmelt. In these regions the seasonal fluctuation of stream flow is likely to change, which can have significant impacts upon water resources. In the western US, for example, peak flows by 2050 are expected to be about one month earlier, significantly affecting hydropower potential: storage facilities will be too small to retain the water that arrives earlier in the season (Barnett et al, 2005). In the Rhine Basin, climate change will result in higher winter discharge due to intensified snowmelt and increased winter precipitation (Middelkoop et al, 2001). In the summer, discharge will lessen due to lower snowmelt and higher evapotranspiration. These changes will have a number of impacts upon water resources. To reduce future flood risks, the water retention capacity in the upstream areas and the discharge capacity of the river channels need to be increased, and there is also the need to improve flood warning systems (Middelkoop et al, 2001). Periods with low flow in summer will cause problems with navigation and the water supply for industry, agriculture and domestic use, as well as for the aquatic ecosystems (e.g. fish and wetlands) that depend upon river water. At the same time, summer water demands are likely to increase due to climate change as a result of higher temperatures (see Chapter 10).

Almost all glaciers around the world are shrinking, and significant parts of all glaciers are projected to melt in the coming century. For example, the glaciers of the Tibetan plateau are projected to decrease by 100,000km^2 by 2035 (IPCC, 2007). Half a billion people in India and 250,000 in China depend upon these glaciers for their water resources (Stern, 2007). The melting of glaciers initially results in increased river runoff, but will eventually cause lower stream flows when the ice has disappeared. So, in these cases, the initial hydrological response to climate change can give a false impression of the future. This initial increase of stream flow and a sudden drop later is predicted for the Himalayan region in particular (Barnet et al, 2005).

Groundwater

Climate change will affect the depth of groundwater tables and the amount of groundwater available through changes in recharge rates. Until now, there has been very little research on the impacts of climate change on groundwater, and it is still very uncertain how changes in climate will affect groundwater. Both changes in average rainfall amount and extremes will have an impact on groundwater recharge rates. In semi-arid regions, only heavy rainfall events result in groundwater recharge. For example in Yemen, recharge rates are very sensitive to changes in rainfall due to the non-linear relation between rainfall and recharge rates. Small changes in rainfall cause large changes in recharge rates especially if there are changes in extreme rainfall events (see Chapter 11).

While increased rainfall variability in semi-arid regions increases recharge rates, in humid regions a higher variation in precipitation could reduce recharge rates because during heavy rainfall events most water is lost through runoff (Bates et al, 2008). Higher temperatures will most likely increase evaporation, which can reduce both recharge and groundwater discharge rates.

Climate change will also increase the *use* of groundwater. If the availability of surface water shrinks, groundwater use usually increases. Where groundwater is already the dominant water resource, the amount of water used could increase due to higher demands as a result of high evaporation rates. For example, in India, 50 per cent of water used for irrigation comes from groundwater, and in some areas groundwater levels are dropping rapidly. If higher groundwater extraction rates are combined with reduced recharge rates, groundwater resources will become depleted relatively quickly.

Sea-level rise

As a result of higher atmospheric temperatures, oceans will become warmer. Due to higher water temperatures, oceans will also expand and sea levels will rise. Global warming will cause significant melting of glaciers, ice-caps and land ice, which will cause an additional rise of sea levels. Due to higher sea levels, there will be reduced protection from extreme storms and flood events because increased sea levels provide a higher base for storm surges. If these higher sea levels are combined with more frequent storms, floods will become more frequent and more severe as well. Sea-level rise potentially has a negative effect on many coastal ecosystems. The impact of sea-level rise on lagoons, mud flats and salt marshes also depends upon sediment transport. If sediment supply can keep up with sea-level rise, the impact can be minimal; but if sea-level rise exceeds the threshold and morphology cannot keep up, irreversible processes can accelerate the impacts of sea-level rise (Van Goor et al, 2003).

In most areas, high sea levels will increase coastal erosion. It is still unclear to what extent sandy shorelines will retreat. As a rough estimate, the model by Bruun (1962) is often used. This model indicates that shorelines will retreat 50 to 200 times the sea-level rise. However, much will depend upon local circumstances. For example, if nearby estuaries or mud flats act as a major sink for sediment to keep up with sea-level rise, coastal erosion could by much higher than calculated by the Bruun model (Van Goor et al, 2003). Higher sea levels will have a significant impact upon coastal zone management. Large parts of the global population live along the coast and several mega-cities, such as Tokyo, Mumbai, New York, Shanghai and Lagos, are located near oceans. Without adaptation, large parts of these cities could be inundated. There is also large-scale economic activity along the coast, which could be affected by sea-level rise. Large deltas, in particular, are sensitive to climate change due to sea-level rise and changes in rainfall and river stream flow. In many of these large deltas, the pressure on available land and the environment is already high and climate change is likely to worsen the situation. Due to rapid population growth, settlements have developed in areas that are much more vulnerable to flooding. These flooding risks are likely to increase due to climate change.

For example, the Nile Delta is highly vulnerable to climate change (Ludwig and Vellinga, 2008). A significant part of Egypt's population lives along the coast. Large cities such as Alexandria, Rosetta City and Port-Said will potentially be affected by sea-level rise as they will become more vulnerable to flooding. Large areas of the most fertile lands in Egypt are located in the coastal delta. Inundation and saltwater intrusion caused by sea-level rise will affect the agricultural activities in the delta. The lakes and wetlands of northern Egypt are responsible for a significant part of the country's fish production. Sea-level rise in combination with higher temperatures can significantly change the ecosystems of these lakes, with a likely reduction in fish production. Rising sea levels can also destroy weak parts of sand belts, which are necessary to protect lagoons and low-lying reclaimed lands. Erosion of the sand belts protecting the coast of Egypt has already increased over the last decades due to sea-level rise in combination with reduced sediment loading due to the construction of the Aswan High Dam. This dam has significantly reduced sediment transport by the lower Nile, which has affected coastal erosion.

Higher sea levels will also have an impact upon the water quality of freshwater aquifers and estuaries. Freshwater recourses in areas with a low elevation near the sea are vulnerable to salinization due to sea-level increase. This can have major impacts upon important drinking water resources and the quality of freshwater ecosystems. Many of these resources are already under threat due to high population pressure in most deltas.

Floods

Global warming will result in the acceleration of the water cycle, which will lead to a higher variability of precipitation. Extreme rainfall events are likely to occur more often and to become more extreme. As a result, floods are likely to become more frequent and potentially more severe in most regions around the world (Milly et al, 2002). Flood frequency and severity will especially increase in regions where more rainfall is predicted; but in areas where a reduction of mean annual rainfall is expected, extreme rainfall events and floods might still increase due to more extreme events. As explained in detail in the case study on Thailand (see Chapter 9), not all floods are necessarily negative, and they are often a part of the normal seasonal cycle. Floods can, for example, be essential for fisheries and agriculture. Floods often become a disaster when they are unusual in timing and/or severity.

Changes in snowmelt patterns and glacier melting can also increase flood risks. Due to the melting of glaciers, new lakes are created in mountainous areas. Accumulation of water in these lakes can cause a sudden discharge of large volumes of water and debris. These glacial lake outburst floods (GLOFs) can result in large-scale disasters, causing death and damage to agricultural systems and infrastructure. Especially in the Himalayas, GLOFs have already increased in number and are a major concern for the future.

Many large urban areas are located in regions that are vulnerable to flooding because they are either located along the coast or along major river systems. One of the problems in these urban regions is that the most vulnerable people with the least access to resources live in areas that are most prone to flooding. Often, slums in major cities in developing countries are built on floodplains where regulated development is prohibited. However, this is not only a problem in the developing world; in New Orleans, the poorer neighbourhoods were also affected the most by the floods caused by Hurricane Katrina in 2005. Due to the fact that many growing urban centres are in areas that are vulnerable to floods, damage caused by floods is predicted to increase rapidly if current flood management policies are not changed.

Droughts and arid zones

On a global level, the area affected by droughts is likely to increase due to a reduction in (summer) rainfall combined with increased evaporation (Sheffield, 2008). Especially in subtropical regions and the Mediterranean, reduced rainfall is predicted to lead to more frequent and more intense droughts. By the end of the 21st century, the land surface affected by extreme droughts at one time could increase by 10-fold to 40-fold (Bates et al, 2008). The regions most affected by increased droughts will be the Mediterranean, Central Asia, Central America, and Southern and West Africa (Sheffield, 2008). The southern parts of the US and Australia – already regularly affected by droughts – are likely to see more frequent and more severe droughts.

Increased droughts will cause significant problems for the water resources of large urban areas. For example, in Australia, almost all major urban centres will face reduced water availability and will therefore need to look for alternatives to satisfy domestic and industrial water demands. In Perth, large reductions in dam inflow are predicted and, as a result, alternative water resources are now being developed (see Chapter 13).

It is not only a reduction in rainfall that influences the number of droughts. Changes in seasonality and variability are also likely to cause an increase in drought frequency around the world. For example, in Northern Europe, no reduction in average rainfall is predicted by the global climate models; yet, summer droughts are likely to increase due to a higher variability in summer rainfall (less days with rainfall) and increased temperatures. These more frequent dry periods in summer, in combination with less runoff from snowmelt in summer, are expected to increase the frequency of low flows of all major river systems in Europe (Middelkoop et al, 2001). These low flows will affect navigation, ecosystems, water quality, and agricultural, industrial and domestic water supply.

All areas that depend upon snowmelt for summer river flows will see more frequent droughts. The combination of less rainfall, reduced stream flow and higher temperatures will put increased pressures on water resources in regions such as the north-western US. Higher temperatures and lower rainfall cause a higher water demand by agriculture, domestic (watering gardens and swimming pools) and industrial use (cooling water). As different pressures come together here, these impacts are likely to be already felt in the near future, and summer water stress and shortages are likely to increase rapidly in several parts of the world. This is especially the case in regions where water demands are increasing due to population growth and the expansion of economic activities – for example, in Perth and Cape Town. In both towns, climate projections indicate lower water availability, while at the same time demand is rapidly growing.

Water quality and health

Climate change is likely to have a negative impact upon water quality in many areas around the world. The most general impact upon water quality will be through higher temperatures of surface water. Algal blooms will occur more frequently in a warmer climate and higher temperatures will also increase microbial activity and bacterial and fungal populations. Especially in developing countries with a lack of proper sanitation, higher temperatures will increase the risks of water-borne diseases. Most diseases spread faster in a warmer climate, increasing the risk of severe outbreaks. In addition to the risks of more frequent and severe outbreaks, a warmer climate could also introduce new diseases into an area. Previously, it might have been too cold for a particular disease to flourish; but a warmer climate can open up areas for a disease. For example, the number of malaria cases is likely to increase in the East African highlands (Hay et al, 2002). This scenario could have severe impacts because the disease might not be recognized by local medical staff and/or medicines may not be available locally because they were not previously needed.

In addition to higher surface temperatures, changed rainfall patterns may also have an impact upon water quality. Higher precipitation intensity results in higher peak runoff. This can cause higher nutrient and pathogen loads into streams, resulting in lower water quality. Mobilization of absorbed nutrients (especially phosphorus) and pollutants can increase due to erosion. Higher runoff is also likely to increase fertilizer and pesticide concentration in streams in areas with high agricultural intensity. Higher stream flows will also result in the dilution of nutrients and pathogens, which improves water quality. More intense rainfall can also result in pollutants reaching groundwater layers sooner. More extreme rainfall events put more pressure on existing sewerage systems. Overflows are likely to occur more often, which will pollute water systems. Overflowing sewerage systems will also increase the risks of spreading diseases, especially in combination with a warmer climate.

Reduced rainfall can also have a negative impact upon water quality because existing pollutions are less diluted, resulting in higher concentrations. In (semi-)arid regions, lower stream flow can increase salinity. For example, in Australia, salinity in the headland water of the Murray-Darling Basin is expected to increase by 13 to 19 per cent, up to 2050 (Pittock, 2003). Human activities in response to a drier and warmer climate will probably increase existing salinity problems (Williams, 2001). For instance, increased water use for irrigation and industry is likely to enhance salinization. Sea-level rise is also likely to increase saltwater intrusion in coastal regions. This will have negative impacts upon agriculture in coastal areas. Yields of salt-intolerant crops are likely to reduce in coastal areas affected by saltwater intrusion, and groundwater use could become more restricted. The water quality of rivers in temperate regions may deteriorate due to climate change as a result of lower summer stream flows. For example, in the Meuse River, in Western Europe, water quality was much lower during droughts with respect to water temperature, eutrophication, major elements and some heavy metals. This decline in water quality was caused by favourable conditions for the development of algae blooms and a lower dilution capacity of point-source effluents (Van Vliet and Zwolsman, 2008).

Environment and natural ecosystems

Climate will have a significant impact upon (semi-)natural ecosystems. Altered precipitation regimes will cause changes in ecosystem water availability. The impact of these changes upon different species is often non-linear and therefore difficult to predict (Bates et al, 2008). Although all ecosystems are threatened by climate changes, aquatic ecosystems have one of the highest numbers of threatened species. Many wetlands are biodiversity hot spots where relatively small changes in rainfall will disturb wetland hydrology, resulting in high extinction rates (Millennium Ecosystem Assessment, 2005).

Coastal wetlands and estuaries are expected to be significantly affected by climate change due to altered river discharge rates and higher sea levels. If river discharges are reduced, the salinity of coastal systems is especially likely to increase. This will have a major impact upon freshwater species and will reduce the amount of water suitable for human consumption.

Human pressures on wetland ecosystems through pollution and excessive water use are already very high and climate change is likely to increase these pressures. Recent developments in relation to defining environmental flows and seeing natural ecosystems as one of the water users in a basin has given a positive boost to protecting natural wetlands and other ecosystems. However, in a changing climate, it is still likely that water levels will be affected first in (semi-)natural systems. Water for irrigation and domestic supply is likely to have a higher priority than water upon which natural ecosystems depend. Changes in seasonality, as well as total water availability, will both have a significant impact upon ecosystems. In many regions around the globe, stream flow and water availability will be reduced during summer months. Many species are not adapted to summer droughts and will disappear. Low streamflow events usually result in much higher pollution concentrations with large negative impacts upon aquatic fauna and flora.

Potential institutional impacts of climate change

Any attempt to map the potential institutional impacts associated with climate change is inherently speculative given the great scientific uncertainties in projecting the nature of changes that will occur in physical systems. Nevertheless, any departure from historical patterns introduces uncertainty and increases the risk that human activities (all of which are based on past experience) will not match future conditions. As a result, there will be consequences for the institutions responsible for dealing with risk and uncertainty. Furthermore, the process of change itself (whether climatic or in other systems) has inherent institutional implications. When changes occur in underlying systems, institutional systems must be able to respond.

These core conceptual points – recognition of changes in uncertainty and risk, combined with the need for flexibility – represent the springboard for analysing the potential institutional impacts of climate change. Some likely impacts may be identified on the basis of clear and tangible relationships. Insurance companies, for example, depend heavily upon the ability to predict probabilities of loss. The ability to calculate insurance risk will reduce as a result of climate change, which will affect the role that insurance institutions play in society. Climate change is also likely to affect other institutional systems that have been developed for managing existing climatic variability and related natural resources. Water rights, management and utility institutions are, for example, highly likely to be affected by climate change.

Insurance

Insurance, the world's largest industry, represents a major part of the institutional landscape for risk management, particularly in industrialized and wealthy regions. Insurance in developing and transition economies is, however, growing rapidly and even now payouts for weather-related events are triple the level of official development aid (Mills, 2005).

While it is impossible to summarize the rapidly growing literature on insurance and climate change here, emerging changes can be said to erode the technical and market insurability of many risks (Mills et al, 2005). As the report by Mills at al points out, climate change is likely to alter the frequency and spatial distribution of events, the variability of losses, and the scaling and covariance of risk vectors (impacts that increase exponentially with, for example, wind speed and the occurrence of single events with multiple consequences). It is also likely to introduce market risks such as correlation between the asset and liability side of insurers' balance sheets, changes in claim patterns and changes in regulatory environments (Mills et al, 2005). In addition, two fundamental impacts upon insurance systems can be expected.

First, insurance depends upon the ability to project the probability and economic impacts associated with events. As the difficulty in projecting event frequencies, magnitudes and economic impacts increases, the viability of the business model underlying insurance systems erodes. In most cases, expectations regarding the frequency and magnitude of extreme weather events are based on historical data. Changes in climate make the historical approach inappropriate, while existing modelling capabilities are unable to produce the types of probabilistic information required to project future risks for specific locations in a quantitative manner. This affects many of the basic tools that the insurance industry uses on a day-to-day basis to communicate risks and set rates. Flood zone maps, for example, are the core tool used to quantify specific flooding risks for specific sites. Climate change alters the assumptions, data and modelling techniques required to develop such maps and, ultimately, their utility as a cornerstone of the industry.

Second, scale and covariance are increasingly major issues (Linnerooth-Bayer et al, 2005; Mills, 2005). When impacts associated with, for example, frequent coastal storms occur in well-defined areas, local insurance systems can be overwhelmed because all insured parties are affected at one time. As a result, multiple simultaneous claims on the insurance pool can easily exceed available resources. To be effective, insurance institutions require a large pool of participants, all of whom are unlikely to be affected simultaneously. If climate change leads to simultaneous impacts across large areas, such as occurred in the Gulf Coast region of the US with Hurricane Katrina, risks are difficult to spread.

Overall, climate change processes are almost certain to have a major impact upon insurability and, thus, upon the role that insurance institutions play in risk-spreading and risk-pooling in society. This is, in fact, already occurring. Insurance companies in the US are limiting the types of coverage that they will provide, limiting the level of coverage and, in some cases, withdrawing completely from high-risk regions (Mills et al, 2005). As a recent report by Munich Re Group (2006) states:

> We are gearing our risk management more than ever to the enormous loss potentials and the changing risk situation. Our products and services are urgently sought after and we will take advantage of this opportunity, but will only accept business at risk-adequate prices and conditions.

A similar report by Lloyds of London (2006), noting the uncertain impacts associated with climate change, states that:

> While some will use this uncertainty as an excuse for inaction, prudent risk managers will take the opposite view. If uncertainty has increased, so has risk, and we must seek to manage it.

Lloyds goes on to emphasize the growing need to 'price risk according to exposure' and to 'underwrite for profit' (Lloyds, 2006). This suggests that the ability of people in vulnerable circumstances to obtain insurance for climate risks from the private sector is likely to decrease. Governments – the insurer of last resort – and individuals will, as a result, increasingly carry the risks associated with climate change.

Water management institutions

Water rights, whether formal or informal, represent the foundation of most water management systems. Most of these institutions are already under stress due to recent increases in water demands from agriculture, industry and domestic use. Current shortages in water supply and mismatches between supply and demand are almost certain to be exacerbated by climate change, particularly if water availability is reduced where flow regimes are changing.

'If you have a water problem, pour water on it and it will go away.' This, the so-called 'California solution' to water problems in the western US, hides a wider reality. In most parts of the world, water rights systems have gradually evolved in what might be called a layered manner – with traditional and individual use practices (which in some cases have achieved the formal status of legal rights) overlain by layers of administrative procedure and regional or basin-level allocation mechanisms. In virtually all cases, such rights systems contain inherent contradictions. Demands on water resources, such as the need to maintain in-stream flows to meet environmental needs as well as the fundamental shifts in economic structures away from agriculture and into other uses, have emerged that were not envisioned at the time that rights systems were developed. Furthermore, rights to water have often been over-allocated. Administrative decisions to apportion stream flows in basins between upper and lower riparian states, for example, may not consider the existence of numerous well owners whose 'right' and ability to extract water (often supported by a separate set of legal and administrative traditions from surface water rights systems) can have a large aggregate impact upon surface flows. This is the case in, for example, the Gangetic Basin, where low-season flows to Bangladesh have been a long-term bone of contention and groundwater extraction in the upper basin is growing explosively. Such contradictions have little practical impact when water availability substantially exceeds demand. As basins approach closure and all water available is allocated, contradictions emerge as points of, often furious, contestation. The easiest solution – and the one that most regions follow when it is technically possible to do so – is to muddle through by adding water or storing it within the system. Where climatic changes reduce overall water availability (the projection for most semi-arid regions), where they reduce dry-season flows (a common projection for snow-fed rivers) or where they increase flow variability, existing contradictions and stresses on water rights systems are likely to come to a head. It will be far harder to 'solve' water problems by 'pouring' more water on them. As a result, changes in climate are likely to have fundamental impacts upon water rights institutions. These impacts are, in turn, likely to ripple upward into virtually all water management institutions. Rights, whether codified or not, describe what uses are recognized as socially legitimate. As a result, they are the starting point that underlies the services that the physical infrastructure is designed to provide and the higher-level institutions that operate such infrastructure. As a result, climate change will affect the foundation upon which virtually all higher-level water management institutions has been constructed.

A detailed review of water rights systems would be a full topic in itself. It is important to note, however, that virtually all such systems contain elements of both flexibility and rigidity that, respectively, are likely to reduce and increase the impacts of climate change.

In most parts of the world, systems for allocating available water supplies have evolved in ways that are explicitly designed to accommodate at least limited amounts of variability. Water rights systems often contain mechanisms for priority based on proportional changes in allocation that rely on water availability. In India, for example, domestic use rights have the highest priority, followed by agriculture and, finally, industrial and commercial uses. In the western US, in contrast, water is allocated based on the prior appropriation doctrine. This essentially states that the 'first in time' is the 'first in right'. As available supplies decline, 'junior' appropriators are cut off to ensure that the rights of more senior appropriators can be met. Rights systems often contain provisions restricting rights to 'beneficial uses' and ensuring that impacts on other right holders are minimized.

Although designed with elements of flexibility that reflect the natural variability of water resource systems, in practice, many water rights systems have high levels of rigidity. In general, the more rigid the systems are, the more they will be affected by climate change. Water transfers between users and uses are often subject to a wide variety of restrictions and conditions. In the US, for example, transfers are often limited to estimates of long-term consumptive use and are also constrained by large 'unknowns' that range from environmental impacts to the unquantified rights of Native Americans or other users. Data availability and a host of other technical issues also combine to reduce the flexibility of rights systems. Beyond this, however, lie political questions and larger processes of social change that, in combination, have tightened water supply availability and increased conflicts over water in many parts of the world. In virtually all parts of the world, agricultural uses dominate. Growing urban populations and industrial uses are, however, creating major strains on institutions for water allocation. These 'centres of political gravity' generally have sufficient social weight to bend water allocation systems in their favour. Their demands are relatively inflexible and, in some cases, are supported by formal water allocation frameworks that match this inflexibility. The *Colorado River Compact*, for example, allocates specific volumes of flow to different states in the basin. These volumes were allocated based on a period of record that is now known to exceed longer-term water availability in the basin. Furthermore, the compact contains no mechanism for reducing allocations in the case of drought or other factors that alter the effective availability of water. As one commentator states:

> Water shortages were not on the minds of compact negotiators; in fact, they seemed to believe that surpluses were more likely. As a result, the compact does not include provisions to deal with shortages due to drought (Gelt, undated).

These water allocation systems will be severely stressed if, due to climate change, water availability is diminished and/or the number of droughts increases.

Successful adaptation to climate change will require high levels of flexibility. As water availability and the dynamics of water resource systems change, patterns of use and allocation will need to alter as well. The inherent contradictions and points of rigidity in water rights systems are, as a result, highly likely to represent major constraints on the ability of regions to adapt successfully to climate change. Furthermore, the impact of changing levels of water availability and flow patterns will contribute to pressure on already stressed water institutions. This may force fundamental changes in such institutions.

Higher-level impacts on agriculture and trade systems

In addition to the above location-specific types of institutional impacts likely to be associated with climate change, higher-level institutional impacts are also probable. If changes in climate and climatic variability increase uncertainty and risk within agricultural systems at either local or regional levels, this is likely to have a major impact upon national and global systems for agricultural production and trade. At a national level, a variety of the institutions that currently support agriculture in most countries will be affected. As with other forms of insurance, the economic viability of crop insurance programmes may, for example, be affected. On a more fundamental level, however, increases in uncertainty, variability and extreme events are likely to result in localized impacts upon agricultural production. This will necessitate increased reliance on regional and global trading systems for maintaining both local economic activity and meeting local food requirements. This may occur gradually as regions gain and lose comparative advantages. In this case, the shifts in global systems for trade are also likely to be gradual. This could enhance existing patterns of trade in 'virtual water' (i.e. grain) such as those that support many Middle Eastern countries. If, however, major regional climatic events (droughts or extreme storms) suddenly affect agricultural production in key regions, the impact upon global institutions could be major. Failure of the monsoon in India or a major storm in China would, for example, affect regional food production and the resulting demand for grain could exceed the capacity of global markets to meet deficits. The probability of this type of sudden impact upon global trading systems has not, that we are aware of, been evaluated in relation to climate change scenarios. If it occurred, the institutional consequences would be major.

References

Barnett, T. P., J. C. Adam and D. P. Lettenmaier (2005) 'Potential impacts of a warming climate on water availability in snow-dominated regions', *Nature*, vol 438, pp303–309

Bates, B. C., Z. W. Kundzewicz, S. Wu and J. P. Palutikof (eds) (2008) *Climate Change and Water*, Technical Paper of the Intergovernmental Panel on Climate Change, IPCC Secretariat, Geneva

Bruun, P. (1962) 'Sea level rise as a cause of shore erosion', *Journal of Water Ways and Harbour Division*, vol 88, pp117–130

de Wit, M. and J. Stankiewicz (2006) 'Changes in surface water supply across Africa with predicted climate change', *Science*, vol 311, pp1917–1921

Gelt, J. (undated) *Sharing Colorado River Water: History, Public Policy and the Colorado River Compact*, http://ag.arizona.edu/AZWATER/arroyo/101comm.html

Hay, S. I., J. Cox, D. J. Rogers, S. E. Randolph, D. I. Stern, G. D. Shanks, M. F. Myers and R. W. Snow (2002) 'Climate change and the resurgence of malaria in the East African highlands', *Nature*, vol 415, pp905–909

IPCC (2007) *Climate Change 2007: Synthesis Report, Fourth Assessment Report*, Cambridge University Press, Cambridge

Kabat, P. and H. Van Schaik (2003) *Climate Changes the Water Rules: How Water Managers Can Cope with Today's Climate Variability and Tomorrow's Climate Change*, Dialogue on Water and Climate, The Netherlands

Linnerooth-Bayer, J., R. Mechler et al (2005) 'Refocusing disaster aid', *Science*, vol 309, pp1044–1046

Lloyds (2006) *Adapt or Bust: 360 Risk Project*, Lloyds, London, p24

Ludwig, F. and P. Vellinga (2008) *Impacts of Climate Change on Water Resource Management in Egypt and The Netherlands*, Alterra, Wageningen UR

Mark, B.G. and G. O. Seltzer (2003) 'Tropical glacier meltwater contribution to stream discharge: Case study in the Cordillera Blanca, Peru', *Journal of Glaciology*, vol 49, pp271–281

Middelkoop, H., K. Daamen, D. Gellens, W. Grabs, J. C. J. Kwadijk, H. Lang, B. Parmet, B. Schadler, J. Schulla and K. Wilke (2001) 'Impact of climate change on hydrological regimes and water resources management in the Rhine Basin', *Climatic Change*, vol 49, pp105–128

Millennium Ecosystem Assessment (2005) *Ecosystems and Human Well-Being – Vol 1: Current State and Trends*, Island Press, Washington, DC

Mills, E. (2005) 'Insurance in a climate of change', *Science*, vol 309, no 5737, pp1040–1044

Mills, E., R. J. Roth et al (2005) *Availability and Affordability of Insurance Under Climate Change: A Growing Challenge for the US*, Ceres, Inc, Investor Network on Climate Risk, Boston

Milly, P. C. D., R. T. Wetherald, K. A. Dunne and T. L. Delworth (2002) 'Increasing risk of great floods in a changing climate', *Nature*, vol 415, pp514–517

Munich Re Group (2006) *Hurricanes – More Intense, More Frequent, More Expensive Insurance in a Time of Changing Risks*, Munich Re and American Re, Munich, Germany

Nicholson, S. (2005) 'On the question of the "recovery" of the rains in the West African Sahel', *Journal of Arid Environments*, vol 63, pp615–641

Oerlemans, J. (2005) 'Extracting a climate signal from 169 glacier records', *Science*, vol 308, pp675–677

Pittock, B. (2003) *Climate Change: An Australian Guide to the Science and Potential Impacts*, Australian Greenhouse Office, Canberra, Australia

Sheffield, J. (2008) *Global Drought in the 20th and 21st Centuries*, PhD thesis, Wageningen University, Wageningen, The Netherlands

Smith, I. (2004) 'An assessment of recent trends in Australian rainfall', *Australian Meteorological Magazine*, vol 53, pp163–173

Stern, N. (2007) *The Economics of Climate Change: The Stern Review*, Cambridge University Press, Cambridge

Usman, M. T. and Reason, C. J. C. (2004) 'Dry spell frequencies and their variability over southern Africa', *Climate Research*, vol 26, pp199–211

Van Goor, M. A., T. J. Zitman, Z. B. Wang and M. J. F. Stive (2003) 'Impact of sea-level rise on the morphological equilibrium state of tidal inlets', *Marine Geology*, vol 202, pp211–227

Van Vliet, M. T. H. and J. J. G. Zwolsman (2008) 'Impact of summer droughts on the water quality of the Meuse River', *Journal of Hydrology*, vol 353, pp1–17

Webster P. J., G. J. Holland, J. A. Curry and H. R. Chang (2005) 'Changes in tropical cyclone number, duration, and intensity in a warming environment', *Science*, vol 309, pp1844–1846

Williams, W. D. (2001) 'Salinization: Unplumbed salt in a parched landscape', *Water Science and Technology*, vol 43, pp85–91

5

Managing Water under Current Climate Variability

Eelco van Beek

Throughout history, people have tried to cope with the variability of their climate. First this was done simply by living in areas in which this variability caused no or few problems (e.g. sufficiently far away from floodplains, or close to springs, lakes or oases with a reliable supply of water). Moreover, people developed sustenance and economic activities that matched their natural conditions – for example, growing crops that were suited to the specific climate in which they were living.

Population pressure and external forces have made people move to areas that are more prone to climate variability. The fertility of floodplains and vicinity of trade routes attracted people to rivers, accepting the risk of occasional floods. To reduce that risk, measures were taken: flood defences and drainage systems were built to prevent flooding, and irrigation systems and drinking water supply systems coped with occasional shortages. Societies and political systems were organized around the need to control, regulate and distribute water for irrigation and food production.

In fact, water management is all about managing climate variability. Climate change and increased climate variability will only transform boundary conditions for water managers. Many water managers consider that such changing boundary conditions will not dramatically influence their basic approach. Others disagree (see the section on 'Changing climate and changing climate variability: Business-as-usual?'). What is clear is that water managers face many problems, of which climate change is only one. Population and economic growth, resulting in an increased demand for water and more pollution, changes in lifestyle and changes in the appreciation of people for nature and ecology are, in some areas, more challenging issues for water managers than changes in climate. It is expected that water managers will have to take changes in climate and climate variability into account in relation to other developments (high population pressure, limited space, etc.): the easy solutions are not available any more. Moreover, the increased safety and reliability of supply that result from good water management will stimulate further development of socio-economic activities in the area. The ultimate result will be that the risk involved will remain at the same level or might even go up, as people tend to accept more risk if the stakes are sufficiently high. They will keep making trade-offs between the socio-economic gain that they get from living and working in a certain area and the risks involved.

This chapter is about how water managers are currently dealing with climate variability, making a distinction in the design of water infrastructure (see the following section) and in the operation of the system (see 'Current/historical practices in *managing* climate variability', page 61). Increased complexity and natural resources limitations have led, during the last decades of the previous century, to the introduction of new approaches, particularly integrated water resources management. These are described in the section on 'Integrated water resources management (IWRM)'. Finally, the last section deals with claims that climate change will require that we make fundamental changes in how we manage our water.

In this chapter, only climate variability and how water managers can deal with this variability is discussed. This implies that in cases of drought, water managers will take care that the water resources system is able to cope with dry years: years in which the available supply is less than average. Scarcity situations in which the demand for water is structurally higher than the average available water will not be discussed. Such scarcity has no direct relation to climate, but is simply the result of an unsustainable combination of socio-economic activities and natural resource assets in a region, particularly as a result of rapid population increase. Climate change can make things worse, but is not the prime responsible cause of scarcity. It might be that such an unsustainable situation occurred somewhat earlier.

As far as specific management measures are concerned, as a general rule, reservoirs provide the most robust, resilient and reliable mechanism for managing water under a variety of conditions and uncertainties. They regulate flood waves and store water for use during dry periods. Non-structural measures (e.g. demand management, agricultural conservation practices, pricing, regulation and relocation) may provide important contributions to water safety and water services in terms of gross quantities of water supply, but not necessarily in terms of system reliability. The choice of packages or portfolio of measures depends upon the degree of social risk tolerance, as well as the complexity of the problem. The permutations for coping with the uncertainties of climate change and variability are limitless – both in the number of strategies and in the combinations of management measures that comprise a strategy. There is no single 'best' strategy. Each depends upon a variety of factors (e.g. economic efficiency, risk reduction, robustness, resiliency or reliability). Moreover, environmental and ecological aspects should be taken into account that, in general, require existing dynamics (variability) to remain the same as much as possible.

Current practices of using climate data for the *design* of water infrastructure

Water management addresses the 'too much' (floods), 'too little' (droughts) and 'too dirty' aspects of water. Too dirty is somewhat less directly related to climate variability and will not be addressed here. It is recognized that indirect relations exist where quality depends upon quantity (flushing, sewage systems, etc.) and is influenced by variability in temperature. However, this section will focus on floods and droughts.

Extreme value analysis

For designing infrastructure for floods and droughts, extreme value analysis is the first and most basic tool to incorporate variability, preferably based on many years of rainfall and river discharge data. The preferred length of the time series is at least 40 years as such a range of years will, in general, contain sufficient dry and wet periods.

Two general approaches are available in extreme value analysis for modelling discharge and precipitation series. The annual maximum series considers only the largest event in each year. The partial duration series (PDS) or peaks-over-threshold (POT) approach includes all 'independent' peaks above a truncation or threshold level. Figure 5.1 illustrates which data points are taken into account in the annual maximum approach (only P1 until P4), while the PDS approach also includes also P1', P3' and P4'. An objection to using annual maximum series is that it employs only the largest event in each year, regardless of whether the second largest event in a year exceeds the largest events of other years. Moreover, the largest annual flood flow in a dry year in some arid or semi-arid regions may be zero, or so small that calling them floods is misleading. When considering rainfall series or pollutant discharge events, one may be interested in modelling all events that occur within a year that exceed some threshold of interest.

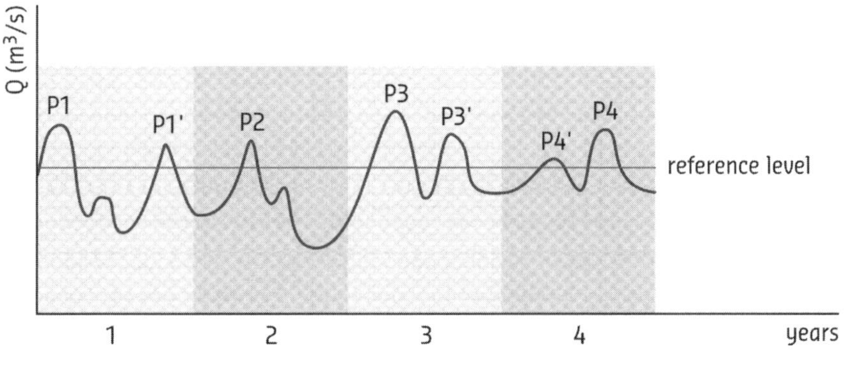

Figure 5.1 *Time series of river discharges*

Source: Eelco van Beek

Use of a partial duration series framework avoids such problems by considering all independent peaks that exceed a specified threshold. Furthermore, one can estimate annual probabilities of exceeding thresholds from the analysis of partial duration series. Arguments in favour of partial duration series are that relatively long and reliable records are often available, and if the arrival rate for peaks over the threshold is large enough, partial duration series analyses should yield more accurate estimates of extreme quintiles than the corresponding annual-maximum frequency analyses.

A drawback of partial duration series analyses is that one must have criteria to identify only independent peaks (and not multiple peaks corresponding to the same event). Thus, such analysis can be more complicated than analyses using only annual maxima. Partial duration models, perhaps with parameters that vary by season, are often used to estimate expected damages from hydrologic events when more than one damage-causing event can occur in a season or within a year.

The peak discharge Q_T exceeded once in average T years ('return period') is called the 'T-years discharge'. The probability of extreme values is called the 'extreme value distribution'. It can be described in different ways. The Gumbel type I for maxima and the Weibull type III for minima are well-known distributions. Other used distributions are log-Gumbel, Pearson and log-Pearson type III distributions. Gumbel type I supposes independent observations of extreme values $X_1, X_2, X_3, \ldots X_n$ (for successive year maxima) to be exponentially distributed. The probability $P' = \exp(-\exp(-y))$ and the reverse $y = -\ln(-\ln(P'))$. The complementary probability $P = 1 - P'$ discharge Q will exceed an observation ($Q > X$) is 1/T and the reverse $P' = 1 - P = 1 - 1/T$. When we arrange the measurements from maximum $m = 1$ until minimum $m = N$ (the number of years available), then the return period $T = (N + 1)/m$ and $P = m/(N + 1)$. If the observations are plotted on a logarithmic scale, the Gumbel I distribution will become a straight line. Figure 5.2 gives an example of such a graph for the Rhine River in The Netherlands.

Various techniques are available for an extreme value analysis. In general, stochastic hydrology considers the chronological sequence of hydrological events (the time series) with the aims of attempting to explain the irregularities of occurrence and, in particular, of forecasting the incidence of outstanding extremes such as floods and droughts. Hydrological and water resources textbooks describe the various techniques available (e.g. Ward, 1967; Maidment, 1993; and Loucks and van Beek, 2005). In cases where insufficient records are available, the statistical procedure can be reversed to generate synthetic time series that are long enough and contain sufficient extremes for the design and management of the water resources system (see, for example, Loucks and van Beek, 2005, Chapter 7). To analyse droughts weekly or even monthly, time steps are sufficient. For floods, daily data is frequently needed.

Designing flood-related infrastructure

The first task ever of 'water managers' was most probably to protect people from floods. Fertile floodplains and deltas are attractive areas in which to live and all four acknowledged 'cradles of civilization' (Yellow River, Euphrates/Tigris, Indus and Nile) are located along rivers and in deltas. The first dikes along the Yellow River were already built in the 21st Century BC.

The oldest design philosophy to determine dike heights is to apply a safety margin on top of the highest water level ever recorded. This was the approach that was used when data and statistical analysis techniques were not yet available. But even today this approach still has its merits: it is simple and easy to communicate to the public, and peak water levels are easy to measure compared to discharge volumes.

A statistical analysis starts with defining the safety level that one would like to achieve. Determining the safety level is often a political decision and is strongly influenced by events. Figure 5.3 illustrates this for The Netherlands. After the major floods of 1953, this safety level was put at 1 every 10,000 years. The costs to achieve this level along the rivers appeared to be very high, which resulted in a more differentiated approach in which the safety level along the rivers was put at 1 every 1250 years. The safety level for the densely populated areas in the west of The Netherlands against flooding from the sea remained at 1 every 10,000 years.

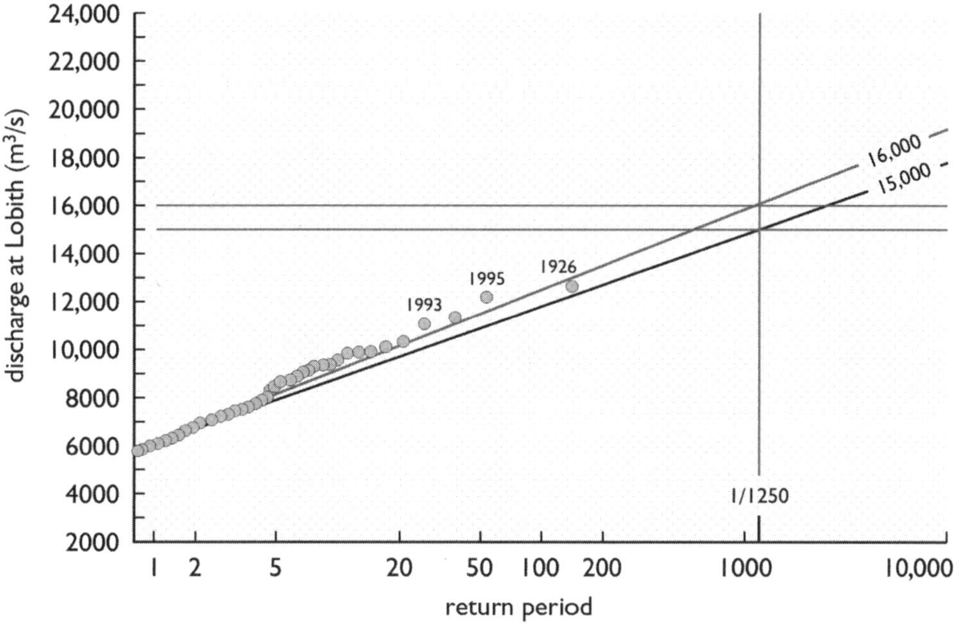

Figure 5.2 *Design discharge for the Rhine River*

Source: Kwadijk et al, 2001

The next step for the design of flood-related infrastructure along rivers is the determination of a design flood (in m³/s) and the related wave form (peak, flat and duration). The design flood is based on the defined flood probability. The 1 every 1250 year safety level for the Rhine River translated until 1993 in a discharge of 15,000m³/s, as illustrated in Figure 5.3. The floods of 1993 and 1995 changed the statistical properties of the data and the design discharge was consequently increased to 16,000m³/s. Safety implies more than just water levels; the stability of dikes should also be considered.

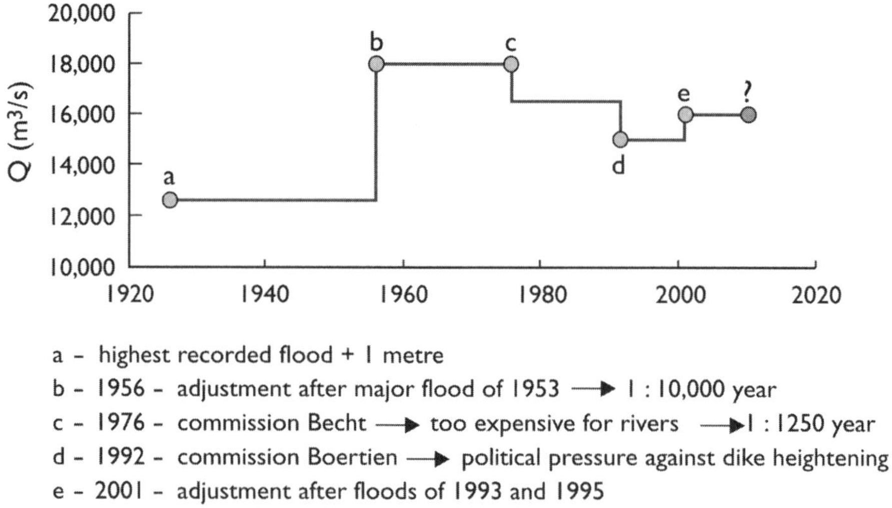

Figure 5.3 *Political adjustments for the safety level of the Rhine River*

Source: Silva et al, 2001

Such a pure statistical approach does not take into account the socio-economic value of the specific area that will be protected by water management measures. In an indirect way, this can be done by differentiating the safety level by applying a higher safety level for densely populated areas and a lower safety level for less populated areas. A more objective approach is to carry out a full flood risk analysis. Flood risk is a function of probability, exposure and vulnerability. Gouldby and Samuels (2005) define flood risk management as a continuous and holistic societal analysis, assessment and reduction of flood risk. Flood risk analysis considers the source, pathway, receptor and consequences involved, as illustrated in Figure 5.4. A flood risk approach is basically a cost-benefit analysis, although quantification of the full benefits remains rather difficult as many benefit elements (reduction in loss of life, disruption of social structure, etc.) are not easy to express in monetary terms.

Once the design discharge is determined, the water managers can design the required flood measures to protect an area. For this flood, engineers can choose between many possible infrastructural measures. These measures can be classified in three main groups:

1 Increase the discharge capacity of the river or drainage system.
2 Protect an area by flood defence structures.
3 Reduce peak flows by retention upstream.

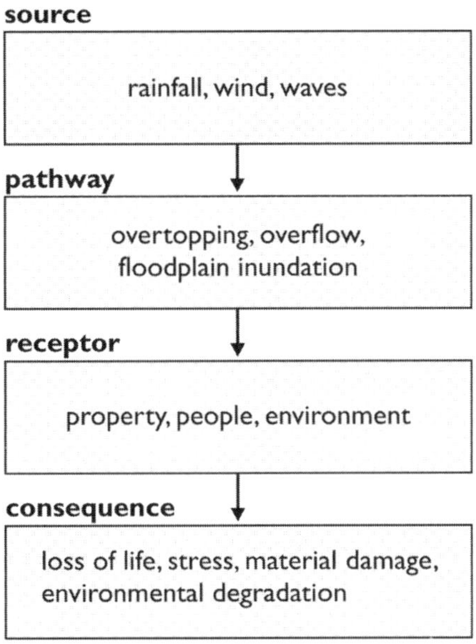

Figure 5.4 *Source, pathway, receptor and consequent steps involved in flood risk management*

Source: Gouldby and Samuels (2005)

In many cases, a combination of these kinds of measures is used. Figure 5.5 provides an overview of the measures that are considered in the Room for the River project in The Netherlands.

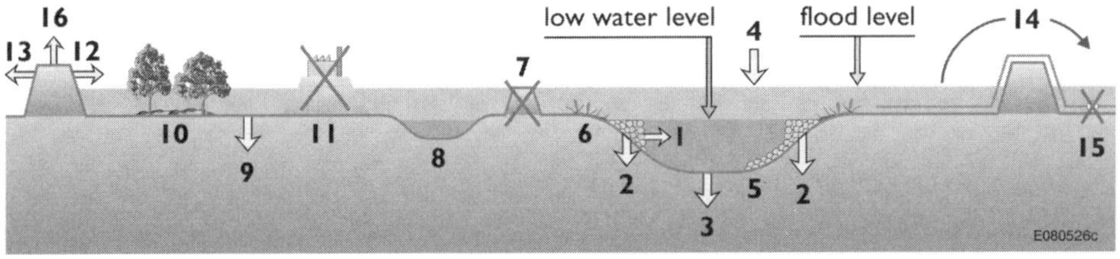

1 - narrowing of main channel
2 - lowering of groyne
3 - dredging
4 - dumping of sediment
5 - permanent layer
6 - natural bank
7 - removing of summer embankment
8 - secondary channel
9 - lowering of flood plain
10 - nature development
11 - removing of high-water free areas
12 - dike reinforcement
13 - dike repositioning
14 - retention
15 - obstruction of lateral flow
16 - dike raising

Figure 5.5 *Measures considered for the Room for the River project in The Netherlands*

Source: Silva et al, 2001

Dikes are most probably the most common flood defence structures. The design defence level of a dike is based on the design discharge and corresponding water level. On top of this a freeboard is inserted, as indicated in Figure 5.6. That freeboard includes three parts: a part to compensate for uncertainties in the statistical hydrological analysis (30cm); a part for wave run-up and engineering uncertainties (= calculated wave run-up with a minimum of 50cm); and the estimated subsidence over the lifetime of the dike. The engineering safety margin accounts for all unrecognized ignorance, as well as all kinds of uncertainties involved in the model calculations.

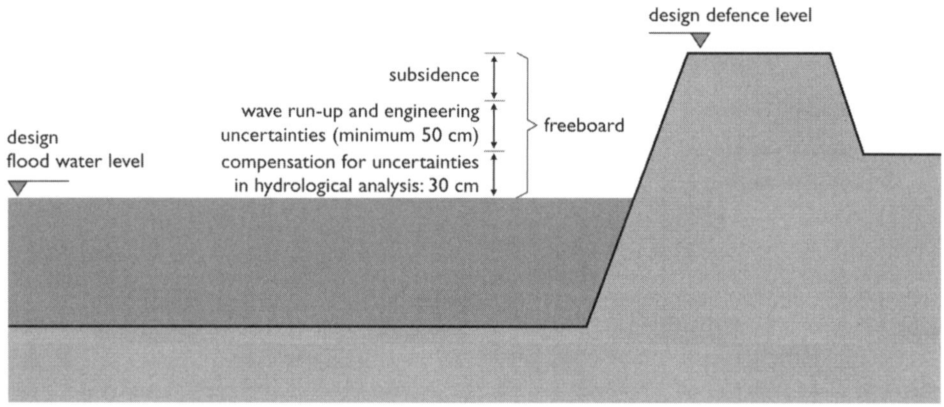

Figure 5.6 *Engineering safety margin for the dike design*

Source: Eelco van Beek

Storing a flood in a reservoir (upstream) or in a retention area (locally) is another much-used measure to deal with the variability in river flows. Reservoirs often have multipurpose functions – for example, by combining the following functions:

- storage for supplying irrigation and drinking water during dry periods;
- flood retention; and
- hydropower production.

Statistical analyses are required to determine the 'design' flood volume that has to be stored. This is done with the same statistical techniques as for the determination of the design discharge and by integrating the flood wave over a certain period of time.

Retention areas are used for peak shaving. Using retention areas requires that good forecasts of the expected flood wave can be made. Starting to use the retention area too early might result in a situation where the retention basin is already full when the top of the flood wave passes by (see Figure 5.7).

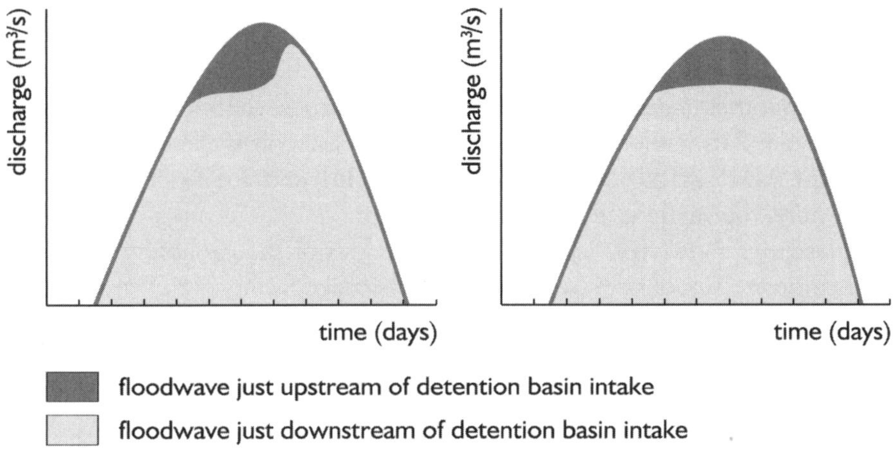

Figure 5.7 *Effect of retention basins on flood levels*

Note: Left panel shows effect of using retention area too early: right panel shows correctly timed use.

Source: Silva et al, 2001

Designing drought-related infrastructure

Droughts have far-reaching effects on humans and their civilizations. Droughts cause crop failures and the death of natural vegetation, livestock, wildlife and people. The World Health Organization (WHO) estimates that droughts and their effects cause half of the deaths worldwide due to all natural disasters (including floods, landslides, earthquakes, etc.). Economic losses from prolonged droughts often exceed those from other more dramatic natural hazards. Yet, droughts often receive less public attention than floods as floods are more spectacular and dramatic. However, floods normally last only a few days up to a few weeks, whereas droughts can last several months up to a few years. Therefore, from a socio-economic point of view, droughts are often much more important than floods.

Moreover, drought issues are often more difficult to solve than floods. Water should be made available during dry periods and this may require large-scale storage and water transfers. These storage and transfer schemes consist mostly of expensive infrastructural works and, in general, have major environmental impacts.

The two main measures to deal with droughts are storage basins and the use of groundwater.

Storage basins

Reservoirs can be designed to:

- store river discharges during the wet period of the year and to make that water available to users during the dry period of that same year; or
- store water during a wet year and make it available during dry years.

The first kind of reservoirs are annual reservoirs and, in general, fill up in the wet period and are (near) empty at the end of the dry period. The capacity of these reservoirs, in general, is less than the mean annual river flow. The second kind of reservoirs provides over-year storage and will have storage capacities well above the mean annual river flow. A good example of an over-year storage reservoir is the Lake Nasser reservoir in Egypt, which at full supply level (combined live and flood storage) can contain about 2.5 times the yearly water demand for Egypt.

The design of a reservoir is based on an analysis of the variability of the stream flow. The ultimate aim of using a reservoir as a storage facility is to provide a certain amount of water (the yield) when this is needed. The safe (firm) yield of a reservoir is the amount of water than can be supplied from the reservoir during a critical dry period. This firm yield is determined by analysing the variable supply to the reservoir in combination with the demand, which is often also variable. The Ripple Method (see Figure 5.8; Ripple, 1883), also called mass diagram analysis, is one of the oldest but still one of the most illustrative methods on how to take the variability of stream flow as a result of climate variability into account.

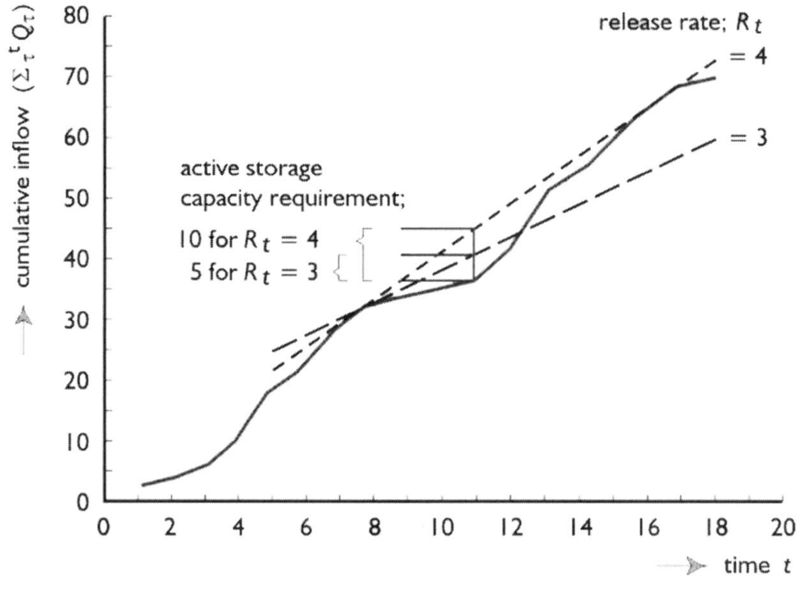

Figure 5.8 *Ripple method to determine safe yield and reservoir size*

Note: For a certain period in time (preferably at least 40 years), the cumulative inflow is plotted versus time. The flat portions of the curve are 'dry' periods, while the steep portions are 'wet'. The slope of the dotted line represents the (constant) yield. The vertical distance represents a volume. For a given demand, the minimum storage (firm yield) is the largest positive deviation between supply and demand. Negative deviations represent water 'spilled' from the reservoir. By changing the slope of the line, the best combination of safe yield and required storage can be found.

Source: Loucks and van Beek, 2005

More advanced methods to determine required reservoir capacities under variable yields are sequent peak analyses and various optimization and simulation approaches using either historical records or synthetic time series of inflow (see, for example, Loucks and van Beek, 2005, Chapters 7 and 11).

Groundwater

Groundwater is, in general, a very reliable supply of water both in terms of quantity and in quality. If the groundwater is used in a sustainable way, this source of water is hardly affected by the normal and current variability of the climate. Vertical drinking-water wells and (nearly horizontal) *qanat* systems have for centuries provided people with a very reliable supply. Qanat systems were already used in the Middle East long before Christ. A qanat with a length of 45km in the Iranian city of Gonabad continues to provide 2700 years of drinking and agricultural water to nearly 40,000 people.

The possible use of groundwater depends strongly upon the natural conditions of an area, such as the depth of the aquifer, porosity, conductivity, infiltration rate, natural drainage situation, etc. Some of these conditions can be influenced positively by increasing infiltration by constructing small dams or infiltration ponds, but also negatively by drainage works. A groundwater system is, in many ways, comparable with a surface water reservoir with the exception that evaporation losses of a groundwater reservoir are nil or very small. Natural conditions permitting, the use and further development of such groundwater reservoirs are preferable to surface water reservoirs. Developing groundwater reservoirs can be a promising option in adapting to further changes in climate variability.

Other design aspects

Designing dikes and reservoirs are just two examples of how water managers deal with climate variability. In nearly all of their design work, they have to take this variability into account. Regional systems such as polders and drainage and irrigation schemes, as well as urban sewerage and drainage systems, should be designed in such a way that they can cope with extreme conditions. The level at which this should be done is a trade-off between the costs involved and the possible impacts if the capacity of the system is too small and flooding and water pollution occur through sewage spills.

Current/historical practices in *managing* climate variability

The operational management of water resources systems is determined by the demands of society (safety and supply) and the need to cope with climate variability. For the professional water resources manager, water management involves the regulation, control, allocation, distribution and efficient use of existing supplies of water, such as in irrigation, power cooling, municipalities and industries; and the provision of water for in-stream uses, such as navigation, hydroelectric power, recreation and environmental flows. Additionally, all levels of government, and especially the private sector and individual stakeholders, are routinely engaged in managing water. Hence, technically, every individual who uses water is a water manager, from the water resource professional to the woman in the village who draws water from a well. Those who pay for its delivery and treatment are also responsible for its efficient use and conservation.

For the purposes of this discussion, all users, including farmers, are considered to be water managers. In terms of water resource systems, both the large-scale, mostly technical, systems as described in the previous section 'Current practices of using climate data for the *design* of water infrastructure' and the small-scale rural systems (including rain-fed agriculture) are taken into account. Addressing the adaptation options that farmers in less developed countries have is particularly critical, owing to the direct impacts that climate variability has on their livelihoods.

Nearly all management decisions related to climate variability boil down to a trade-off between maximizing the output and the risk of failure. Using Egypt again as an example, Figure 5.9 shows the trade-off that Egyptian water managers are making between the additional benefits of increasing the yearly release from Lake Nasser above the present 55.5 billion m^3 (BCM) through the High Aswan Dam (HAD) and the reliability of this yearly release. The figure shows that increasing the yearly release will, on average, result in a higher crop production; but the chance that once in a while some drought damage will occur will also increase. A release of about 56.4 billion m^3 would provide the maximum additional benefits – up to 65 million Egyptian pounds (MLE) per year. After this, the expected drought damage will outweigh the benefits of additional crop production. Hence, from an economic point of view, the release should be increased. However, Egyptian water managers stick to a release of 55.5 billion m^3 as they value the reliability of the system more than the economic benefits.

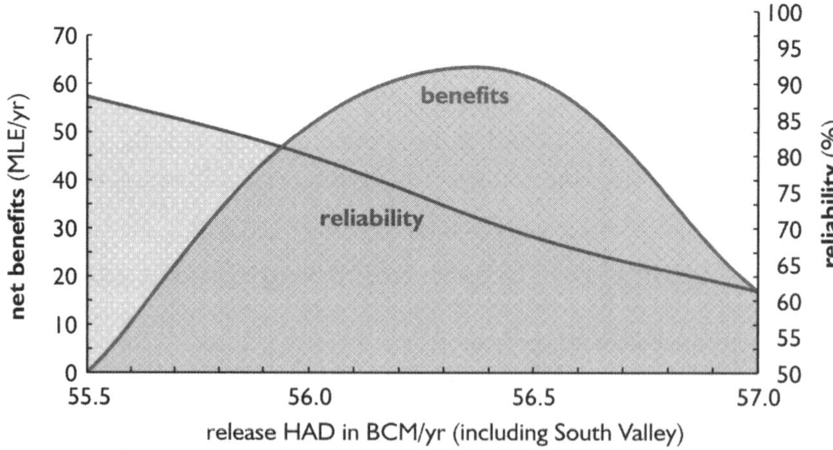

Figure 5.9 *Benefits versus reliability of supply from Lake Nasser*

Source: MWRI, 2008a

What are the management options that are available to water managers in dealing with floods and droughts resulting from climate variability? Some of these options are described below and include forecasting and warning, setting priorities, and operating the various infrastructure elements of the system, such as reservoirs.

Forecasting floods and droughts

The basic objective of developing an operational flood-and-drought warning system is to provide timely warning such that measures can be taken to reduce the impact of the event. Flood warning systems are, at present, much more developed than drought warning systems, mainly because there are more options to take action after such warning, such as evacuation, temporary relocation or implementation of flood control strategies. An operational flood warning system combined with these simple measures provides a cost-effective way of reducing flood risk, and may help to avoid large investments in traditional engineering flood control measures, such as raising dikes or building flood control dams.

A difference between flood-and-drought forecasting systems also exists in the timescale addressed. Flood forecasting systems have lead times of a few days and up to a week, while drought warning systems aim to forecast these droughts one or more months ahead.

To be effective, the ability to provide timely warnings must be complemented by the awareness and preparedness of those at risk. A good example of how losses can be reduced was shown in the Meuse Basin. Losses due to flooding in The Netherlands during December 1993 greatly exceeded those of January 1995, despite the two events being very comparable in magnitude. The reduction was partly attributed to the provision of an early warning and subsequent response (Wind et al, 1999).

Operational flood warning systems have been developed or are under development in many river basins throughout the world. All of these rely on the detection of floods through hydro-meteorological observation networks, and the use of observation data is a primary element of a flood warning system. To increase the potential utility of the flood warning service through extension of the lead time with which a flood event can be predicted, the more state-of-the-art systems also incorporate some form of (model-based) flood forecasting. A wide range of techniques may be employed to provide flood forecasting capabilities, ranging from data-driven modelling techniques such as neural networks to complex networks of conceptual and physical models. The necessity for different types of techniques is very much determined by the given requirement of delivering a forecast at a certain point with a minimum lead time. The lead time defines how much warning can reliably be given of imminent flooding. The lead time at which effective warning can be delivered is clearly dependent upon the lag times between precipitation falling and the flood peak reaching the point of interest. Warning lead times therefore vary greatly even in a single basin; for example, in the Rhine catchment, warning lead times in the upper basin may only be of the order of 24 hours, while in the lower basin, forecasts at lead times of up to four days can be provided using only hydro-meteorological observations and a rainfall runoff and routing model (Sprokkereef, 2001). The requirement to provide effective warning, together with the objective of increasing lead time and accuracy, poses new challenges in the implementation of flood warning systems.

Through developments in meteorological forecasting and the ability to link these with flood forecasting systems, the requirements in handling large amounts of data have increased greatly. Realization of the importance of considering the reliability of flood warnings (Krzysztofowicz et al, 2003) has equally led to the requirement of dealing with uncertainty. These uncertainties may be due to uncertainties in the meteorological forecasts, reflected through ensemble forecasting, while the models used may contain uncertainties due to uncertain model parameters (Beven et al, 2000). In particular, ensemble forecasts, such as those provided by the European Centre for Medium-Range Weather Forecasts (ECMWF) Ensemble Prediction System (EPS) or the American National Weather Service (NWS) Ensemble Streamflow Prediction system, have made the uncertainty in precipitation forecasting more explicit. EPS creates 50 ensemble members for a ten-day lead time weather forecast through perturbation of the initial conditions. Each ensemble member can then be used to derive precipitation and temperature boundary conditions for the hydrological models of a catchment. Figure 5.10 shows the results of a series of ensemble forecasts for the forecasting location at Lobith on the Rhine River leading up to the January 1995 event. The figure shows that the ensemble weather prediction has almost no influence on the first two days of the forecast: at these short lead times, flows at Lobith are dominated by water already in the main river. At increased lead times, the influence of the ensemble spread starts to dominate the discharge prediction.

As mentioned above, drought warning systems are much less developed. To take action, they require a lead time of up to several months and weather predictions for such timescales are highly uncertain. In certain river basins, forecasts can be made based on snowfall in winter. Progress is also made by linking the forecasts to El Niño Southern Oscillation (ENSO) events. ENSO events seem to have a strong influence on regions in the lower latitudes, especially in the equatorial Pacific and bordering tropical areas. India and Eastern and Southern Africa also show a strong correlation between ENSO events and a lack of rainfall that brings drought. Understanding these teleconnections will help to develop better drought forecasting and warning systems.

An example of applying such teleconnections are the studies that have been conducted by Dawod and El-Rafy (2002) to explain the Nile flows in relation to sea surface temperatures (SSTs) and jet streams. It was found that the annual natural Nile flow at Lake Nasser could be well predicted in advance from the following equation ($r = 0.82$):

$$Q_{Nile}\ (\text{billion m}^3/\text{year}) = 299.1 - 10.75\ P_1 + 7.92\ P_2 + 7.26\ I_1 - 7.38\ I_2 - 10.36\ A$$

where:

- Q_{Nile} = natural flow at Lake Nasser (billion m^3/year);
- P_1 = SST of the Pacific Ocean in June at 16°–21°S and 125°–135°W ($r = -0.40$) (°C);

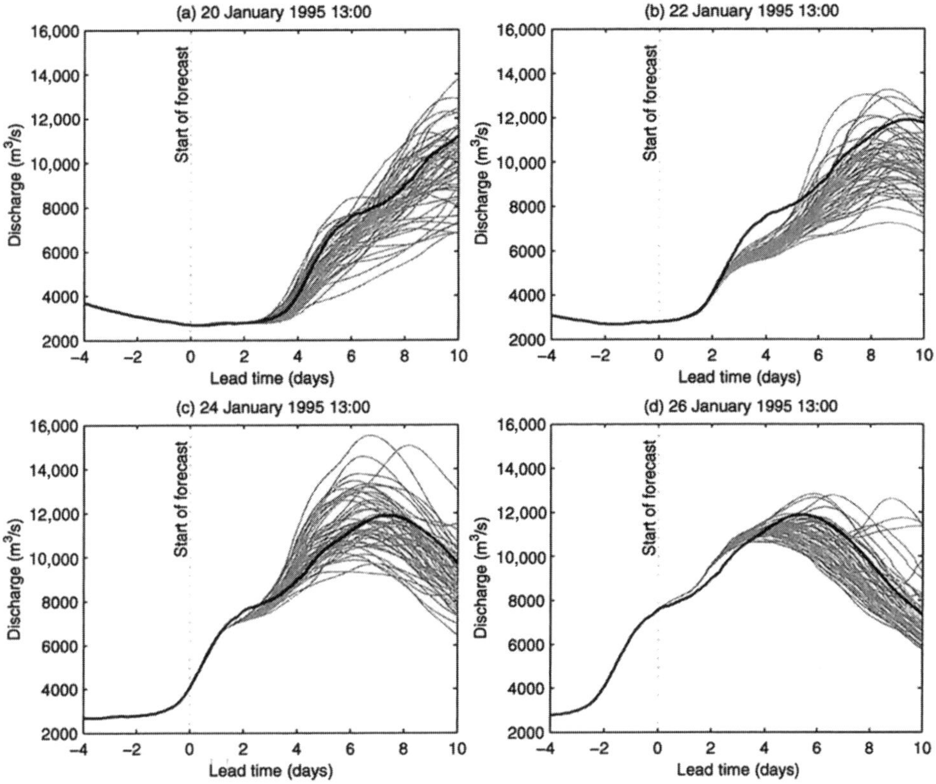

Figure 5.10 *Ensemble forecasts for the January 1995 event on the Rhine River at Lobith: Forecasts shown are issued at two-day intervals starting 20 January 1995 at 1.00 pm*

Note: Grey lines show the results of ensemble runs, depicting updated (simulated) discharge before the start of the forecast and ensemble forecast results after the start of forecast. The black line is the observed discharge.

Source: adapted from Werner et al (2004)

- P_2 = SST of the Pacific Ocean in June at 20°–27°N and 138°–160°E (r = 0.55);
- I_1 = SST of the Indian Ocean during January to March at 15°–25°N and 62°–70°E (r = 0.50);
- I_2 = SST of the Indian Ocean in June at 35°–40°S and 56°–70°E (r = –0.55);
- A = SST of the Atlantic Ocean in January at 22°–35°N and 72°–85°W (r = –0.40);
- r = correlation coefficient.

Using this relation, by the end of June, a fair prediction can be made of the Nile flow in the next hydrological year. The performance of this equation is shown in Figure 5.11.

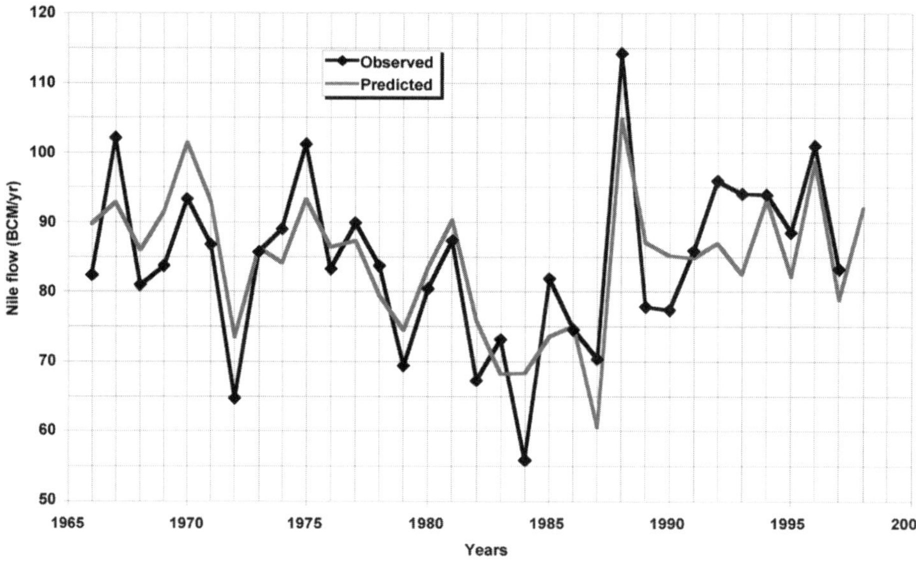

Figure 5.11 *Predicted and observed inflow in Lake Nasser*

Source: MWRI, 2008b

Priority-setting in drought events

In most cases, a water manager has few options for operational measures in the event of drought. The manager cannot influence precipitation, and if there is no storage facility available, there is little to be done to prevent low flows. A large part of occurring water shortages cannot be solved because the water cannot be conveyed to the right place in the right quantities at reasonable costs. The extent of economic damage often justifies only a limited investment in measures. Moreover, solutions in one sector will have adverse effects in other sectors. For example, increasing the extraction of groundwater through deep wells for irrigation may cause shallow groundwater wells used for drinking water in rural areas to fall dry. Consequently, there must be a balance between sectoral interests and suboptimal solutions for each sector.

In the event of water shortages, available water will have to be distributed as adequately as possible. National or regional committees should be in place to lay down the rules for such distribution. In The Netherlands, the distribution of water during water shortages is based on the so-called water supply priority series (see Figure 5.12). The highest priority is assigned to functions relating to population safety and the prevention of irreversible damage. The second priority is given to public utilities, such as drinking water and electricity production. The third priority is assigned to water-use functions that are relatively small and have a large economic value. All other categories of water use have a lower priority.

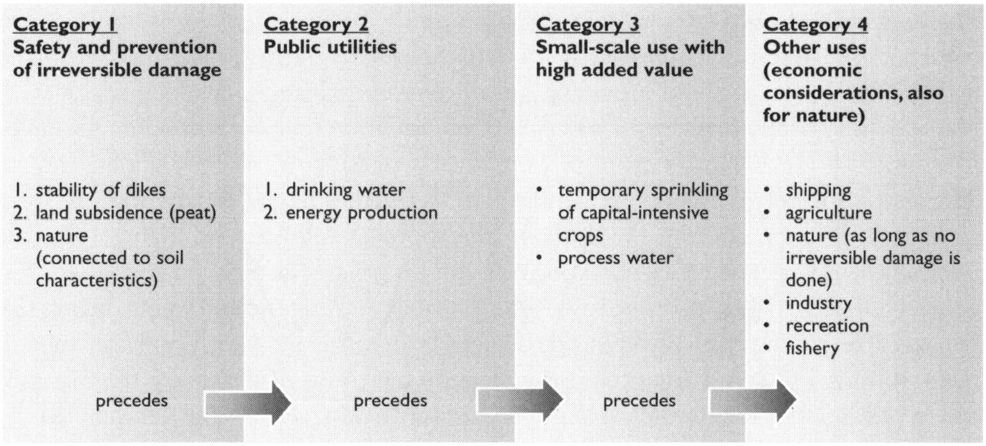

Figure 5.12 *Priority-setting in drought situations in The Netherlands*

Source: Ministerie van Verkeer en Waterstaat, no date

Reservoir management

Reservoirs are examples of the most powerful measures that can be taken to deal with climate variability since they address floods as well as droughts. Moreover, reservoirs support the generation of hydropower and, in combination with locks, improve the conditions for navigation. Nevertheless, reservoirs can also have negative social and environmental impacts. The management of reservoirs has to take all of these aspects into account.

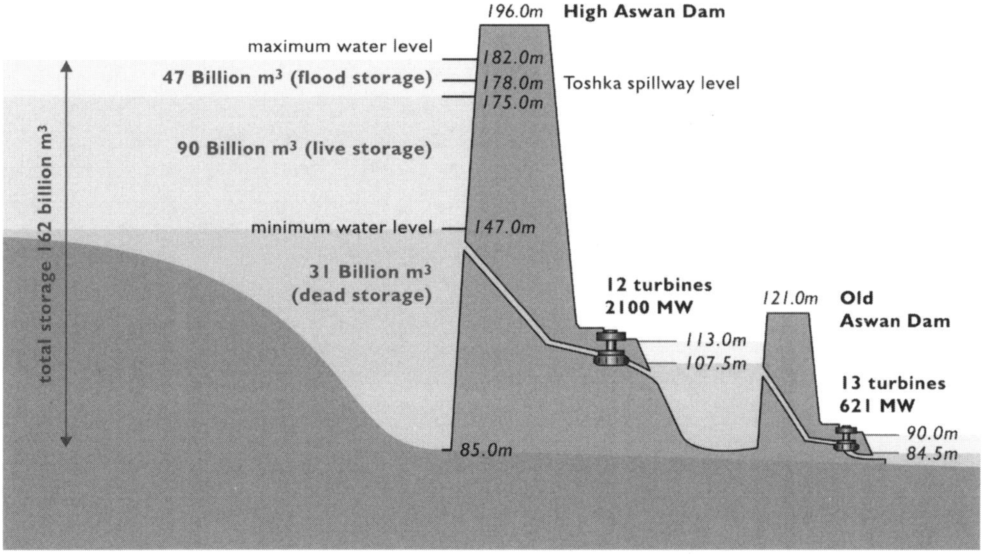

Figure 5.13 *High Aswan Dam, Lake Nasser*

Source: MWRI, 2008a

The operation of reservoirs is often based on predefined rule curves, as illustrated in Figure 5.14. These rule curves accommodate the reservoir functions of flood control, hydropower generation and providing water to users (irrigation, drinking water, etc.). Figure 5.14 typically describes a water resource situation that has a flooding season of December to March and a crop-growing season of May to October. Flood control demands sufficient reservoir space at the start of the flooding season in order to store or at least to attenuate the flood wave. Hydropower requires the highest possible reservoir level. Consumers of irrigation and drinking water also like to keep the reservoir as full as possible in order to ensure that they have sufficient water during dry periods. The trade-off involved in operating a reservoir is between flood control and water supply. The rule curves tell the manager what to do. If necessary, the manager has to release additional water in order to have sufficient storage for flooding. In the case of shortage, it is important to cut down on the supply in order to ensure a continuous and sufficient water supply for high-value use, such as drinking water.

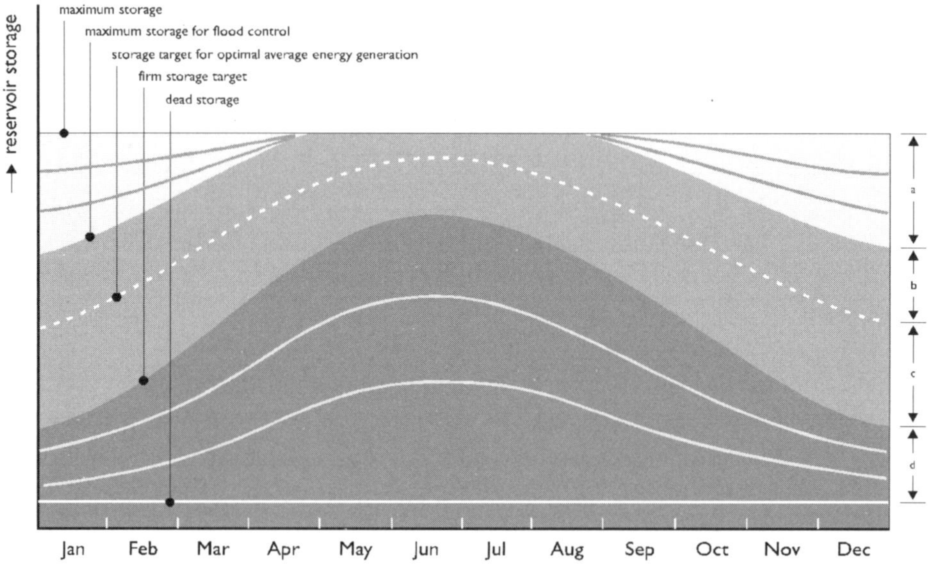

Action for each storage zone:

a – firm target draughts + extra energy generation (as much as possible) + forced spilling if necessary
b – firm target draughts + extra energy generation to avoid spilling later on
c – firm target draughts only
d – firm target draughts – possible reduction for firm storage preservation (hedging)

Figure 5.14 *Surface water reservoir rule curves and associated operation*

Notes:

1 The flood control curve indicates the maximum storage in order to provide for floods. If flood waters are contained instead of overrunning, flooding downstream will be reduced.

2 The storage curve for maximum average energy generation presents the optimum balance over time between creating head and avoiding spillage.

3 The firm storage curve indicates the amount of water that should be kept in storage in order to satisfy downstream demands throughout a critical dry period.

Source: Eelco van Beek

If the reservoir level drops below the firm storage curve (as depicted in Figure 5.14), then the reservoir target release is reduced (hedged). Various hedging methods can be applied, for example, based on storage (supply oriented) and based on water allocation priority (demand oriented). Box 5.1 explains the operation and hedging rules as applied to Lake Nasser.

Box 5.1 Operation strategy for the High Aswan Dam

Lake Nasser has an active storage volume of nearly 140 billion m^3. This means that the lake can contain almost twice the annual yield of the River Nile. The purpose of this massive reserve is to ensure over-year storage as opposed to annual storage. With over-year storage, it is possible to give the same reservoir yield every year because differences in inflow can be buffered by the active storage. Depending upon the variability of the inflow and the level of allowed uncertainty, the release is determined.

Figure 5.15 shows the average inflow in Nasser in relation to the water demand in Egypt. The lowest lake level will occur when the two lines cross (actually, it will be somewhat later as the inflow of Lake Nasser includes the evaporation of the lake). Figure 5.15 illustrates that the reservoir is able to fully control the Nile flows; the high flows during August and September are completely eliminated and maximum discharges are now limited to 270 million m^3 per day (i.e. less than one third of the earlier peak values).

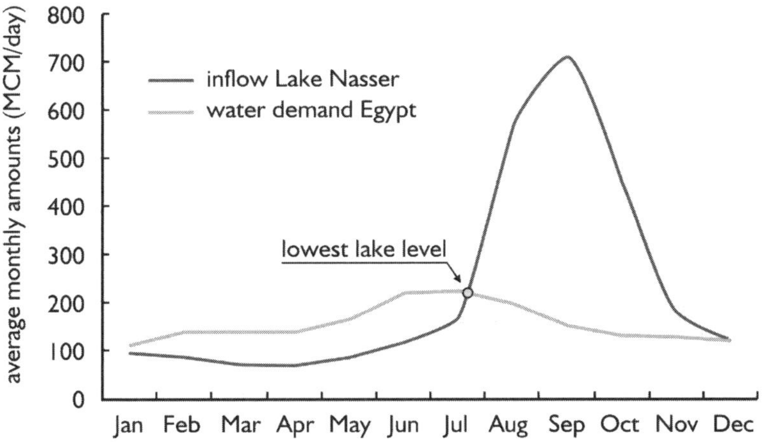

Figure 5.15 *Average inflow in Nasser in relation to the water demand in Egypt*

Source: MWRI, 2008a

Although no explicit rule curve is used in the operation of the High Aswan Dam, the following formal rules are applied:

- By 1 August the reservoir level should not be higher than 175m to allow sufficient room to receive the coming flood. The date of 1 August is defined as the start of the hydrological year in Egypt.

- The release from the dam is determined by the agricultural demands, which are fixed at an annual amount of 55.5 billion m^3. A maximum amount of 230 million m^3/day is released in May, whereas in January, when maintenance work is carried out on the irrigation system, a minimum amount of 80 million m^3/day is discharged to satisfy various requirements.
- The reservoir level should not exceed 182m, with an upper limit of 183m above mean sea level in emergency situations.
- If the reservoir level exceeds a height of 178m, water from the reservoir is spilled over to the Toshka depression.
- In the case of low reservoir levels, releases are reduced and a sliding scale is applied to Egyptian and Sudanese water demands (since 1968, this has only applied once). These sliding scales are:
  - If storage $\geq$ 60 billion m^3, then the full share is used.
  - If 55 $\leq$ storage $<$ 60 billion m^3, a reduction of 5 per cent is to be applied.
  - If 50 $\leq$ storage $<$ 55 billion m^3, a reduction of 10 per cent is to be applied.
  - If storage $<$ 50 billion m^3, a reduction of 15 per cent is to be applied.

Indigenous coping strategies for dealing with droughts

Climate variability is not new and people have learned to live with it. Drought situations occur in a spectrum of arid to dry sub-humid areas. Experience of these areas shows a low total annual precipitation combined with high rates of potential evaporation and an extreme spatial and temporal variability in precipitation. Over the centuries, societies living in these areas have developed a broad range of mechanisms to cope with climatic variability. In agrarian communities based on rain-fed agriculture, this has focused on local development of social, economic and biophysical management strategies in order to bridge droughts and dry spells. The indigenous knowledge base of climatic coping strategies certainly dates back at least 7000 years.

Indigenous strategies to cope with climatic variability vary between different geographical locations and between social, religious and cultural settings, as well as between livelihood cores (e.g. between agro-pastoral communities depending upon livestock-raising compared to sedentary farming communities depending primarily upon crop production). It is thus impossible to give a generic overview of indigenous coping mechanisms.

Climate change is likely to increase variability, and one would expect the proliferation of current and the revival of old indigenous coping mechanisms. In general, however, the situation seems to be the reverse. Population pressure, degradation of land and water resources, and migration have, in large parts of the water-scarce environments, resulted in a deterioration (and, in many cases, complete loss) of indigenous coping mechanisms.

From the birth of agriculture until recently in tropical environments (often until the late 19th century or even early 20th century), farming systems were based on shifting cultivation, which depended upon spatial rotation of cropland and long fallow periods. In environments with large spatial and temporal rainfall variability, this production strategy was strategically designed to spread risks in space and in time. In the Sahel, fallow-based rain-fed farming has essentially disappeared under escalating pressure from population growth, and farmers depend for their livelihoods upon continuous cultivation on small (far too small) parcels of land. Granaries were used as cereal banks to store surplus grain from 'wet' rainy seasons for use during dry years, in accordance with Joseph's advice to the Egyptian pharaoh in the Old Testament (save the surplus from the seven good years to cope with the seven dry years that follow). This management strategy, dating back several millennia, formed the backbone of many farming systems in climatically variable environments until modern times.

In West Asia and North Africa, coping strategies dealing with climatic variability and water scarcity date back at least to 5000 BC. In Mesopotamia, southern Jordan and the Negev Desert, water-harvesting systems collecting surface water from intensive rainstorms for use during droughts and dry spells, both for agricultural and domestic purposes, were probably developed simultaneously with the introduction of sedentary societies. Agarwal and Narain (1997) mention that water harvesting dates back three millennia BC. These indigenous coping strategies died out during the 20th century as a result of the modernization of water management during the hydraulic era of irrigation developments. Interestingly, these coping strategies are reviving in pace with the realization by local farming communities that governments are not able to provide security from climatic variability.

Integrated water resources management (IWRM)

Box 5.2 Definition of integrated water resources management (IWRM)

IWRM is a *process* that promotes the coordinated development and management of water, land and related resources in order to maximize the resultant *economic and social welfare* in an equitable manner without compromising the *sustainability of vital ecosystems.*

The approaches described in previous sections form the technical base for existing water systems. The design and management of new systems will also largely be based upon these kinds of techniques, which have developed over time. At present, more detailed analysis is possible for dealing with climate variability, although, essentially, the approach has not changed. What has altered dramatically is how we apply these technical approaches. During the 1970s and 1980s, it became clear that the classical top-down and supply-driven approach did not yield the promised results. Another way to apply water management had to be found. This has led to the development of the concept of integrated water resources management (IWRM). IWRM should be seen as a response to the increased pressure on our water resources systems due to a growing population and socio-economic developments. Water shortages and deteriorating water quality have forced the world's countries to reconsider their options with respect to managing their water resources, in developed and developing countries alike. As a result, 'classical' water resources management has been undergoing a drastic change worldwide, moving from a mainly supply-oriented, sector-focused and engineering-based approach towards a demand-oriented, multi-sectoral approach. IWRM is not a technique: it is a way of thinking and is subject to considerable debate. The consensus about the implications of IWRM are best reflected in the Dublin Principles of 1992 (see GWP, 2000) that have been universally accepted as the basis for IWRM. The concept of IWRM makes us move away from a top-down 'water master planning', which focuses on water availability and development, towards 'comprehensive water policy planning', which addresses the interaction between different sub-sectors, seeks to establish priorities, considers institutional requirements, and deals with the building of capacity.

Box 5.3 Dublin Principles

- Water is a finite, vulnerable and essential resource, essential to sustain life, development and the environment.
- Water resources development and management should be based on a participatory approach, involving users, planners and policy-makers at all levels.
- Women play a central role in the provision, management and safeguarding of water.
- Water has an economic value in all of its competing uses and should be recognized as an economic good.

A key-aspect of IWRM is that the management and development of resources should take place in interaction with users, uses (the socio-economic system) and institutions. IWRM applied in this way considers the use of the resources in relation to social and economic activities and functions. These also determine the need for laws and regulations for the sustainable use of water resources. Infrastructure made available in relation to regulatory measures and mechanisms will allow for effective use of the resource, taking due account of environmental carrying capacity.

Compared to 'classical' water resources management, which has focused on economic efficiency, IWRM incorporates and pays special attention to two other overriding criteria: social equity and environmental and ecological sustainability (see Figure 5.16). Social equity addresses the basic right of all people to have access to water of adequate quantity and quality for the sustenance of human well-being. Environmental and ecological sustainability should ensure that the present use of the resource is managed in a way that does not undermine the life-support system, thereby compromising use by future generations of the same resource.

The concepts of IWRM do not specifically address climate change and climate variability. However, the holistic view taken in IWRM does facilitate finding solutions

Figure 5.16 *General framework and overriding criteria for IWRM*

Source: GWP, 2000

to cope with this variability. Climate change and climate variability affect all water-related sectors and influence the supply and demand side involved. This link is, for example, shown in Table 5.1, which gives the Intergovernmental Panel on Climate Change (IPCC) adaptation recommendations for water resource managers. Note that there are, indeed, no major changes in coping with climate variability and climate change compared to what is being done (or should be done) in IWRM already. Determining the necessity of these measures will vary depending upon whether or not (and to what degree) climate variability and change are taken into account (i.e. the amount of variability and the level of an unexpected event).

Changing climate and changing climate variability: Business-as-usual?

The previous sections seem to imply that it is business-as-usual for water managers. Water managers will design and manage their systems based on statistical analysis of monitoring data. If new data becomes available, they will revise their designs. Since much of the infrastructure involved has a relatively short life span of between 5 and 50 years, the new design parameters can be taken into account in the next generation of the infrastructure. The increase of the design discharge in the Rhine from 15,000m^3/s to 16,000m^3/s and the corresponding increase in the design level of the dikes along the river is an example of this (see also Figure 5.3). Stakhiv (1998) brings the following question to the fore:

Table 5.1 *IPCC recommendations for water resources managers*

Supply side		Demand side	
Option	**Comments**	**Option**	**Comments**
Municipal water supply			
• Increase reservoir capacity	Expensive; potential environmental impacts	• Incentives to use less (e.g. through pricing)	Possibly limited opportunity; needs institutional framework
• Extract more from rivers or groundwater	Potential environmental impacts	• Legally enforceable water-use standards (e.g. for appliances)	Potential political impact; usually cost-inefficient
• Alter system operating rules	Possibly limited opportunity	• Increase use of grey water	Potentially expensive
• Inter-basin transfers	Expensive; potential environmental impacts	• Reduce leakage	Potentially expensive to reduce to very low levels especially in old systems
• Desalinization	Expensive (high energy use)	• Development of non-water-based sanitation systems	Possibly too technically advanced for wide application
		• Seasonal forecasting	Increasingly feasible
Industrial and power station cooling			
• Increase source capacity	Expensive	• Increased water-use efficiency and water recycling	Possibly expensive to upgrade
• Use of low-grade water	Increasing used		
Hydropower generation			
• Increase reservoir capacity	Expensive; potential environmental impacts; may not be feasible	• Increasing efficiency of turbines; encourage energy efficiency	Possibly expensive to upgrade
Navigation			
• Build weirs and locks	Expensive; potential environmental impacts	• Alter ship size and frequency	Smaller ships, more trips; increased emissions and costs
Pollution control			
• Enhance treatment works	Potentially expensive	• Reduce volume of effluents to treat (e.g. charging discharges)	Requires management of diffuse sources of pollution
		• Catchment management to reduce polluting runoff	Requires buy-in from farmers, e.g. incentives

Table 5.1 *(continued)*

Supply side		Demand side	
Option	**Comments**	**Option**	**Comments**
Flood management			
• Increase flood protection (levees, reservoirs	Expensive; potential environmental impacts	• Improved flood warning and dissemination	Technical limitations in flash-flood areas and unknown effectiveness
• Catchment source control to reduce peak discharges	More effective for small than large floods	• Curb floodplain development	Potential major socio-political problems
Irrigation			
• Increase irrigation source capacity	Expensive; potential environmental impacts	• Increase irrigation-use efficiency	By technology or increasing prices
		• Increase drought-tolerant varieties	Genetic engineering is controversial
		• Change crop patterns	Change to crops which need less or no irrigation

Source: adapted from IPCC (2001, Table 4–13)

> The question is whether the current methods of water resource development and management, based on the assumption of a stationary climate, can be suitably employed to accommodate the uncertainties of a non-stationary climate. Several authors, notably Fiering and Matalas (1990), Rogers and Fiering (1990) and particularly Matalas (1997) believe that the framework of stochastic (synthetic) hydrology, that is widely used in project planning, can accommodate the uncertainties in water supplies induced by global warming.

Of course, the managers acknowledge that the length of their time series is limited and that there might be events that are not captured in these time series. For this reason, they always apply a kind of additional safety margin, such as the extra freeboard in dike design (see Figure 5.6). The basic assumption underlying the statistical approach is that the natural systems fluctuate within a fixed range of variability and that this range does not change. The probability curves of the times series (rainfall, river discharges, etc.) are assumed to be stationary.

On the other hand, an increasing number of scientists dispute this approach. Gleick et al (2000), for example, argue that sole reliance on traditional management responses is a mistake:

> First, climate changes are likely to produce – in some places and at some times – hydrologic conditions and extremes of a different nature than current systems were designed to manage; second, climate changes may produce similar kinds of

> variability but outside of the range for which current infrastructure was designed and built; third, relying solely on traditional methods assumes that sufficient time and information will be available before the onset of large or irreversible climate impacts to permit managers to respond appropriately; and, fourth, this approach assumes that no special efforts or plans are required to protect against surprises or uncertainties.

An important argument by Gleick is that applying only traditional methods may lead to severe impacts that may have to be mitigated or prevented by cost-effective actions taken today. Although this is true for all kinds of future developments, this argument seems to be specifically applicable in the case of climate change and climate variability.

An important aspect that should be taken into account is that in some basins small changes in rainfall and rainfall pattern can result in big changes in river discharges. The discharge of a basin is the difference between two significant numbers (rainfall and evaporation). It is estimated that a 10 per cent increase in rainfall in the equatorial lake area and in Ethiopia will result in a 40 per cent increase in the annual flow in the Nile. On the other hand, a 10 per cent decrease in rainfall will also reduce the annual flow by 40 per cent, which will be disastrous for Egypt. Similar examples can be given on important reductions of groundwater recharge as a result of moderate changes in rainfall. These changes can cause situations that are currently sustainable (supply in equilibrium with demand) to become unsustainable, even if the demand does not increase.

Some scientists have started to question the fundamental assumption of stationarity. Milly et al (2008) assert that stationarity is dead and should no longer serve as a central default assumption in water-resource risk assessment and planning. Their main arguments are the substantial anthropogenic changes of the Earth's climate that are altering the means and the extremes of precipitation, evapotranspiration and resulting river discharge.

But if this is true, what should water managers do instead? Milly et al (2008) suggest the use of multiple climate models, driven by multiple climate-forcing scenarios to develop new probability density functions.

The overall conclusion is that water managers will, indeed, face some serious challenges. Pressure on water systems will increase, as will societal demands on possible solutions, reflected in the social equity and environmental sustainability criteria of IWRM. In addition, water managers will have to work with the uncertainties involved in possible changes in climate and climate variability. Easy solutions are no longer available: water managers will have to look for other options that are most likely complicated and expensive to develop.

References

Agarwal, A. and S. Narain (1997) *Dying Wisdom: Rise, Fall and Potential of India's Traditional Water Harvesting Systems: State of India's Environment 4 – A Citizens' Report*, Centre for Science and Environment (CSE) New Delhi, India, p398

Beven, K. J., R. Romanowicz and B. Hankin (2000) 'Mapping the probability of flood inundation (even in real time)', in M. Lees and P. Walsh (eds) *Flood Forecasting: What Does Current Research Offer the Practitioner?*, British Hydrological Society Occasional Paper no 12, pp56–63

Dawod, M. A. A. and M. A. El-Rafy (2002) 'Towards long range forecast of the Nile flood', in *Proceedings of the Fourth Conference: Meteorology and Sustainable Development*, Meteorologist Specialist Association, Cairo, Egypt

Fiering, M. B. and N. C. Matalas (1990) 'Decision-making under uncertainty', in P. E. Waggoner (ed) *Climate Change and US Water Resources*, John Wiley & Sons, New York, pp75–84

Gouldby, B. and P. Samuels (2000) *Language of Risk*, Floodsite, Wallingford

Gleick, P. H. et al. (2000) *Water: The Potential Consequences of Climate Variability and Change for the Water Resources of the United States*, Report for the US Global Change Research Program, September

GWP (2000) *Integrated Water Resources Management*, TAC Background Papers no 4, Global Water Partnership, Stockholm, Sweden

IPCC (2001) 'Hydrology and water resources', in *Climate Change 2001: Impacts, Adaptation and Vulnerability*, Report of Working Group II of the Intergovernmental Panel on Climate Change, Cambridge University Press, Cambridge, UK

Krzysztofowicz, R. (1995) 'Recent advances associated with flood forecast and warning systems', *Reviews of Geophysics*, vol 33 supplement

Kwadijk, J., N. van Gemert, M. van Asselt, W. van Deursen, H. Middelkoop, H. Buiteveld, M. Haasnoot and J. Rotmans (2001) 'Maatgevende afvoeren, onzekerheden en wereldbeelden', *Stromingen*, vol 7, no 2

Loucks, D. P. and E. van Beek (2005) *Water Resources Systems Planning and Management*, UNESCO

Maidment, D. R. (1993) *Handbook of Hydrology*, McGraw-Hill Company, McGraw-Hill, New York, NY

Matalas, N. C. (1997) 'Stochastic hydrology in the context of climate change', in K. D. Frederick, D. C. Major and E. Z. Stakhiv (eds) *Climate Change and Water Resources Planning Criteria*, Kluwer Academic Publishers, Dordrecht, The Netherlands

Milly, P. C. D. et al (2008) 'Stationarity is dead: Whither water management', *Science*, vol 319, pp573–574

Ministerie van Verkeer en Waterstaat (no date) 'Verdringingsreeks', Ministerie van Verkeer en Waterstaat website, www.verkeerenwaterstaat.nl/onderwerpen/water/droogte/_verdringingsreeks/

MWRI (2008a) *Lake Nasser Flood and Drought Control Project, Main Report,* Ministry of Water Resources and Irrigation, Egypt

MWRI (2008b) *Lake Nasser Flood and Drought Control Project, Volume VIII, Hydrology*, Ministry of Water Resources and Irrigation, Egypt

Ripple, W. (1883) 'Capacity of storage reservoirs for water supply', *Minutes of Proceedings of the Institution of Civil Engineers*, vol 71, pp270–278

Rogers, P. P. and M. B. Fiering (1990) 'From flow to storage', in P. E. Waggoner (ed) *Climate Change and US Water Resources*, John Wiley & Sons, New York, NY, pp207–221

Silva,W., F. Klijn and J. P. M. Dijkman (2001), *Room for the Rhine Branches in the Netherlands: What the Research Has Taught Us*, WL | Delft Hydraulics, Delft, The Netherlands

Sprokkereef, E. (2001) *Extension of the Flood Forecasting Model FloRIJN*, NCR Publication 12-2001

Stakhiv, E. Z. (1998) 'Policy implications of climate change impacts on water resource management', *Water Policy*, vol 1, pp159–175

Ward, R. C. (1967) *Principles of Hydrology*, third edition, McGraw-Hill, Maidenhead, UK

Werner, M. G. F., P. Reggiani, A. De Roo, P. B. Bates and E. Sprokkereef (2004) 'Flood forecasting and warning at the river basin and at the European scale', *Natural Hazards*, vol 36, nos 1–2, September, pp25–42

Wind, H., T. Nierop, C. De Blois and J. De Kok (1999) 'Analysis of flood damages from the 1993 and 1995 Meuse floods', *Water Resources Research*, vol 35, pp3459–3465

6

Using Seasonal Climate Forecasts for Water Management

Fulco Ludwig

Recognizing and understanding historical and current climate variability

The previous chapter described how many water management decisions related to climate variability focus on a trade-off between maximizing the output and minimizing the risk of failure. In Chapter 2, the different timescales at which climate varies are discussed. There is seasonal, annual and sometimes decadal variation in rainfall that should be considered. Most water management systems are set up to manage seasonal variation in water availability. Annual and decadal variation in climate and water availability are more difficult to manage due to the lower predictability, understanding and experience with these longer timescales. A better understanding and enhanced predictability of climate variability can potentially improve the trade-off between maximizing output and minimizing failure. This chapter focuses on the use of seasonal climate forecasts as a tool to improve water management and how the analysis of historical climate data can improve the understanding of climate variability and its implications for water management.

An important first step in coping with or adapting to climate variability in a particular region is to understand the historical climate and to take stock of all available data relevant for water resources management in a particular region. Traditionally, historical data on temperature, precipitation and, occasionally, stream flow or other indicators of water levels are used for decision-making (Hartmann, 2005). Instrumental records are generally the most reliable source of historical climate information. In North America and Europe, proper data records on temperature and precipitation are typically available for at least 100 years. However, access to such data, as in the case of Europe, is still a significant problem. In developing countries, dependable historical data records tend to be less available and are often of lower quality and of poor spatial resolution. The way in which data is used often determines how reliable the records need to be. For example, if data is used to determine average temperatures over an extended timespan, the presence of minor gaps in the historical data is often insignificant. The timespan of the database and the quality of the data become more relevant when determining rare extreme events.

When basing forecasts and decisions on *historical* climatic data in the absence of climate change scenarios, one important assumption is made: stationarity (Milly et al, 2008). One assumes that the climate of the future will be identical to the one of the past. Examples from the early 20th century have demonstrated the limitations of assuming stationarity. The allocation of water in the Colorado River in 1922 was based on data from the previous ten years (Frederick and Kneese, 1990). However, later observations showed that these years were unusually wet and therefore too much water was allocated to the different states, resulting in eventual conflict.

If data for longer periods is available, it is often difficult to define which part of the data set is relevant for forecasting and making decisions, particularly when long-term fluctuations and trends are present within the data set. An example for stream flow of major water sources around Perth, Australia, demonstrates that using different sets of data can result in different management strategies. In this case, an average annual stream flow of almost 0.30km^3/year was observed over the entire historical record, whereas only 0.16km^3/year was observed over the last 30 years (IOCI, 2002; Power et al, 2005). To supply the long-term water needs for the City of Perth with the latter observation would require additional sources of water. This example demonstrates that when long-term climatic data is available, not only should the average climate be analysed, but it is also necessary to perform trend analyses or to compare different sets of data. While most trends are probably an indicator of the direction of the future climate, care should be taken in extrapolating all trends. Some trends could be part of a 'natural' cycle and may not be the result of anthropogenic climate change.

Historical data should not only be used to analyse how variable the climate is at different timescales. Depending upon the application, historical data can be used to analyse the frequency and duration of dry and wet periods. For analyses of the average climate, less data is needed and only the most recent data should be included. All available data should be used when calculating the frequency of rare extreme events. A next step in the analyses should be to study whether unusually dry and/or wet periods are linked to climate variability indicators such as the El Niño Southern Oscillation (ENSO) and La Niña phenomena or the phases of the Decadal Pacific Oscillation (Hamlet and Lettenmaier, 2000). If this is the case, then ENSO indicators can be used for future water management. Occasionally, special meteorological conditions occur that require a different management approach. For example, the concurrence of exceptionally dry periods and very high temperatures can increase water stress, necessitating different management approaches than for isolated warm or dry conditions. In conclusion, analysing historical climate data is the critical starting point for managing climate variability and understanding local climate patterns. Furthermore, it should not be assumed that the climate of the future is identical to the past (Milly et al, 2008).

Using seasonal climate forecasts

The use of seasonal climate forecasts is still limited in the water sector. In many areas of the world, however, the use of seasonal climate forecasts is likely to be an important tool for improving water management. Where weather forecasts usually provide an amount of rainfall or temperature expected in a short-term timeframe, seasonal forecasts are almost always probabilistic. For example, a probability is given for above- or below-seasonal average rainfall (see also 'Predictability of climate variability at the seasonal timescale' in Chapter 2 of this book). Depending upon the forecast, the management approach can be adapted to the newly forecasted climate. For example, if a forecast predicts greater-than-average rainfall, the likelihood that the threshold for too much water is reached can increase, and measures can be taken to reduce this risk of excess water. If a storage basin has an increased risk of flooding using the new forecasts, water could be released as a precautionary measure to reduce the likelihood of flooding. However, there is a new risk associated with this because seasonal climate forecasts are never perfect and it is still possible that the actual rainfall is lower than average. In this case, the water released from dams could later result in water shortages.

Seasonal forecasts can also be used for opportunities to improve water productivity. Water allocations could be increased if above-average rainfall is expected or groundwater use can be reduced if more surface water becomes available. If these kinds of measures are taken on a continuous basis, the average risk will be reduced and there is a lower probability of exceeding critical thresholds. Seasonal forecasts can also be used to modify water pricing or to introduce water-use restrictions (Chiew et al, 2003; Brown et al, 2006).

When to use seasonal forecasts

Seasonal forecasts are not appropriate under all circumstances. Their utility depends upon a specific set of factors. The future climate for the coming season has to be predictable in the region of interest. The forecast should be comprehensible and trustworthy to the people who use or translate it. Most importantly, the critical requirement of seasonal forecasts is the possibility of adjusting management to accommodate the forecast.

The climate must be predictable

Before establishing a seasonal forecasting framework, the first step should be to assess the level of predictability of the climate of the region of interest. As discussed in 'Detection of trends in a fluctuating signal' in Chapter 2, the quality of seasonal forecasts is not universal around the globe, and it is easier to predict climatic trends in areas strongly influenced by, for example, the El Niño Southern Oscillation. Therefore, seasonal forecasts require a sufficient skill in order to be useful for a particular application (Lemos et al, 2002). The level and type of skill is determined by the application for which it is intended to be used. For 'low cost, low risk' measures, a lower skill is needed, whereas for more expensive and riskier applications, a higher skill is required. Also, in areas where there is a great potential for the use of climate forecasts, not every forecast is useful. It is sometimes the case that only average climatology is predicted or that forecasts are only reliable during specific seasons.

Forecast information is available, understood and trusted by decision-makers

If climate forecasts can potentially be used for water resources management in a region, the availability and accessibility of the information may be a critical constraint (Pagano et al, 2002). Accessibility of information is probably a larger problem in the developing world; however, in developed countries, the format of the forecast and/or the way in which it is communicated can also limit the use of seasonal forecasts (Rayner et al, 2005). The availability of a well-communicated forecast is the responsibility of both the forecast agency (usually the national meteorological service) and the end users in the water sector. The end users must explain to the forecast agency what kind of information is useful for them, and the agency must be willing to adjust their forecast format and/or the way in which it is communicated. For example, analogue year forecasts, such as the Southern Oscillation Index (SOI) phase system, are much more useful for farmers than probabilistic forecasts that give the probability of above-average rainfall. However, probabilistic forecasts are often more accurate. Sometimes an intermediate organization can be used to 'translate' and combine the different forecasts into a format useful to the end users. For example, the International Research Institute for Climate and Society (IRI) uses forecasts from different agencies and produces forecasts for use in developing countries.

For optimal use of seasonal climate forecasts, it is important that people within the organization who drive decision-making understand and trust the forecast. Most hydrologists and managers in the water sector are not used to working with probabilities and need information such as cubic metres of stream flow, dam levels or amount of rainfall. Skill scores are also difficult to understand. It is essential to understand and to think in terms of probabilities if seasonal forecasts are to be used properly. Different farmer programmes that promote the use of seasonal forecasts have demonstrated that professionals can be taught how to use seasonal forecasts with relatively little training and support (Ash et al, 2007). Although there are currently many more programmes that focus on using seasonal forecasts oriented towards the farming community than towards water managers, it should not be difficult to set up similar programmes for the water sector.

Skill and format of forecast can be translated into actual measures

The most challenging step in using climate forecasts is taking action in response to the forecast. A framework has to be developed within the organization on how to incorporate the seasonal forecast within the decision-making process. Ideally, an action plan will be drawn up. For example, in case 'X', forecasted, measure 'Y' will be taken. Sometimes already existing decision-support systems can be used to incorporate the climate forecast (e.g the use of daily weather data in combination with models to predict future stream flow). Data from historical analogue years (based on the forecast) can be used to make distribution curves of likely future stream flow. If these distribution curves show significantly higher- or lower-than-average stream flow, appropriate measures can be taken. For example, water allocated for different irrigation sites can be increased or reduced (Smith, 2005).

Examples of the possible use of seasonal climate forecasts in water resources management

Nearly all studies on the use of seasonal climate forecasts in water management focus on statistical methods based on the analogue year approach. Historical climate records are divided into different types of years or seasons based on prevailing ocean and/or atmospheric conditions (Meinke and Stone, 2005). In most cases, the analogue years are linked to the different phases of the El Niño Southern Oscillation or to the Pacific Decadal Oscillation (PDO) (Hamlet and Lettenmaier, 2000). One of the main benefits of using the analogue year approach is that parts of the historical climate data can be used to predict future conditions. This links to the historical approach of water management where (all) historical data are used to design water management systems. Still rarely used are numerical methods for seasonal climate forecasts that, for example, incorporate hydrological models coupled with regional climate models (RCMs). RCMs still have significant biases in rainfall fields and normally need a statistical correction of their outputs. The biases in global climate model (GCM) and RCM outputs also result in limited trust of these kinds of forecasts by water managers.

Currently, the potential to use seasonal climate forecasts in water resource management is identified mostly in regions where the climate is strongly affected by ENSO, such as North America, Brazil, Australia and the Philippines. For example, north-east Brazil has a strongly seasonal climate where rainfall is affected by ENSO. In north-east Brazil, water from the wet season (January to June) is stored in reservoirs to be used for irrigation, domestic and industrial purposes during the dry season (July to December) (Broad et al, 2007). Due to annual variation in rainfall in this region, flow into reservoirs significantly varies from year to year. Reservoirs are historically managed while assuming a zero inflow during the next season. If stream-flow prediction could be improved, on average, more water could be allocated to the different users. In north-east Brazil, a NINO3.4 index (an indicator for the ENSO state) and a North Atlantic dipole index was used to forecast stream flows of rivers and flows into reservoirs (Souza and Lall, 2003). Some interesting lessons were learned from this study, which probably apply to other regions as well (Broad et al, 2007). First of all, the irrigators using water from the reservoirs supported the use of forecasts to release more water in the current season. These farmers tend to maximize the current harvest and can use the extra water, if available. In contrast, the technicians from the water agencies in charge of releasing water were much more reluctant in using forecasts. These water managers felt that they would not be rewarded if more water were released by using forecasts, but they would be punished if water shortages occurred due to the use of these forecasts. In terms of the value of the forecast, it was shown that the less water available relative to demand, the higher the value of using forecasts. If there is no water stress, there is no need for using forecasts because enough water is available anyway. However, in times of shortages, using forecasts has more value. In north-east Brazil, the value of forecasts will probably increase in the future due to increased water demand as a result of development and population growth.

Seasonal climate forecasts can also be used to manage domestic water use. For example, in Australia, domestic water in most cities comes from reservoirs, and depending upon the amount of water being stored, municipalities have imposed water restrictions on the use of water. Restrictions mostly focus on the use of domestic sprinkler systems. Initially, sprinkler systems can only be used, for example, two days a week; and during periods of more severe water shortages, the use of water for gardening can be completely banned. Filling swimming pools and washing cars are activities that can also be banned as part of water restrictions. In order to decide whether restrictions are introduced, water boards generally use historical climate data in combination with the amount of water currently stored in the reservoirs. In Australia, the amount of rainfall and stream flow is strongly linked to El Niño (Chiew et al, 1998). This demonstrates that regulations imposed by water boards could be optimized by using seasonal forecasts. The average Southern Oscillation Index or sea surface temperatures (SSTs) could be used to forecast future rainfall (Chiew et al, 2003). If the forecast predicts lower-than-average rainfall, restrictions can be introduced even sooner. By introducing less severe restrictions earlier, more severe restriction can be avoided later. If more-than-average rainfall is predicted, then restrictions may be introduced later, and unnecessary restrictions will be avoided.

There is generally a potential to use seasonal climate forecasts to improve water allocation for irrigation in semi-arid regions with variable climates (Chiew et al, 2003; Ritchie et al, 2004). Water managers tend to be conservative in the amount of irrigation water that they allocate at the beginning of the season. Initial allocations are usually relatively low and based on historical data, and it is often the case that the allocation increases during the season. By using seasonal forecasts, allocation mechanisms could be improved. If above-average rainfall is predicted, higher allocations could awarded than in the case of dry forecasts. For example, Ritchie et al (2004) suggest using the SOI phase system (Stone et al, 1996) to improve water allocation in the Murray-Darling Basin, Australia.

While the use of seasonal climate forecasts in agriculture has rapidly increased over the last decade (Meinke and Stone, 2005), its use is still very limited within the area of water resources management. Several studies indicate that water managers tend to be conservative and avoid risks in the use of seasonal forecasts (Rayner et al, 2005). Most farmers in semi-arid regions are accustomed to taking risks and see seasonal forecasts as a tool for improving risk management. Most water managers, on the other hand, are not used to short-term risk management and tend to be more conservative when allocating water to different users.

Linking the management of climate variability and adapting to climate change

Managing climate variability and adapting to climate change are often seen as two separate issues; however, there are important links between the two. Improved management of (current) climate variability is usually a good first step towards climate change adaptation. Where water is not properly managed under current climate variability, it is likely that climate change will aggravate water-related problems. If seasonal forecasts or other information on climate variability are used to manage the water, trends in climate are probably observed relatively quickly. In contrast, management systems based on the 'average' year or climatology are more likely to ignore climate trends because recent climate information is not used as part of the decision-making process. Organizations that have a structure to manage climate variability and use recent climate information are usually more flexible and are thus better prepared to adapt to climate change. For example, if water allocations and/or water restrictions are based on seasonal climate information and are not fixed, it is much easier to adapt the allocations to climate change.

Another important link between managing climate variability and change is the impact of a changing climate on seasonal forecasts. As discussed above, most seasonal climate forecasts are based on statistical methods and use the analogue year approach. This approach assumes that an El Niño year in the future will result in similar rainfall patterns compared to El Niño years during the early part of the 20th century. Basically, most statistical approaches assume that there is no climate change.

In the near future, dynamic climate modelling will probably provide forecasts with higher skill than statistical approaches (Coelho et al, 2006). One of the advantages of using output from coupled ocean–atmosphere general circulation models (OAGCMs) is that the impacts of climate change are automatically taken into account in the projections. By using outputs from climate models, it is also easier to manage climate variability at different timescales. Model outputs are usually available at timescales from one month up to the next century. When decision and management systems are modified in order to use this kind of climate information, it is much easier to adapt water management to climate change. The main problem, however, is that it is much more difficult to use climate outputs for water management than the analogue year approach. The main challenge for water managers and climate scientists is, thus, to develop methods and tools that facilitate the use of climate outputs in water management decision-making (Fowler et al, 2007).

References

Ash, A., P. McIntosh, B. Cullen, P. Carberry and M. S. Smith (2007) 'Constraints and opportunities in applying seasonal climate forecasts in agriculture', *Australian Journal of Agricultural Research*, vol 58, pp952–965

Broad, K., A. Pfaff, R. Taddei, A. Sankarasubramanian U. Lall and F. de Assis de Souza Filho (2007) 'Climate, stream flow prediction and water management in northeast Brazil: Societal trends and forecast value', *Climatic Change*, vol 84, pp217–239

Brown, C., P. Rogers and U. Lall (2006) 'Demand management of groundwater with monsoon forecasting', *Agricultural Systems*, vol 90, pp293–311

Chiew, F. H. A., T. C. Piechota, J. A. Dracup and T. A. McMahon (1998) 'El Niño Southern Oscillation and Australian rainfall, streamflow and drought: Links and potential for forecasting', *Journal of Hydrology*, vol 204, pp138–149

Chiew, F. H. S., S. L. Zhou and T. A. McMahon (2003) 'Use of seasonal streamflow forecasts in water resources management', *Journal of Hydrology*, vol 270, pp135–144

Coelho, C. A. S., D. B. Stephenson, F. J. Doblas-Reyes, M. Balmaseda, A. Guetter and G. J. van Oldenborgh (2006) 'A Bayesian approach for multi-model downscaling: Seasonal forecasting of regional rainfall and river flows in South America', *Meteorological Applications*, vol 13, pp73–82

Frederick, K. D. and A. V. Kneese (1990) 'Reallocation by markets and prices', in P. E. Waggoner (ed) *Climate Change and US Water Resources*, John Wiley & Sons, New York, NY, pp395–419

Fowler, H. J., S. Blenkinsop and C. Tebaldi (2007) 'Linking climate change modelling to impacts studies: Recent advances in downscaling techniques for hydrological modelling', *International Journal of Climatology*, vol 27, pp1547–1578

Hamlet. A. F. and D. P. Lettenmaier (2000) 'Long-range climate forecasting and its use for water management in the pacific Northwest region of North America', *Journal of Hydroinformatics*, vol 2, pp163–182

Hartmann, H. C. (2005) 'Use of climate information in water resources management', in *Encyclopedia of Hydrological Sciences*, John Wiley, Chichester, UK

IOCI (Indian Ocean Climate Initiative) (2002) *Indian Ocean Climate Initiative: Climate change in South West Western Australia*, www.ioci.org.au/publications/pdf/IOCI_Technical Report02.pdf

Lemos, M. C., T. J. Finan, R. W. Fox, D. R. Nelson and J. Tucker (2002) 'The use of seasonal climate forecasting in policymaking: Lessons from northeast Brazil', *Climatic Change*, vol 55, pp479–507

Meinke, H. and R. Stone (2005) 'Seasonal and inter-annual climate forecasting: The new tool for increasing preparedness to climate variability and change in agricultural planning and operations', *Climatic Change*, vol 70, pp221–253

Milly, P. C. D., J. Betancourt, M. Falkenmark, R. M. Hirsch, Z. W. Kundzewicz, D. P. Lettenmaier and R. J. Stouffer (2008) 'Climate change – stationarity is dead: Whither water management?', *Science*, vol 319, pp573–574

Pagano, T. C., H. C. Hartmann and S. Sorooshian (2002) 'Factors affecting seasonal forecast use in Arizona water management: A case study of the 1997–1998 El Niño', *Climate Research*, vol 21, pp259–269

Power, S., B. Sadler and N. Nicholls (2005) 'The influence of climate science on water management in Western Australia: Lessons for climate scientists', *Bulletin of the American Meteorological Society*, vol 86, pp839–844

Rayner S., D. Lach and H. Ingram (2005) 'Weather forecasts are for wimps: Why water resource managers do not use climate forecasts', *Climatic Change*, vol 69, pp197–227

Ritchie, J. W., C. Zammit and D. Beal (2004) 'Can seasonal climate forecasting assist in catchment water management decision-making? A case study of the Border Rivers catchment in Australia', *Agriculture Ecosystems and Environment*, vol 104, pp553–565

Smith, I. N. (2005) 'Assessing the skill and value of seasonal climate predictions', in A. Zerger and R. M. Argent (eds) *MODSIM 2005 International Congress on Modelling and Simulation. Modelling and Simulation Society of Australia and New Zealand*, Melbourne, pp1703–1708

Stone, R. C., G. L. Hammer and T. Marcussen (1996) 'Prediction of global rainfall probabilities using phases of the Southern Oscillation Index', *Nature*, vol 384, pp252–255

Souza, F. A. and U. Lall (2003) 'Seasonal to interannual ensemble streamflow forecasts for Ceara, Brazil: Applications of a multivariate, semiparametric algorithm', *Water Resources Research*, vol 39

7

Adapting to Climate Change in the Water Sector

Jeroen Aerts and Peter Droogers

The Intergovernmental Panel on Climate Change's *Fourth Assessment Report* (IPCC, 2007) defines adaptation practices as 'actual adjustments, or changes in decision environments, which might ultimately enhance resilience or reduce vulnerability to observed or expected changes in climate'. This seems, at first, a new issue; but adaptation is not new to water management. Water managers throughout history have a long record of adapting to the impacts of weather and climate through a range of practices such as irrigation, drainage and flood protection strategies (e.g. Aerts and Droogers, 2004; Adger et al, 2007). In terms of governance and institutional settings, water management also has a track record, such as the establishment of the Dutch water boards several centuries ago.

Long-term changes, however, such as climate change and socio-economic trends, pose a new challenge to water management as they are inherently uncertain (IPCC, 2007). This makes it difficult to translate these trends into quantifiable boundary conditions on the basis of which concrete water management strategies and measures for daily operational water management can be developed (Aerts et al, 2008a).

As a result, the new element in adapting to climate change in water management is an unknown future. In decision-making, this means that you may have to invest in water infrastructure based upon future (unknown and even unsure) risks. The aspect of long-term changes, uncertainty and possible feedback mechanisms between different trends and their effects leaves two fundamental issues related to adaptation in water management:

1 How much adaptation is needed to cope with future climate change?
2 What is required to develop and implement adaptation strategies given the aspect of uncertainty?

The first aspect has been addressed in the previous chapters. This chapter focuses on the second issue and provides an overview of the main challenges that need to be tackled in water management and adaptation.

Dealing with future uncertainty in water management

The future is inherently uncertain; hence, climate change presents another set of complex conditions for water managers seeking to identify effective interventions in the water system. On the one hand, developing responses to these impacts that can be well specified is, at least conceptually, relatively straightforward. For example, where accurate estimates of changes in stream runoff affect hydropower generation, distribution effects can be compensated for by changing reservoir operating rules. Similar changes can be devised for most non-catastrophic changes in climate and water resource systems as long as their nature can be accurately projected.

It is far more conceptually difficult to plan for uncertainty, variability and risk. These are, however, probably the most important consequences of climate change since climate change projections are often inconsistent and lack accuracy at the regional and local scales. Moreover, relatively short series of historical data can no longer be assumed to represent, however imperfectly, future conditions.

The aspect of uncertainty has driven climate adaptation research into new – adaptive – approaches in water management that anticipate and enable water managers to cope with future uncertainties. Hence, at its core, adaptation is about flexibility: the ability, vision and resources required to shift water management strategies so that they become resilient under a wide range of future conditions.

New approaches for dealing with future uncertainties in water management have been introduced (e.g. Gleick, 2003). For example, the development of flood insurance, flood risk-mapping systems and general risk management approaches that specifically address the probability of certain future trends are commonly used in spatial planning research and are gaining increasing attention in water management (e.g. Burby et al, 1999). Furthermore, in the social sciences, the concept of adaptive (water) management has been introduced, which aims at more institutional flexibility and provides stakeholders with a central role in an iterative 'social learning process' (Folke et al, 2002; Pahl-Wostl et al, 2007).

Numerous studies examine the vulnerability of the water system to climate change and also highlight the aspect of uncertainty in this context (e.g. O'Brien and Sculpher, 2000; Smit et al, 2000; Adger et al, 2007; Füssel and Klein, 2006). The IPCC (2007) shows how vulnerability and adaptation relate to one another (see Figure 7.1). If we apply this concept to the water system, the figure shows that the current water system has an adaptability (coping range), which is set up and designed according to current climate conditions and historical information. Climate change, however, will enhance both climate variability and mean climatic parameters; hence, the water system becomes more *vulnerable* as future climatic effects are projected beyond the current coping range. Through extra adaptation measures (both physical and political), the coping range can be increased as well as the threshold above which the system becomes vulnerable.

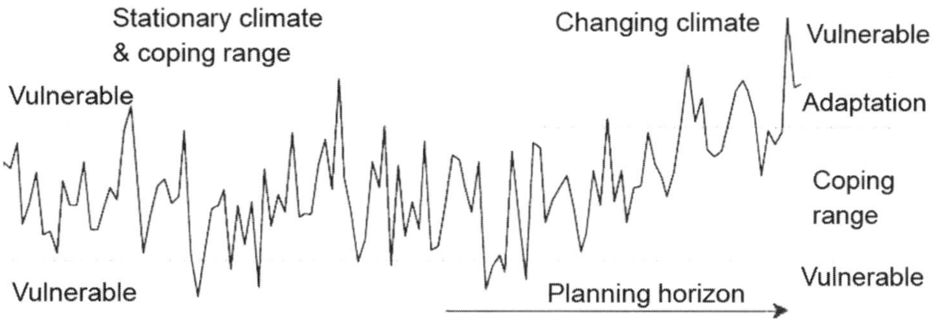

Figure 7.1 *Adaptability (coping range) of the water system under current climatic conditions*

Note: Climate change will enhance the range of effects and, hence, enlarge the necessary coping range; additional adaptation measures are needed to maintain current acceptable vulnerability levels

Source: IPCC (2007)

Box 7.1 Dealing with uncertainty in the case studies

Perth Water Corporation, responsible for serving 1.5 million people, experienced real water shortages at the beginning of this century with shallow aquifer production well fields taken off line to reduce abstraction from environmentally sensitive areas. A new study on metropolitan water supply for Perth, Western Australia, did not use climate change projections as the general impression by decision-makers is still that climate change projections are too uncertain. The study assesses future droughts on the basis of some severe droughts in the past. The main conclusions, however, were that at least some of the droughts that occurred in the past could be the result of the enhanced greenhouse effect and for future planning, only a qualitative 'hotter and drier' statement is considered as the principal planning strategy.

Similar to the Perth case, in Germany, awareness about climate change was fuelled after the major flood in 2002 and the drought in 2003. Regarding the use of climate change scenarios, it was stated that the uncertainty in hydrological responses to changing climate is higher than the uncertainty in climate input. Moreover, while climate change is considered as an important challenge for future water management, until today, the major impacts on regional water resources are due to changes in socio-economic systems.

It is interesting that a number of climate change assessment studies do not rely on the 'impact-analysis' approach following the steps of:

- select one or a set of climate change projections;
- evaluate the impact of these projections;
- define adaptation measures; and
- determine the effectiveness of these measures.

A typical example of where this classical approach is not followed is described in the Thailand case (see Chapter 9). The focus of this case is on flooding, and a combined approach of using past trends (in impact and adaptation) and general climate change statements (higher frequency of intense rainfall events, sea-level rise and decreased inland rainfall) was employed. Based on these qualitative evaluations, a coping and response strategy at the policy-maker level was devised.

In cases where climate change projections were actually used, this was achieved in a rather straightforward way. For the Berg River Basin in South Africa, climate change impact and coping assessments were undertaken for only one story line and only one global climate model (GCM). The use of only one deterministic climate change scenario was advocated by the fact that downscaled stochastic climate scenarios do not currently exist for the region.

Furthermore, it appears that the impact of climate change on aquifers is a relatively unexplored area. A complicating factor in this respect is that groundwater recharge is still considered as the final outcome of other complex hydrological processes. In Yemen, however, where groundwater is the main resource for drinking water, great concerns about the sustainability of the resource are emerging. In this area, the extent and pace to which groundwater resources are being threatened are unknown. The uncertainties of changes in precipitation in the region are substantial: GCMs could not provide unambiguous results. However, all GCMs are consistent in expected increases in temperature leading to higher evaporation and, thus, lower recharge rates. In addition to this remains the uncertainty that people's behaviour and politicians' policies can be a larger challenge to the overexploitation of scarce groundwater resources in Yemen, rather than climate change itself.

What are the requirements for adaptation in water management?

The IPCC and other bodies point to the importance of building adaptive capacity and resilience in (water) management practices in order to respond to future uncertainties. *Resilience* can be defined as the ability to absorb disturbances. *Adaptability* is the capacity of a socio–ecological system (SES) to manage resilience, also referred to as 'adaptive capacity'. Systems with high adaptive capacity are able to reconfigure themselves after a shock due to an extreme event.

Although the above definitions capture many of the features of resilient and adaptive 'natural systems' (the water system), they do not emphasize *agency*: the ability of water managers and stakeholders to take proactive action and to shift strategies in response to perceived or projected changes. Hence, there is a fundamental challenge linking broad concepts of system dynamics with the day-to-day world of water management and responses to increases in risk and uncertainty associated with climate change.

The key to managing uncertainty and promoting flexibility is *risk management*, which needs to be addressed within current water management. The practical implications of concepts regarding risk management for this day-to-day world are far from fully defined; but various research results show that the following elements are key to more adaptive water management:

- *Flexibility and robustness.* Dealing with uncertainty in water management policies is crucial in reducing vulnerability. The development of water management strategies and infrastructure that have high levels of flexibility – or robustness – will almost certainly contribute to both resilience and adaptive capacity as climatic and, consequently, water resource conditions change. This can be done by risk-pooling mechanisms, emphasizing diversification and considering a variety of future scenarios under which alternative solutions are evaluated (Figge, 2004; Aerts et al, 2008b). Many water resource decisions, particularly those involving the construction of large-scale infrastructure, are essentially irreversible and need to be assessed against more flexible strategies.
- *Cross-sectoral cooperation.* More attention will need to be paid to related sectors, notably finance and insurance, regional economic development and livelihoods. Water management should not only focus on managing the probability of events, but also on reducing its consequences. This implies that improved cooperation between water management and spatial planning is important.
- *The ability to learn.* As change proceeds, the ability to learn (i.e. to draw on experience and analysis in the formulation of new strategies rather than 'reinvent the wheel') is widely emphasized in the resilience literature as central to the ability to adapt. This means that there is a need to develop educational approaches for climate and water specialists that are capable of evolving as new information and perspectives emerge. Stakeholder involvement within participatory processes is also seen as key to stimulating adaptive capacity (Pahl-Wostl et al, 2007)
- *Governance.* The ability of systems and populations to recognize change and to respond to it are central to adaptive capacity. This is as much about *agency* as it is about the structural or technical flexibility of the water system. Recent research, for example, demonstrates that effective social responses to floods and droughts often have little to do with water *per se*, but with institutional structures and effective governance (Moench and Dixit, 2004; Aerts et al, 2008a).

Risk management and IWRM

Although almost all forms of potential management intervention have been discussed at one time or another under the integrated water resources management (IWRM) framework, risk management or adaptation to change and uncertainty are still rarely addressed. Furthermore, many of the water-linked but non-water-focused sets of intervention (such as economic diversification and insurance for risk management) that are likely to be central to adaptation have not been a major focus of attention in IWRM activities. Instead, the emphasis of most work on IWRM, as in the definition, is on water development and management to maximize economic and social welfare.

IWRM has been defined as:

> A process that promotes the coordinated development and management of water, land and related resources in order to maximize the resultant economic

> and social welfare in an equitable manner without compromising the sustainability of vital ecosystems. (GWP, 2000)

One of the greatest risks that global society faces is, in fact, the likelihood that most decision-making will occur incrementally without recognition of the broad, and often irreversible, strategic choices being made. As the impacts of climatic variability and change become evident, local areas are likely to demand investment in protective infrastructure or water supply. If such investments are made without wider evaluation of alternative solutions or their long-term sustainability, then relatively inflexible, often unsustainable, hard infrastructure-led approaches are likely to dominate.

On a conceptual level, therefore, adaptation will require conceptual frameworks that, while retaining many of the elements that have been developed as part of IWRM, have a significantly different focus. Water management strategies will need to evolve in ways that place a much greater emphasis on risk, uncertainty and the ability to respond to change and inevitable surprises. This will require very tangible interventions to control risks using a combination of adaptive management approaches, such as *diversification* (Aerts et al, 2008b) of measures and *risk-pooling* mechanisms (e.g. insurance), along with strategies for living with water (e.g. wetland restoration and mangrove rehabilitation for flood protection), rather than attempting to control water according to purely cost-efficiency rules that apply under current climate conditions.

Box 7.2 Institutional aspects and adaptation in the case studies

Most case studies emphasize the important role that institutions play in climate change policies. It is not always clear what 'institutions' refer to. In the case study for the Elbe in Germany (see Chapter 15), a useful definition of institutions is applied: 'set of rules, decision-making procedures and programmes that define social practices, assign roles to the participants in these practices, and guide interactions among the occupants of individual roles'. The case study defines basic elements of institutional adaptation:

- knowledge and information;
- polycentric governance;
- participation;
- sectoral integration;
- flexibility;
- openness for experimentation; and
- political willingness.

The Umgeni Water Utility in South Africa (see Chapter 12) decided that, in terms of institutional aspects regarding climate change, two issues should be covered:

1. more accurate evaluation of impact and adaptation; and
2. awareness-raising for all clients.

The first aspect is mainly covered by collaboration with universities and research institutions. By developing improved downscaling techniques combined with hydrological impact models, Umgeni Water is trying to achieve better predictions of threats and potential adaptation measures. In addition, creating awareness among its clients, the water service authorities and end users is a very high priority despite uncertainty regarding the impacts of climate change.

History also shows that institutional reforms are often triggered after a weather-related disaster. Typical examples from Australia show that after a dry period where a ban on sprinkler use was effective, sudden changes were imposed, such as new water acts, the establishment of a new water agency and the construction of a desalination plant. Similarly, in The Netherlands, two near-flooding events have completely changed the policy regarding safety, where a purely technical approach has been replaced by a more spatial-planning approach.

For Thailand, the priority has been to respond to the challenges posed by climate change in relation to flood and disaster management. Table 7.1 provides an example of how the Thailand case addresses each of the four key elements of adaptive water management.

Table 7.1 *The four key elements of adaptive water management in Thailand*

		Key elements in adaptation strategies and policies			
Case study	**Issue**	**Flexibility and robustness**	**Cross-sectoral solutions**	**Social learning and participation**	**Governance**
Thailand	*Floods*	Diversification Use of disaster cycle Address uncertainty in planning	Space for water Wetland restoration Relocation of settlements	Enabling local communities Strengthen link between knowledge institutes and practitioners	Establishment of Department of Disaster Prevention (2002)

A focus on risk management within IWRM is an essential starting point. Risk management implies the evaluation of alternative courses of action, attempting to balance strategies and recognizing when irreversible decisions are being made. Giving risk management *the* central place in IWRM should shift the attention of professional communities away from a narrow focus on water or climate impacts *per se* and towards the much wider array of strategic pathways that are necessary to respond flexibly to climate change. In a practical sense, it implies the development of specific mechanisms to bring together the institutions and organizations that society has developed for dealing with risk and uncertainty together with the institutions and organizations that society has developed for managing water. It also implies developing the capacities within communities and institutions that will encourage an explicit focus on risk and the establishment mechanisms to support flexibility and ability change as conditions evolve.

Water management and spatial planning

Research shows that societal change and economic development are mainly responsible for increasing losses due to climate-related disasters (Bouwer et al, 2007). By 2015, loss potentials among the world's ten largest cities, most of which are in developing countries, are projected to increase from 22 per cent (Tokyo) to 88 per cent (Jakarta). A repeat of the July 2005 floods in Mumbai in 2015 could cause 80 per cent higher losses and affect 20 per cent more people, independent of climate change.

This shows that water management strategies should not only focus on managing (or lessening) the probability of an event (e.g. floods and droughts), but should also address mitigating its consequences. For example, decisions to build hard infrastructure to protect regions from floods, storms and sea-level changes or to supply water in arid zones are turning points. Once an area is protected, people will count on that protection and will have little incentive to take steps to further reduce vulnerability within protected areas. Furthermore, as population and investment grow in protected areas, political pressure to maintain protection at whatever cost will extend as well. The combination of spatial planning and water management plays a crucial role in this respect.

Table 7.2 *Increase in mega-city disaster loss potential from 2005 to 2015: Ranking is by population in 2015*

	Population estimates (million)			Estimated GDP (US$ billion at 2005 purchasing power parity)		
City	2005	2015	Change (%)	2005	2015	Change (%)
Tokyo, Japan	35.2	35.5	0.8	1191	1452	22
Mumbai, India	18.2	21.9	20.2	126	226	79
Mexico City, Mexico	19.4	21.6	11.1	315	489	55
São Paulo, Brazil	18.3	20.5	12.0	225	336	49
New York, USA	18.7	19.9	6.2	1133	1408	24
Delhi, India	15.0	18.6	23.6	93	170	82
Shanghai, China	14.5	17.2	18.8	94	167	77
Dhaka, Bangladesh	12.4	16.8	35.5	52	94	81
Jakarta, Indonesia	13.2	16.8	27.3	98	184	88

Source: Bouwer et al (2007)

This dynamic also holds true for water supply in arid areas. Once water supplies are ensured, people have little direct incentive to diversify into low water-intensity forms of livelihood. Such incentives can be created through water pricing and other economic or regulatory mechanisms – but the political difficulty in implementing such measures should not be underestimated.

Decisions in adaptation options

In many cases, climate change may necessitate choices between alternative strategic approaches to water management, while in others it will require shifts or changes in the mix of strategies employed. Such strategic issues involved are best illustrated in relation to the choices likely to be faced in vulnerable regions. Because climate change impacts on the water sector are likely to be particularly pronounced in coastal areas, large river basins and arid zones, we will focus on these.

Coastal zones

Where sea-level rises and extreme storms are concerned, water managers can, in a broad sense, either attempt to fully protect large areas by using structural interventions in order to maintain current land-use and development trends, or they can concentrate on structural protective measures in smaller areas while leaving large sections of land open to either permanent or intermittent inundation. These alternatives are, of course, not mutually exclusive in an absolute sense: most coastal regions already provide different levels of structural protection in different areas. As broad strategic approaches, however, the emphasis on one or the other of the alternatives is fundamentally different.

Approaches in large coastal regions that rely on infrastructure for protection will engender a series of essentially irreversible decisions. Large-scale engineering works with long construction lead times and high levels of investment will be required. More importantly, once regions have attained some level of protection against storms and sea-level rise, high levels of investment by individuals, corporations and other entities are almost certain to occur within the 'protected' areas. This will, in turn, create political and land-use conditions that are far harder to reverse than even the investments in the protective works themselves. Establishing a minimum level of protection will, in effect, commit society to development pathways that assume and require such protection to be maintained however climate conditions evolve.

Where risks are concerned, investments in structural protection will almost certainly reduce the impact that moderate storms or sea-level changes have on buildings and economic activity in coastal areas. Risk, however, is a function that depends upon both the probability of an event and its consequences. When structural interventions reduce the frequency with which storms and sea-level rises inundate coastal areas, but those same structural interventions catalyse increased investment, then aggregate risks will tend to increase. If protective structures fail, then the consequences in terms of lost investment will be far larger. Furthermore, although flood frequency may decline, the ultimate probability of inundation occurring as sea levels continue to rise may well increase. As a result, unless hydrologists and engineers can both specifically and accurately project climate conditions, approaches that rely primarily on structural approaches to protecting large coastal areas will carry high levels of apparent risk.

Such approaches will also catalyse patterns of development and investment that have low levels of flexibility and are difficult to adapt if climate conditions do not evolve as anticipated. At society level, people may assume that protection is ensured and they will build and make other investments within 'protected' areas on that assumption. Furthermore, because they are acting on the assumption that risks are low, they are likely to have little incentive to invest time, energy or finances in diversification or other courses of action that could reduce vulnerability at the level of households, businesses or other local organizations. Finally, because large-scale structural protection measures can only be constructed and maintained by equally large-scale technical and professional organizations, such organizations and their perspectives will tend to dominate water and climate management debates. In addition to the irreversible nature of the infrastructure itself, this institutional dimension may have implications for society's ability to 'iterate' – that is, for regions to evaluate and incrementally adjust strategies as experience accrues and conditions evolve.

In contrast to strategies that rely primarily on structural protection, approaches that confine protective investments to small, particularly high-value, areas (such as urban and town centres) and emphasize a mix of techniques for adjusting to rather than controlling the consequences of sea-level changes and coastal storms would generate very different development and risk trajectories. Such strategies would encourage land-use patterns (such as the maintenance of wetlands, coastal marshes and agricultural areas) that 'allow water to spread' and absorb the impact of storms. They would also involve a focus on flood- and storm-'adapted' infrastructure that is designed to ameliorate the impact of inundation on economic activity, environmental values, housing and so on. This type of infrastructure would, almost certainly, involve much more distributed and individually smaller patterns of investment than would be the case with large structural protection works. It would consist of changes, for example, in the design of buildings (raised or floating) and protective works (reductions in scale that are sufficient to reduce the force of storms, the concentration of large flood flows or depth of flooding, while not actually eliminating inundation). It might also involve investments in early warning, communications and transport systems that allow for movement of people and goods out of the path of storms when they occur.

This second strategic approach is likely to catalyse very different coastal development pathways and patterns of risk from approaches that rely on large structural protective measures. Where coastal development patterns are concerned, reliance on more 'adapted infrastructure' and reduced protection from regular events is likely to encourage patterns of development where high-value investments are concentrated in areas with available protection, while investments in other areas are reduced or are of a nature that is not affected by intermittent inundation. Housing and major industrial activities would tend to be more concentrated while more extensive land use (environmental, agricultural, etc.) would occupy less protected areas. Where risk is concerned, because individual investments are likely to be both smaller scale and require less lead time, flexibility and reversibility will be higher. The probability of inundation within any given period may be higher – but the consequences would be, by design, much lower. Net risk should, as a result, be lower. Whoever bears the risk may also change because full protection (or the illusion of protection) is never made as an explicit priority except in the highest-value areas; investments in less protected areas are also less likely to qualify for insurance. Individuals and organizations interested in making such investments will, as a result, be forced to both recognize and absorb much more of the risk that they entail. The political and social dynamics may be quite different between strategic approaches. The mix of individually smaller and more reversible investments combined with greater exposure to more frequent, but arguably smaller, risk vectors may generate conditions that encourage greater levels of social involvement in climate- and water-response activities. This could build social and political dynamics that, in effect, result in continuous re-evaluation or iteration regarding the effectiveness of individual interventions and wider strategic approaches within regions. The political dynamics would, as a result, probably be different from the highly centralizing tendencies inherent in approaches that rely primarily on structural protective measures.

Although the above contrasts in the implications of different strategic approaches for coastal areas are far from comprehensive, they illustrate the fundamental nature of the alternatives involved. Decisions to protect areas (whether made incrementally or proactively as part of an overall coastal protection plan) catalyse patterns of development that are politically and economically inflexible and difficult to reverse. More adapted approaches have, in contrast, greater flexibility. Where risks are concerned, the balance between approaches influences the nature of exposure, the degree to which different groups are aware, and who is likely to bear the risk. Choices early in the development process regarding the relative balance between approaches shape long-term strategic options. Overall, although different mixes of protective and adaptive measures are likely to be used in any given situation, the choice between approaches does represent true alternatives that are likely to generate very different patterns of development and risk as climate change proceeds.

Large river basins

The contrast between strategic approaches for dealing with climatic variability in large river basins has very similar elements to the contrast discussed above for coastal areas. Where flood control is concerned, for example, fundamental distinctions exist between strategies that rely primarily on structural measures (dams, embankments and diversions) to control flows and strategies that allow water to spread, but mitigate the impacts of flooding by discouraging the development of vulnerable activities and investments in floodplains, strengthening early warning and encouraging drainage. Unless future climatic conditions can be specified with a high degree of precision, any attempt to identify future flow volumes and, hence, required embankments will carry a high degree of uncertainty. Strategies for 'living with water' may be much more robust in relation to the uncertainties associated with climatic change than control-based strategies. In addition, new types of early warning information could enable populations in floodplains to move valuable assets out of harm's way and take other actions to reduce flood impacts on relatively short notice.

Box 7.3 Adaptation strategies in the case studies

In general, one can distinguish between adaptation (also referred to as coping or measures) in infrastructural and non-infrastructural actions. A typical example of a non-infrastructural adaptation measure has been followed in The Netherlands. The traditional approach of flood protection by elevating dikes has been replaced by the so-called principle of 'room for water'. This change in thinking, for many seen as a paradigm shift, has been worked out especially for the main rivers where a €2.4 billion project, referred to as Room for the River, created controlled flooding areas along the river. These areas are used for agriculture during normal to low river discharges, while during periods of high flows these areas are used as a buffer to protect the more vulnerable sites from flooding. Landowners are being compensated for potential damage; and, in some cases, flood-prone areas are purchased by water boards responsible for water management.

Adaptation to potential water shortage in the drinking water sector has always been considered as a straightforward solution: either increasing the supply or encouraging water conservation. Often, the latter has been seen as the cheapest and the most environmentally friendly method. In Australia, measures to reduce water consumption have been very effective so that a further reduction in consumption in the future would be harder to achieve. Management of climate risk and uncertainty should therefore include other measures as well. One of the risk-avoiding strategies is to have diversified sources of water that are climatically and hydrologically distinct. Perth Water Corporation, responsible for drinking water supply in Western Australia, has implemented this principle by obtaining drinking water from groundwater, surface water and a desalination plant.

This principle has already led to concrete adaptation strategies by Perth Water Corporation, including the following five measures to enhance supply:

1. seawater desalination;
2. recycling of treated wastewater;
3. managed aquifer replenishment;
4. thinning of selected trees in forests; and
5. water trading.

Reducing water consumption has been mainly achieved by temporarily banning sprinklers; but long-term population growth and economic development are expected to increase demand substantially over the coming decades, so these five concrete actions have to be further implemented.

The case of Thailand describes probably one of the most effective adaptation measures: economic development in conjunction with decreasing social vulnerability. Highly developed and populated countries, on the contrary, consider the extent to which a small disaster might result in enormous economic damage, such as in The Netherlands (see Chapter 10). Moreover, people in more developed countries are not 'used' to disasters and are less prepared to overcome a potential disaster. This contradiction is so far a somewhat unexploited field of research. Infrastructural projects are, in general, still considered the best adaptation measures. A recent example launched by disaster experts and politicians was a proposal to build an 80km-long wall to protect Bangkok and two surrounding provinces. It is interesting that alternatives were considered later as well, such as making more space for water by restoring multifunctional seasonal wetlands, and directing settlements further away from low-lying coastal areas.

The case study on the Berg River in South Africa (see Chapter 14) concentrated on:

- increasing water storage capacity by constructing a reservoir; and
- introducing water markets.

A quantitative so-called 'hydro-economic' modelling approach was developed to assess the impact of climate change on the 'welfare' that water provides expressed as hypothetical monetary revenues. The study shows that for the Berg River Basin (15,000ha of irrigated land), potential damage will be about US$800 million annually for the near future and will reach a level of US$1.5 billion at the end of this century.

The contrast between structural and more adapted approaches to flood control also has implications for environmental management. Maintaining 'space' for water to spread is often equivalent to maintaining riparian zones and wetlands and would, as a result, provide the diverse ecosystem niches necessary for species to adapt as climatic conditions evolve. Structural approaches tend, in contrast, to limit wetland and riparian zones. As a result, maintenance of environmental values would require much more proactive, directed interventions if such values are to be maintained under infrastructure-led strategic approaches.

Finally, there is a major difference between structural and more adapted approaches to flood control in the context of major transboundary basins. Structural approaches often involve interventions, such as the construction of dams and diversions that have major implications for other riparian countries. Close coordination on infrastructure investments in order to minimize the negative impacts for upstream and downstream riparian countries is, as a result, essential in most large transboundary basins. Where more adapted approaches to flood management are concerned, the types of coordination involved can be quite different. Early warning systems, for example, can require sharing of flood data in ways that have mutual benefits to upstream and downstream riparian regions. A transboundary river basin committee with representatives from all riparian countries may facilitate this process.

The flood case illustrates clear strategic differences in options for responding to the impacts of climate change on flood control in large basins. Similar strategic differences may also emerge in relation to the wide array of low flow, water quality and water temperature concerns likely to emerge as a consequence of climate change. Water quality, for example, can be controlled either through approaches that focus on treatment prior to delivery for domestic, industrial or other uses, or it can be managed through watershed-level interventions that emphasize land use, the nature of vegetative cover and avoidance of pollution.

Box 7.4 The Ganges Basin

In the Ganges Basin in South Asia, a significant component of dry-season flow is supported by snow and glacial melt. Climate projections suggest that such flows are likely to decline as the snow melts earlier and glaciers retreat with climate change. As a result, overall dry-season flows are also likely to decline. The likelihood that low flows will decline suggests that targeted courses of action to maintain supplies for the hundreds of millions of users living there will be of critical importance. Sites for surface reservoirs are few and most are located upstream in the upper riparian country, Nepal. Reservoirs are, furthermore, extremely vulnerable even to existing patterns of variability – sudden storms can mobilize huge flows that result in extensive sedimentation within a few hours. The Khulkani reservoir in the Nepal Himalaya, for example, had a design lifetime of 100 years, but was virtually filled when 14.47 million cubic metres of sediment were mobilized by a single cloudburst. Fortunately, the Ganges Basin overlies one of the world's largest aquifers. Although water quality problems (particularly arsenic) are a significant issue in this aquifer, it represents a massive resource that, if utilized with care, could serve as a buffer source of water supply. Strategic decisions need to be made, as a result, regarding the relative emphasis on developing surface water sources to meet low-season demands as opposed to relying on groundwater development. Encouraging further groundwater development would probably be an irreversible decision with major consequences of its own. Once catalysed, groundwater development has proved virtually impossible to regulate in most parts of India.

Another strategic decision in the Ganges Basin involves the relative emphasis on flood control structures, particularly embankments, as opposed to approaches that focus more on 'living with floods'. The role of embankments as a core structural mechanism for flood control in the Ganges Basin has been the subject of intense debate for decades. Even without considering the impacts of climate change, the region faces a fundamental strategic choice across different alternative solutions. It can either:

- attempt to protect large areas from flooding using structural measures, such as embankments; or
- improve the ability of local populations to live with floods through a combination of:
    - small protected areas – ring dikes around urban areas combined with raised villages;
    - improvements in drainage; and
    - early warning and flood mitigation.

As embankments are built, flood protection goes up in the embanked areas, but flood flows may concentrate in smaller areas and may undermine the viability of the techniques for living with floods in unprotected areas. On the other hand, emphasizing the role of climate and water risks would result in a strategic focus on activities that improve flexibility and, hence, the ability of local populations to live with water. Operationally, this would lead to investments in early warning and distributed investments in flood protection (ring dikes or raising villages, flood planning, the development of flood-adapted agricultural systems, etc.). Such investments could be further supported by a portfolio of other operational interventions that include economic diversification to reduce the flood vulnerability of income streams, insurance systems to pool risks and spatial planning.

It is important to emphasize again that differences are, in some ways, not absolute alternatives, but more a question of strategic balance. Attempting to manage the water quality impacts of climate change at a basin level would, for example, not eliminate treatment needs at points of diversion. In contrast, the absence of watershed-level management would force users to rely on treatment.

Once upstream and downstream riparian countries have developed relationships that reflect incremental (and often competitive) individual decisions to construct infrastructure for protective and water development purposes, establishing the trust necessary to move towards data-sharing, common early warning systems and common strategies at a basin level is also likely to be difficult. Overall, as a result, any approach for responding to the impacts of climate change on large basins will require a combination of 'hard infrastructure' and softer 'adaptive' strategies. However, the decisions made regarding relative emphasis have major implications for basin-level relationships, flexibility and the ability to iterate.

Drought and arid zones

Projections from the IPCC indicate that increases in drought frequency and overall declines in precipitation can be expected in many regions, particularly mid-continent zones, as a consequence of climate change. Broadly, approaches for responding to declines in water availability where they occur can focus either on supplying the water required to meet existing and emerging needs, or they can emphasize shaping 'needs' to match water availability. As with hard structural measures for flood control, supply-focused measures for delivering water tend to require large long-term investments in infrastructure that entail significant levels of uncertainty with regard to water availability, demand and a host of other factors. Techniques for shaping 'needs' to water availability, in contrast, often involve a myriad of much more distributed and individually smaller interventions (such as the shift to drip irrigation or different crops; Tonhasca and Byrne, 1994) to improve the economic and technical efficiency of water use. They can also involve shifts in the nature of economics and other activities to displace water-intensive activities. This type of change can range from micro-level interventions to economy-level shifts (such as the increasing reliance of countries in the Middle East on grain imports (virtual water) as opposed to domestic production.

Most water management debates over the past decade have emphasized the difference between supply- and demand-side management in response to water scarcity. These have tended to focus on the contrast between supplying more water and improving the technical efficiency within *existing uses*. They have not emphasized the role of shifts in economic and other activities and how this would require altering the water sector.

Attempting to respond to anticipated declines in water availability associated with climate change through supply- and efficiency-focused measures and use shifting would involve different strategic approaches. Supply-dominated strategies require the ability to predict conditions well in advance so that long-term investments in infrastructure can be evaluated and, where technically and economically viable, made to meet projected needs. Due to the large engineering nature of such infrastructure, this would probably require leadership from fairly centralized governmental and private-sector organizations. Demand-side management strategies would involve very different implementation arrangements, including regulation and economic pricing in order to create incentives for end users to adopt efficient technologies. Finally, high-level policy decisions regarding, for example, regional economic development strategies (the emphasis on agriculture and other water-intensive activities as opposed to low water-demand forms of economic activity) could play a major role. Such high-level decisions would create an environment that enables and catalyses numerous micro-level decisions within households, businesses and other economic units regarding the specific courses of action that they will take in response to water scarcity.

In many ways, the contrast between strategies that are supply led and those that emphasize demand-side management and strategy shifting in the drought context are similar to the contrasts already discussed for coastal and flood-prone regions. Where the uncertainties associated with climate change are concerned, each of the above strategies has very different implications. Infrastructure-led supply-side strategies require long-term advanced planning and investment. They carry a high level of risk if conditions do not match those anticipated. Demand-side management and use-shifting strategies are much more flexible since many (though not all) of the investments can be made rapidly at local levels as conditions dictate. Institutionally, supply-led strategies are likely to require and encourage reliance on fairly large, centralized institutions, while those that emphasize use shifting and demand-side management are likely to require institutions that operate on a much greater diversity of levels from the national policy environment down to local areas.

The complexity of taking directed action to respond to climate change in such pluralistic institutional environments is important to recognize. Demand-side management requires institutional arrangements that enable the transmission of technologies (drip irrigation) and behavioural changes (turning taps off) to millions of end users. The development of institutions capable of accomplishing this in a directed manner is complex, particularly under the stressed conditions common in the context of many developing countries. It has, for example, proved impossible to exert much control over groundwater pumping in most of the world despite steadily increasing recognition at state, national and global levels over the last three decades that this is essential (Burke and Moench, 2000). The institutional difficulty of demand-side led approaches should not, as a result, be underestimated. Responding to water scarcity by use shifting is also complex. National-level decisions, for example, to allow unrestricted imports and exports of grain can create contexts in which water users are forced to shift livelihood strategies as climatic conditions evolve. Relying on 'virtual water' imports could, as a result, enable highly flexible responses to climate change. Reducing reliance on local resources would, however, require the establishment of transport, communication, finance and production systems at local levels that have access to, and are able to produce, products as part of global markets.

Risk management in water management policies

In most countries, water management is comprised of a range of stakeholders, including private companies, community-based watershed management groups, individual farmers and water users, irrigation system managers, municipal utilities, and government organizations at state, national and international levels. Integration, where it occurs, emerges from the common training that different professional cadres receive (e.g. as in the importance of integration under the growing global emphasis on IWRM) and through direct activities where legal, regulatory or operational considerations force interaction.

The risk management landscape is populated by an equally diverse host of actors. These include organizations involved in disaster response and risk reduction, such as police and fire departments, the International Red Cross and other non-governmental organizations (NGOs), national and state governments, and, globally, the International Strategy for Disaster Reduction (ISDR) and the United Nations system. They also include organizations involved in advanced warning applications (such as the weather service) and those explicitly involved in risk management, such as financial institutions and the insurance industry. Finally, and this is essential to recognize, much risk management occurs within communities as households and individuals actively strategize and manage their assets to respond to the opportunities and constraints that they perceive. Actions taken at the household level to diversify livelihood strategies, to build climate-resistant structures, and enter into regional/global labour markets through commuting and migration are often some of the major existing forms of adaptation to floods, droughts and other forms of climatic variability (Moench and Dixit, 2004).

In this inherently diverse context, mechanisms for operationalizing risk management in IWRM in relation to climate impacts can be divided into two broad complementary avenues:

1 The first avenue for operationalizing risk management involves the development of a global discipline and toolkit followed by projects and training of professionals. It would involve the introduction of a major focus on risk into IWRM objectives, concepts, strategies, tools and activities. It would also require effective engagement between actors involved in water management and new communities of actors who, at a minimum, include climate specialists, groups that already specialize in risk (finance, insurance and disaster management) and economic development specialists.
2 The second avenue for operationalizing risk management could be described as an enabling one. It emphasizes the role that individual actors (individuals, households, businesses and other actors) play in recognizing and responding to risk. This approach recognizes that exposure to risk and the impact of events are shaped, in essence, by the behaviour of different actors within national and global contexts. When people have access to education, transport, communication and

financial or other assets, combined with freedom of mobility and the right to organize, they are in a far better position to recognize and respond to risks than when such factors are absent. Think of conditions under which individuals can diversify livelihood strategies, such as building climate-adapted infrastructure (e.g. flood-proof or insulated houses) or moving when threatened. The ability to manage and adapt to climate risks as they become evident is not equivalent to economic development or income alone. Instead, it can be seen as depending upon the capacities, assets and information that individuals have access to, their freedom of mobility and the balance between markets, civil society (the right to organize) and governmental forms of organization.

In practical terms, it means the integration of a major focus on climate risk within water management paradigms, training and implementation activities. Examples of these are outlined in more detail below.

Global mechanisms

At a global level, three mechanisms already exist that could either be built upon directly or used as operational 'models' to integrate risk management in IWRM. These are:

1. *The Global Water Partnership (GWP).* The GWP was formed, in essence, to develop, promote and disseminate IWRM concepts through support for a combination of research, pilot implementation, training, technical support and, importantly, network development activities. At a global level, formation of the GWP was intended to bring together the very diverse array of academic and applied work on water management and to promote its synthesis into a conceptually integrated approach that could then be applied through the actions of numerous local water managers in the course of their daily work. It represents, in essence, an attempt to create a global discipline that integrates many much more specialized components.
2. *The Hyogo Framework for Action 2005–2015: Building the Resilience of Nations and Communities to Disasters.* This globally negotiated framework identifies a broad set of 'general considerations' (conceptual elements) that need to be incorporated to reduce disaster risk. These are the basis for identified priority areas for action, along with much more specifically identified 'key activities' associated with each priority.
3. *The National Adaptation Programmes of Action (NAPAs).* Following the publication of the *Third Assessment Report* of the IPCC, guidelines were established at the Seventh Conference of the Parties (COP 7) for producing National Adaptation Programmes of Action in least developed countries (LDCs). The NAPA process is, in essence, the primary process at a global level for developing applied responses to climate change. It provides a structure for LDC governments

to plan specific courses of action to support adaptation that would then be financed either through existing national resources or external sources such as the Global Environment Facilty (GEF), multilateral banks or bilateral donors.

The approach used by the GWP could be used to, in essence, create the global networks and set of capacities needed to bring existing and new sets of actors together for the development of a common professional discipline and the capacities necessary to support implementation. This should include members of the water management, as well as private- and public-sector risk management, disaster response and development communities. A framework for action such as that prepared in relation to disaster management would help to identify specific responsibilities and priorities for action that could be used as a starting point for the wide variety of implementation efforts that will ultimately be required. Finally, planning structures combined with specific financing mechanisms could be used to drive the broad professional perspectives and priorities towards the identification of specific directed implementation activities in particular national or local contexts.

Regional mechanisms

At the regional levels, the implementation approaches for climate and water risk management can build upon a variety of mechanisms that already exist in related fields. These include:

- *Risk management and planning processes.* Key to promoting risk management in ongoing and new planning processes is stakeholder involvement. Risk reduction will require the involvement of a larger array of actors than those commonly involved in water management debates. In addition to communities and water-related professionals, actors should also include those involved in disaster management, insurance and regional economic development. Overall, stakeholder-based adaptive planning processes need to be developed to insert risk- and adaptation-related perspectives into the daily activities of utilities, planning departments, irrigation organizations, environmental organizations and other entities active at the local level.
- *Institutional and legal frameworks.* Institutional frameworks and organizations are needed to enable local stakeholders to implement adaptation strategies. Institutions of this type already exist in some countries. In The Netherlands, a legal framework exists ('*watertoets*') that requires new infrastructure and housing projects to be evaluated on their capacity to temporarily store water during extreme precipitation events. In other countries, water boards have been set up to develop institutional arrangements under the IWRM principles. Considering risk management within these activities would encourage addressing long-term uncertainty in water management practices.
- *Educational activities and the integration of professional disciplines.* The building of

professional capacities through universities and widespread education is a core mechanism that has been used in the water sector and other fields to support long-term implementation. New tool kits that enable evaluation and communication of climate- and water-related risks into planning and implementation activities would support this process.

Conclusions

Conceptually, water management approaches need to place much greater emphasis on risk and uncertainty. Strategic decisions in water management are of fundamental importance because they are often irreversible and shape long-term development paths (and, therefore, the nature of risk) within regions. As a result, these strategies must be evaluated on their robustness under a variety of possible futures.

Key to increased resilience of water management strategies is incorporating risk management practices within IWRM. This can be done by increasing flexibility through risk-pooling, diversification of measures, improved interaction with spatial planning, and stakeholder participation.

At a global level, the core challenge is to develop the combination of professional capabilities and implementation experiences that are necessary to give water and climate risk management a central role in the water sector. This is, in many ways, a similar challenge to the one faced by early proponents of IWRM concepts. Operationally, it could be achieved by developing a professional discipline (perhaps using the GWP approach as a model), establishing common global frameworks and developing programmatic approaches that enable the implementation of extensive pilot projects. At regional levels, adaptation requires the creation of institutional environments that *enable* local populations to take appropriate action in response to the specific needs within their areas combined with more directed implementation activities.

Creating such mechanisms within society depends at least as much upon enabling conditions as it does upon interventions that focus directly on specific climate or water risks. Operational interventions to support such enabling conditions are closely related to wider processes of economic development and existing objectives, such as the Millennium Development Goals (MDGs). The GWP could play an important role on the global scale to facilitate this process.

General educational levels within society improve the ability of individuals, households and communities to learn and shift strategies (change occupations). Education may also increase understanding and agency (i.e. willingness/ability to act). Similarly, improvements in transport, finance and communication systems increase the ability of communities to respond flexibly by shifting strategies, migrating or accessing external resources in response to climate, water or other surprises. Such improvements also enable individuals, households and businesses to pool risks by diversifying assets, activities and sources of income.

References

Adger, W. N., S. Agrawala, M. M. Q. Mirza, C. Conde, K. O'Brien, J. Pulhin, R. Pulwarty, B. Smit and K. Takahashi (2007) 'Assessment of adaptation practices, options, constraints and capacity', in M. L. Parry, O. F. Canziani, J. P. Palutikof, P. J. van der Linden and C. E. Hanson (eds) *Climate Change 2007: Impacts, Adaptation and Vulnerability, Contribution of Working Group II to the Fourth Assessment Report of the Intergovernmental Panel on Climate Change*, Cambridge University Press, Cambridge, UK, pp717–743

Aerts, J. and P. Droogers (2004) *Climate Change in Contrasting River Basins: Adaptation Strategies for Water, Food and Environment*, CABI, London

Aerts, J., T. Sprong and B. Bannink (2008a) *Rapport Aandacht voor Veiligheid*, www.adaptation.nl, accessed April 2008

Aerts, J. C. J. H., W. Botzen, A. Van der Veen, J. Krykrow and S. Werners (2008b) 'Portfolio management for developing flood protection measures', *Ecology and Society*, vol 13, no 1, p41, www.ecologyandsociety.org/vol13/iss1/art41/

Bouwer, L. M., R. P. Crompton, E. Faust, P. Höppe and R. A. Pielke Jr. (2007) 'Confronting disaster losses', *Science*, vol 318, 2 November, p753

Burby, R. J., T. Beatley, P. R. Berke, R. E. Deyle, F. French, S. P. Godschalk, E. J. Kaiser, J. D. Kartez, P. J. May, R. Olshansky, R. G. Paterson and R. H. Platt (1999) 'Unleashing the power of planning to create disaster resistant communities', *Journal of the American Planning Association*, vol 65, pp247–258

Burke, J. J. and M. Moench (2000) *Groundwater and Society, Resources, Tensions and Opportunities: Themes in Groundwater Management for the 21st Century*, United Nations, New York, NY

Figge, F. (2004) 'Bio-folio: Applying portfolio theory to biodiversity', *Biodiversity and Conservation*, vol 13, pp827–849

Folke, C., S. Carpenter, T. Elmqvist, L. Gunderson, C. S. Holling and B. Walker (2002) 'Resilience and sustainable development: Building adaptive capacity in a world of transformations', *Ambio*, vol 31, pp437–440

Füssel, H. M. and R. J. T. Klein (2006) 'Climate change vulnerability assessments: An evolution of conceptual thinking', *Climatic Change*, vol 75, pp301–329

Gleick, P. (2003) 'Global freshwater resources: Soft-path solutions for the 21st century', *Science*, vol 302, pp1524–1528

GWP (Global Water Partnership) (2000) *Integrated Water Resources Management*, TAC Background Papers no 4, Stockholm, Sweden, www.gwpforum.org/gwp/library/Tacno4.pdf

IPCC (Intergovernmental Panel on Climate Change) (2007) *Climate Change 2007: Synthesis Report, Fourth Assessment Report*, 'Impacts, adaptation and vulnerability', Cambridge University Press, Cambridge, www.ipcc.ch/ipccreports/ar4-wg2.htm

Moench M. and A. Dixit (2004) *Adaptive Capacity and Livelihood Resilience: Adaptive Strategies for Responding to Floods and Droughts in South Asia*, Institute for Social and Environmental Transition, Boulder, CO

O'Brien, B. J. and M. J. Sculpher (2000) 'Building uncertainty into cost-effectiveness rankings, portfolio risk–return tradeoffs and implications for decision rules', *Medical Care*, vol 38, p460

Pahl-Wostl, C., H. Sendzimir, P. Jeffrey, J. Aerts, G. Berkamp and K. Cross (2007) 'Managing change towards adaptive water management through social learning', *Ecology and Society*, vol 12, no 2, p30

Smit, B., I. Burton, R. Klein and J. Wandel (2000) 'An anatomy of adaptation to climate change and variability', *Climatic Change*, vol 45, pp223–251

Tonhasca, A. and D. N. Byrne (1994) 'The effects of crop diversification on herbivorous insects: A meta-analysis approach', *Ecological Entomology*, vol 19, no 3, pp239–244

8

Climate-proofing

Jeroen Veraart and Marloes Bakker

Introduction

Every day, key decisions are being made about future (infrastructural) investments related to water management and land use across the globe. Expected changes in climate and socio-economic water demands require that water managers reconsider their strategies that aim to minimize flood risks and optimize water supply. Coping with climate variability has been part of water resources management for ages; but the use of systematically collected climate information and daily weather forecasts stems from a more recent date. The usefulness of daily weather forecasts is widely acknowledged in various sectors. However, the operational use of (seasonal) climatic forecasts (see Chapter 6) and climate change scenarios (see Chapter 3) in water management is still limited.

Since the 1970s, several governments have proposed programmes and technologies designed to weather-proof or drought-proof their countries in order to cope with climate variability (see Box 8.1). These programmes can be regarded as the earliest forms of 'climate-proofing', a term currently becoming a buzz phrase. The phrase started to appear in Australian and American policy documents about ten years ago. In the scientific literature, it was probably first described by Glantz (2003), followed by elaborations and alternative interpretations by, for example, Kabat et al (2005) and Hay et al (2005). The concept of 'climate-proofing' could be interpreted in three different ways:

1. a policy objective or an additional standard, a set of risk thresholds or criteria for (sustainable) water management and land use;
2. a decision support system for interventions in water management and land use focused on climate change and climate variability, taking into account the uncertainties that come with a changing climate; or
3. a new planning paradigm for water management, natural resources management and spatial planning, taking into account the future claims on natural resources, risks, opportunities and associated uncertainties.

Moreover, assessing and dealing with future uncertain risks (see Chapter 7) is central to all interpretations of climate-proofing.

Box 8.1 Drought-proofing and weather-proofing: The precursors of climate-proofing

The history of successful agriculture in the Canadian prairies has been punctuated by episodes of drought. During the 1970s, following the recurrence of severe drought in the Canadian prairies, the government launched a programme to 'drought-proof' the prairies. Drought-proofing measures included changes in land-use practices, such as leaving stubble and crop residue in the ground after harvest. Expectations for successfully drought-proofing this region, however, were soon undermined by nature, as droughts and crop losses continued to reappear in the region. Today, Canadians in the region are more specific in their activities – for example, in calling for the drought-proofing of farm water supplies. Different interpretations still surround the concept of drought-proofing; however, it is still being proposed by United Nations agencies as well as by various national governments, such as recently in Australia and India.

In late 1999, the US Weather Research Programme launched a national computing system for forecasting purposes in order to weather-proof economic activities. But within a matter of days, a forecast of light snow for the Washington, DC, area proved wrong when a major winter storm developed, depositing 30cm of snow in the metropolitan area. In March 2001, a storm of major proportions – referred to by some forecasters as a potential 'storm of the century' – had been forecast for the lower half of the north-eastern US. The forecast prompted people and government to take precautionary measures. Stores were emptied of shovels, salt, mechanical snow-removing devices and the like. The track of the storm unexpectedly shifted more than 160km to the north. Afterwards, the governor of New Jersey threatened to sue the National Weather Service for the adverse costly impacts of what he viewed as a grossly 'erroneous' forecast.

Source: Glantz (2006)

Climate-proofing in water management: Debates and paradigms

This book discusses in detail two planning paradigms. The approach described in Chapter 5 will, throughout the remainder of this chapter, be referred to as the *foundational water management paradigm*, while in Chapter 7, *adaptive management* is discussed. Both chapters mention the paradigm of *integrated water resources management* (IWRM). Next to these named planning paradigms, many others exist within policy sciences and spatial planning, such as the distinction between state-led systems and multilayered governance of water resources, or the division of tasks between the public and private sectors. Keeping this in mind, climate-proofing will be placed in the context of the foundational water management paradigm and adaptive management in this section. The section on 'Conceptualizing climate-proofing' describes how climate-proofing could be conceptualized in water management and spatial planning. The final section summarizes the main conclusions and explains the difficulties of the conceptualization of climate-proofing.

The foundational water management paradigm and climate-proofing

Many expert communities involved in water management tend to construct, assess and approach climate risks based upon the stationarity principle, as illustrated in Chapter 5. The basic assumption underlying this stationarity principle is that natural systems fluctuate within a fixed range of variability and that this range does not change (Milly et al, 2008). Water managers supporting this paradigm favour the formulation and revision of design rules and management criteria based upon the statistical analysis of monitoring data and the methodology of stochastic (synthetic) hydrology. Several experts within this paradigm, notably Fiering and Matalas (1990) and particularly Matalas (1997), state that this approach can also accommodate the uncertainties in water management induced by global warming with the operational assumption that stationarity is as meaningful as the assumption of non-stationarity.

Nearly all land-use and water management decisions related to climate variability within this paradigm will result in a trade-off between maximizing water supply and/or economic income and the risk of failure (see Chapter 5). Water managers who use this paradigm in their daily activities tend to favour structural adaptation measures over non-structural measures. The rationale behind this is that they perceive these measures as more robust, resilient and reliable based on the uncertainties as compared to structural measures.

Usually (incremental) no-regret decisions (i.e. measures whose benefits equal or exceed their cost to society) are taken into consideration within this paradigm. These decisions do not suffer from errors of caution and their implementation improves welfare relative to the reference case. In adaptation policies for climate change, the 'error of precaution' could be seen as welfare consequences (Smith and Lenhart, 1996; see also the Berg River Basin case study in Chapter 14). For example, reservoirs are the preferred measure to cope with drought, rather then water demand policies (see Chapter 5 and the Perth, Australia, case study in Chapter 13).

Design floods (Chapter 5) are calculated based on a maximum of accepted probability, and if the maximum accepted probability is exceeded, engineering solutions are largely sought, such as constructing or heightening dikes. The occurrence of such a flood is derived from historical and/or synthetic time-series analyses. In order to take unknown flood risk into account, safety margins are added to design rules for infrastructural measures. Should evidence of increased variability continue (non-stationarity) to become more substantial, water managers have the option to increase the safety margin. In this paradigm, water managers recognize increased vulnerability to extreme weather events by acknowledging the fact that the population living in flood- and drought-prone areas and associated economic capital will increase. This is the main argument used that legitimizes the consideration of coping strategies to climate variability. In addition, but of less importance within this paradigm, water managers have to cope with the uncertainties linked to possible changes in climate.

The adaptive management paradigm and climate-proofing

Alternatively, other experts state that climate changes are likely to produce – in some places and at some times – hydrologic conditions and extremes of a different nature than current systems were designed to manage (Gleick, 1998; see also Chapter 7). In addition, social scientists call for more institutional flexibility and stakeholder participation within water management with a central role in an iterative 'social learning process' as a strategy in complex problem situations, such as climate change (Folke et al, 2002; Pahl-Wostl and Hare, 2004). Because the first ideas about adaptive management were developed by ecologists during the 1970s (Holling, 1978), it is often also reviewed as a paradigm that addresses a widely perceived need to give more prominence to ecological imperatives, such as the concept of 'living with floods' in river basins. In addition, the notion of possible discontinuities and turning points, caused by disturbances and surprises in ecology and society, is further emphasized. These discontinuities are frequently caused by small incremental changes in the drivers that trigger a rapid and large response. The dissolution of the Soviet Union, the Berlin Wall coming down and putting a man on the moon are examples of rapid non-linear surprises within society (Costanza, 2000). Water systems and ecosystems are also exposed to gradual changes in climate, nutrient-loading, habitat fragmentation or biotic exploitation. One might assume that these systems respond to gradual change in a smooth way. However, studies on lakes, coral reefs, oceans, forests and arid lands have shown that smooth changes can be interrupted by sudden switches to a contrasting state, such as shallow lakes suddenly turning from a 'clear water state' into a 'turbid state' or rapid irreversible coral bleaching (Scheffer et al, 2001).

Within this paradigm, attempts are sought to improve the management of scientific uncertainty and ignorance (unknown bounds of the set of potential outcomes and unknown probabilities) in situations where both regulatory action and inaction can have costly but unforeseeable impacts or surprises (Pahl-Wostl, 2002; van der Sluijs, 2007). For example, if one forgets to maintain a dike system or work with fixed risk thresholds, the cost of inaction and regulatory action, respectively, could be an unforeseeable flood.

Scientific experts and stakeholders who approach water management from the adaptive management paradigm tend to prioritize non-structural measures, such as institutional change (e.g. flood risk zoning in spatial planning) or behavioural change (e.g. via water demand policies), and argue that structural measures are far less flexible and often result in irreversible environmental impacts. Within this paradigm, more decisions are taken based upon the precautionary principle (i.e. the absence of full scientific certainty is not used as an argument to postpone decisions that could prevent unproven threats of serious or irreversible harm).

Conceptualizing climate-proofing

As said before, the process of conceptualizing weather- and climate-proofing started in the 1970s (see Box 8.1). Recent programmes initiated in Europe include initiatives in The Netherlands (Box 8.2), the UK and Sweden, but also in other continents – for example, initiatives by the World Bank in Asia (Hay et al, 2005). It is clear that no society, rich or poor, is able to fully insulate its people and human activities from climate- and weather-related anomalies (Kabat et al, 2005; Glantz, 2006). However, the alternative is less clear, resulting in policy-makers searching for design rules, climate risk thresholds and management criteria.

Here, we view climate-proofing as a decision-making process (see Figure 8.1) in land and water management, in which both risks and opportunities of climate change are taken into account in line with the definition of Kabat et al (2005). The opportunities involve, but are not limited to, technological, institutional and societal innovations. In order to meet sustainability objectives as well, the adaptation strategies should not lead to additional greenhouse gas emissions compared to the business-as-usual scenario. The concept of climate-proofing uses a combination of infrastructural and institutional measures in order to adapt to future climate change. Within the climate-proofing approach, risk is seen as a social construct that is not only determined by the probability of exposure and the potential amount of damage (see Chapter 7), but also by elements such as the voluntariness of exposure, the expected benefits of taking the risk and the imaginableness of the consequences. Risk experience and, as a result, the determination of (risk) thresholds are also influenced by factors such as personal experience, access to scientific information and media attention. As a result, in addition to the definition of Kabat et al (2005), we state that climate-proofing is a policy objective that should not be presented in a set of fixed risk threshold(s) derived from science, or supported with a single decision-support system. It also poses the question of who is responsible for determining climate risk thresholds.

Crucial steps in evaluating adaptation policies and associated risks and opportunities are:

- exploring the future by vision-building on climate-proofing;
- learning adaptation to climate change by doing; and
- the design of tailor-made interfaces between science, society and policy.

These steps will be explained in the sections below.

Exploring the future by vision-building in climate-proofing

The future is a moving target: divining its characteristics is always tough. Methodologies to explore the impact of policies on the future have a long history, going back as far as the Greeks and their Oracle of Delphi. Systematic studies, such as vision-building and scenario analysis, have been applied on a large scale in policy-making ever since World War II, but also by private companies – for example, Shell in strategic management in oil exploration and refineries. During the 1970s, the Club of Rome started using newly developed quantitative modelling tools for trend analysis to assess the extent of future environmental problems (Meadows, 1992). Currently, a

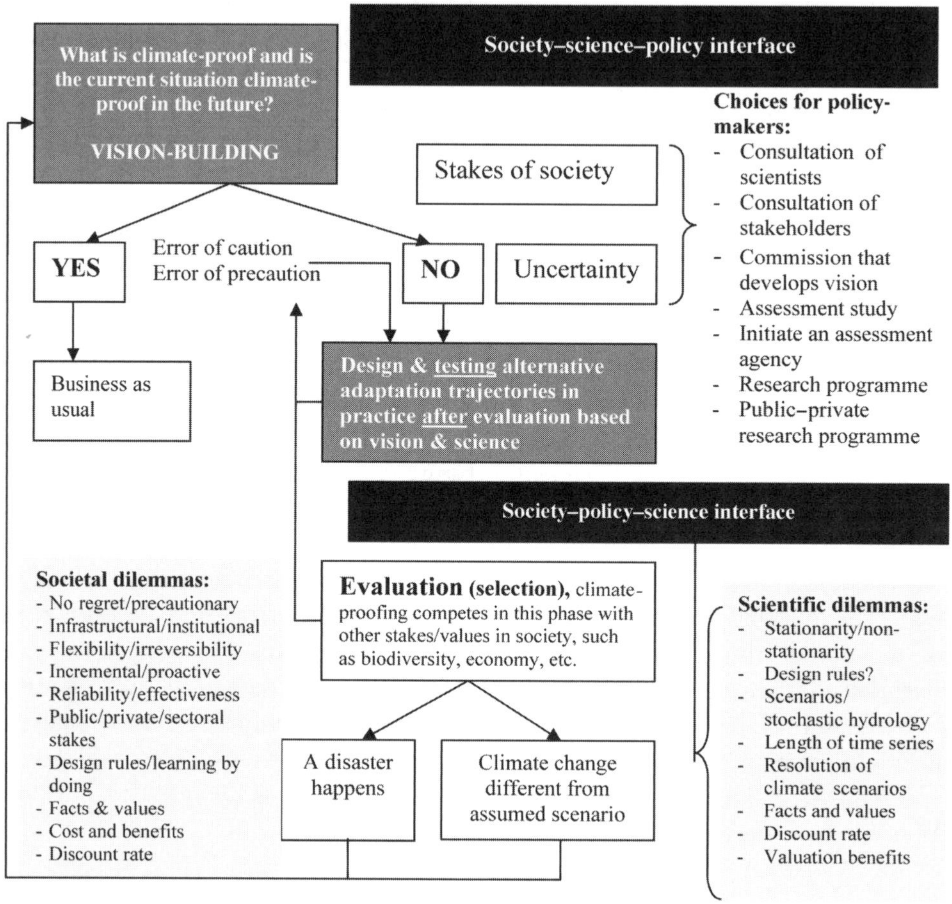

Figure 8.1 *The decision-making process regarding climate change adaptation strategies*

Source: Jeroen Veraart and Marloes Bakker

wide array of methods is available to explore the future. Distinctions can be made (Ruijgh-van der Ploeg and Verhallen, 2002) between:

- formal methods based on a mathematical (global) approach; and
- normative methods based on expert knowledge, including (local) stakeholder participation.

Formal scenarios provide plausible descriptions of how the future could develop, based on a coherent and internally consistent set of assumptions ('scenario logic') about key relationships and driving forces, such as economic growth, energy consumption or emission of greenhouse gases. A scenario is part of a set of scenarios, which together span the range of likely future developments. This is also the approach followed by the Intergovernmental Panel on Climate Change (IPCC). Normative methods go one step further and aim to create a shared vision of a sustainable and desirable society, thereby involving values. The development of these scenarios relies much more on qualitative information and personal perceptions than on analytical methods (Ruijgh-van der Ploeg and Verhallen, 2002).

The climate-proofing approach should start with vision-building by scientists, policy-makers and other representatives of the stakeholders in the involved catchment or region. Vision-building could be done by formulating narrative stories for the future or designing maps of the future. The desired future could also be visualized or described by architects, film-makers, journalists and product designers. At this stage, a normative method is preferred because in designing adaptation strategies, practical knowledge is equal to scientific knowledge in judging how effective and desirable adaptation options are. In addition, the preferred future is, by definition, a societal value. Experience also shows that the use of pre-described 'formal' socio-economic scenarios, in particular, is frequently debated in politically charged atmospheres and hampers the process of designing adaptation trajectories in dialogue with scientists. On the other hand, quantified information about future climate change and related impacts on natural resources and economic sectors are also often requested by decision-makers. Questions such as: 'What is climate-proof?' or 'Is our region vulnerable to climate change?' are often posed at this stage. Within the climate-proofing approach, the aim is to develop a vision that is a co-production of policy, societal values and scientific expertise. The design of the process is complex and requires delicate tuning between these three entities, and preferably chaired by someone from the society or policy domain. Expert judgement derived from formal climate scenario methods and impact assessments can be included, but should not dominate the process of vision-building. At a later stage, when societal preferences for the future are clear, the identified adaptation trajectories should be evaluated using formal scientific methods in order to assess the impact of the designed adaptation trajectory compared to the business-as-usual scenario.

Evaluation of adaptation trajectories

The most important difference between the climate-proofing decision-making pathway (see Figure 8.1) and the 'foundational water management paradigm' is that the (scientific) evaluation phase of alternative adaptation options is done later in the process after the initiations of practical pilots (learning adaptation by doing). These 'adaptation trajectories' consist of a set of cross-sectoral (structural and non-structural) measures for a defined system – for example, a catchment, river delta or country. This approach has many components that fall within the scope of integrated water resources management (IWRM) (see Chapter 5). The difference is the way in which knowledge transfer (society–policy–science interfaces), future risks (partly unknown) and current uncertainties are treated within the decision-making process.

This book describes a selected number of scientific tools that can be used within the interface of climate science and water management, particularly climate forecasts, climate scenarios, time-series analysis and risk assessments (see Table 8.1). In addition, some socio-economic decision-support models are discussed within the case study of the Berg River Basin, South Africa (see Chapter 14). For additional reading, we refer the reader to Toth (2000) and Dessai and van der Sluijs (2007).

Table 8.1 *Scientific tools addressed in this book*

Tool	Remarks regarding application in water management (pros and cons)
Time-series analysis of climate variables (Chapter 2) Stochastic (synthetic) hydrology (Chapter 5) Extreme value analysis (Chapter 5)	• Predictability: nature and amplitude of climate variability strongly vary on spatial and temporal scales. • Trend detection: signal-to-noise ratio, as well as the availability of (long-term) homogeneous observational records, determines detection of a trend.
Climate scenarios (Chapter 3)	• Climate projections: first 50 years of uncertainties in the initial conditions are more important than uncertainties in external forcings (GHG emissions). • Different approach of using climate scenarios in transboundary rivers (i.e. the Rhine). • Tailoring climate information: lessons learned from the climate effects atlas.
Climate forecasts (Chapters 2 and 6)	• Correlations between sea surface temperatures and ENSO, PDO, NAO and IOD as a basis for seasonal forecasts (ENSO is the best). • Land–atmosphere interactions, particularly soil moisture, as a basis for seasonal forecasts. • Developments in decadal forecasts (10 to 15 years).
Climate models (GCM/RCM) (Chapter 3)	• Predictability is determined by initial conditions and external forcings. • Model scenario projections: from general circulation models to Earth system models. • Uncertainty associated with imperfect modelling systems: can be covered by multi-models.
Risk assessments (Chapters 7 and 9: Case study on Thailand)	• Empirical correlations between water management and (mitigated) impacts are based on current (climatic) conditions. • Learn from other risks (e.g. nuclear power). • Risk = P × potential damage, whereas perceived risk = f (P) = damage, media attention, imaginableness of a disaster, risk distribution, voluntariness of exposure, etc.). • How to deal with/assess the impacts of adaptation measures outside the water management arena? • The case study on flood and disaster management in Thailand (Chapter 9) describes societal uncertainties regarding mitigating flood risks.
Cost-benefit analysis (Chapter 14: Case study on Berg River Basin)	• Deterministic scenarios instead of stochastic scenarios. • Identification of parameter values based upon empirical data. • Coupling of climate scenarios with hydrological models, and with economic models. • The principles of 'caution' and 'precaution' are defined by mathematical expressions. • Use of different possible future water supply and demand scenarios (dams and markets). • Discount rate of 6 per cent.

Transparency

Since the evaluation of adaptation trajectories is inevitably a normative process, 'transparency' is an important criterion in the selection and/or design of decision support tools for climate policies. Scientific information on climate risks (and opportunities) is not always consistently used within the internal decision-making process due to conflicting stakes. Decision-makers and stakeholders may have a tendency to rank adaptation options to cope with, for instance, drought risks, solely with reference to one of the stakes (e.g. agriculture), ignoring all other differences between different adaptation options (van der Heide et al, 2008). Hence, lexicographic preferences are defined as preferences without trade-offs because of the respondent's attitudes or fundamental beliefs, and are an indication of the strategic behaviour of the respondent (van der Heide et al, 2008).

An element of subjectivity cannot be excluded in scientific expert judgement regarding the presentation of uncertainties. For example, water and climate scientists may have dilemmas in accounting uncertainties in making climate scenarios and related impacts spatially explicit at the local (provincial) level (see Chapter 3). If water and climate scientists refuse to make climate risk maps for zoning at a provincial level, they may lose credibility. But if they do not communicate uncertainties and simply make the risk zoning maps, they may lose legitimacy as well. It is therefore important to give special attention to disclaimers and to develop benchmarks that provide some indication of the amount of associated uncertainty. In the reporting and presentation phase of the evaluation, it is also important to give attention to the linguistic aspects of expressing probabilities and risks. For example, the IPCC has developed verbal equivalents for probability intervals: in the sentence 'Drought-affected areas are likely to increase in extent', *likely* means 'with a 66 to 90 per cent probability'.

In order to visualize the uncertainty in expert judgement, particularly if those judgements are not supported by empirical observations or model analyses, we propose involving several experts of each discipline during the evaluation of adaptation strategies. Often, due to financial and time constraints, this is not done. However, it is worthwhile investigating alternative methods to get multiple expert judgements – for example, by internet questionnaires. The expert could be asked about the uncertainty range in their judgement and their perception of the probability that their expert judgement might be wrong. For example, the question: 'What will be the necessary safety margin for dike heights in order to cope with future sea-level rise?' is given to a group of 20 scientific experts. One expert will say a safety margin of 50cm with a lower and upper bound of 25cm and 100cm, and the probability that he or she will be wrong is less than 10 per cent. Another expert estimates a safety margin of 25cm with a lower and upper bound of 10cm and 45cm, with a probability estimation of 50 per cent. Following a Bayesian data analysis, prior beliefs in the form of the probability statements are multiplied with likelihood functions to obtain a joint probability distribution function (PDF). The entropy measure of a PDF is maximum if all estimated safety margins are equally likely to be mentioned by the experts (= maximum uncertainty). The entropy could be seen as a benchmark of uncertainty or degree of disagreement among experts about the effectiveness of a certain adaptation strategy.

Boundary organizations (science–policy–society interfaces)

Climate-proofing requires, like other environmental problems, clearly (re)defined and negotiated boundaries between science and policy (Janasoff, 2004; Tuinstra, 2006). Problem-defining, policies and research agendas need to be mutually constructed in boundary organizations, which may also lie outside the traditional domain of water resources management. Boundary organizations between science and policy as such are not new. Examples exist in the field of air quality policies (Tuinstra, 2006) and the field of IWRM and climate change (e.g. the Co-operative Programme on Water and Climate, or CPWC). Other examples of boundary organizations are national environmental/economic assessment agencies, public–private research programmes, and a participatory research project or a temporary commission set up by the government (see Box 8.2). Boundary organizations that also include representatives from society, in addition to science and policy, are not as common (Turton et al, 2007). The Trialogue Model from Turton et al (2007) assumes three interfaces (processes) – namely, society and science; government and society; and government and science. Knowledge transfer within the interfaces can be supported by decision support tools, but usually also contain other communication strategies, including the media and the internet.

The use of 'science', 'society' and 'government' in three entities is, of course, an oversimplification. When designing interfaces between these entities, it is therefore important to take into account the historical context and user characteristics of existing policy and scientific networks already in place that need to cooperate within the boundary organization. The individuals (or institutes) that take leadership within the policy–science interface will strongly influence the choices in response to the mentioned dilemmas (see Figure 8.1).

Not only the national government, but all levels of policy institutions, such as water boards or municipalities, are often routinely engaged in the process of climate-proofing water management and land use. Water managers are typically considered to be formally trained professionals and involved in some institutionally organized component of water development, delivery or regulation, with the responsibility and accountability for the decisions that are made (Kabat et al, 2003). In state-led planning systems, interfaces between the three entities are less complex; one boundary organization might be enough. However, in multilayered water management governance systems, you have to deal with national, regional and local levels of government. In these circumstances, cooperation between many regional boundary organizations in combination with a national or a transboundary river catchment boundary organization is necessary. This will increase in complexity due to upscaling and downscaling issues at both the scientific and societal levels (problem shifting).

The scientists working on this subject are from a myriad of disciplines: climate sciences, hydrology and governance studies, to name but a few. Within the same discipline, different paradigms may exist. In addition, each country has its own design of scientific networks that may include universities, research programmes, applied research institutes and bodies that take care of review and scientific quality. Cultural

Box 8.2 Boundary organizations mentioned in the case studies in Part II

Climate change and spatial planning in The Netherlands

In The Netherlands, many key decisions about future (infrastructural) investments are being taken in the field of spatial planning and water management. The senate of the Dutch Parliament called for incorporating climate change risks and opportunities within these decisions in 2005 (*motie Lemstra*). In response, policy-makers developed a boundary organization called Adaptation Programme for Spatial Planning and Climate (ARK) that includes two interfaces:

1 national policy–regional policy/society; and
2 national policy–science.

The management of the boundary organization falls under the Ministry of Environment and Spatial Planning (VROM), supported by an interdepartmental steering group and national associations for the provinces and municipalities. The science–policy interface is developed through the initiation of a new public–private research programme (Knowledge for Climate) linked to an existing public–private research programme (Climate Changes Spatial Planning). Both platforms are managed by scientific institutes. Parallel to these more long-term interfaces, commissions are also set up in The Netherlands, such as the Delta Commission (2007). This commission consisted of both scientists and policy-makers and had the one-year task to develop a new vision for future water management up to 2200. This is an example of knowledge management in practice, which does not necessarily follow the proposed scheme as presented in Figure 8.1. This exemplifies how difficult knowledge management is. Due to the power of existing networks in policy and science, it is not easy to design the desired science–policy–society interfaces that are necessary to be fully equipped for adaptive management approaches such as climate-proofing. However, designing science–policy–society interfaces is also 'learning by doing'.

Drought management in Australia

The development of the Indian Ocean Climate Initiative (IOCI) could be seen as an institutionalization of the science–policy interface between water managers and climate scientists in Australia. Overall, the research outputs such as climate risk assessments and climate scenarios from this partnership have provided acknowledged guidance for the *State Water Strategy* and the *State Greenhouse Strategy*. However, underpinning decision-making with information obtained from the latest developments in climate science is still in its infancy. The sequence of debating decisions by the water corporation and its predecessors was largely driven by observed dam inflow reductions due to multi-decadal droughts.

aspects and even law determine the character of these networks and the attitude of scientists towards policy-makers and their willingness to participate in these types of boundary organizations.

Each individual is an element of society; however, it is clear that it is impossible for each individual to participate in a boundary organization. Selection of societal representatives within a boundary organization is therefore an important, but also a difficult, step within the design phase due to the political dimension, especially in state-led planning systems. This is, for example, exemplified in the case study of Thailand (see Chapter 9), where the empowerment of local communities in water management and disaster-reduction policies is called for.

It is therefore important to take into account the historical context and user characteristics of existing policy institutions, societal governance networks and scientific infrastructure that are already in place and which need to cooperate within the boundary organization. The designer should be aware of the initial objectives of the involved institutions, how these organizations currently deal with uncertainty, how they frame climate-proofing, how they deal with dissenting opinions, and their familiarity with stakeholder participation. These initial settings will differ from region to region. In African countries, water management is more organized in informal institutions than in Europe or the US. In addition, economic settings and the influence of economic sectors (agriculture, multinationals, services, etc.) on policies and their climate sensitivity differ.

Conclusions

Climate-proofing can be seen as adaptation trajectories: (cyclic) operational steps towards achieving a progressively more protected society by increasing the resilience and adaptive capacity of both the physical and societal systems in question. Developing visions on climate-proofing and adaptation trajectories to respond to climate change is strongly recommended. Within the climate-proofing approach – primarily, though not exclusively – normative methodologies for exploring the future are used in the phase of vision-building. After vision-building, practical test cases (learning by doing) are initiated and an evaluation phase follows. This evaluation phase of adaptation strategies is supported by formal quantitative scenario methods and decision-support systems. The use of formal scenario methods in a later stage of the decision making process is different from the current approach in water management and climate policy.

In most cases, the dilemmas for scientists and policy-makers (see Figure 8.1) within the decision-making pathway of water management can be boiled down to choices between the 'adaptive management paradigm' and the 'foundational water management paradigm'. In practice, a combination of individual scientists and policy-makers is involved in the decision-making process. As a result, a mixture of proactive and conservative arguments from both paradigms is used, relevant to each region, country or catchment. The individuals (or institutes) that take leadership within the policy–science interface will strongly influence the choices in response to the described dilemmas (see Figure 8.1). The definition of climate-proofing depends upon whether the local or national situation of the decision-making process is dominated by people using the 'adaptive management paradigm' or the 'foundational water management paradigm'. The need for climate-proofing is acknowledged in both existing paradigms, but will lead to different risk thresholds. Our described preferred climate-proofing approach is mostly in line with the adaptive management paradigm and can be seen as an explicit example of this paradigm. While it is not a new planning paradigm, it does build upon existing theories regarding decision-making in uncertainty.

The conceptualization of 'climate-proofing' is a process of joint learning, where some expert and policy communities are taking the lead in this process, while others are still struggling. Training and educating both scientists and water professionals is therefore important in order to select and use the most appropriate decision-support tools, such as (tailored) climate scenarios (Chapters 2 and 3). In addition, training of water professionals, scientists and other representatives of society with the intention to design tailor-made boundary organizations is vital. The successes of boundary organizations are usually judged upon credibility, legitimacy and social relevance (Tuinstra, 2006). It is therefore important to take the historical context and user characteristics of existing policy and scientific networks already in place into account. As a result, no blueprint exists. Finally, it is important to design a knowledge transfer procedure and decision support tools where water managers can identify the effectiveness of adaptation strategies for themselves in dialogue with climate scientists, rather than provide information on costs and benefits beforehand. The entity that is responsible for evaluating and selecting the decision support tools is preferably a boundary organization interfacing science, policy and society.

Acknowledgements

We thank Michael van der Valk (Co-operative Programme on Water and Climate) for providing comments on the contents of earlier drafts of this chapter. Much of the presented work is a spin-off from the following research programmes:

- Climate Changes Spatial Planning (www.climatechangesspatialplanning.nl)
- Co-operative Programme on Water and Climate (www.waterandclimate.org)
- NEWATER project (www.newater.info)

References

Costanza, R. (2000) 'Visions of alternative (unpredictable) futures and their use in policy analysis', *Conservation Ecology*, vol 4, no 1, p5, www.consecol.org/vol4/iss1/art5

Dessai, S. and J. P. van der Sluijs (2007) *Uncertainty and Climate Change Adaptation – A Scoping Study*, Task for The Netherlands Environmental Assessment Agency, The Netherlands

Fiering, M. B. and N. C. Matalas (1990) 'Decision making under uncertainty', in P. E. Waggoner (ed) *Climate Change and US Water Resources*, John Wiley & Sons, New York, NY, pp75–84

Folke C., S. Carpenter, T. Elmqvust, L. Gundersin, C. S. Holling and B. Walker (2002). 'Resilience and sustainable development: Building adaptive capacity in a world of transformations', *Ambio*, vol 31, pp437–440

Glantz, M. (2003) *Climate Affairs: A Primer*, Island Press, Washington, DC

Glantz, M. (2006) *Weather- and Climate-Proofing: Dreaming the Impossible Dream*, www.fragilecologies.com/jan20_06.html

Gleick, P. H. (1998) *The World's Water 1998–1999: The Biennial Report on Freshwater Resources*, Island Press, Washington, DC

Hay, J., R. Warrick, C. Cheatham, T. Manarangi-Trott, J. Konno and P. Hartley (2005) *Climate Proofing: A Risk-based Approach to Adaptation*, Asian Development Bank, the Philippines

Holling, C. S. (1978) *Adaptive Environmental Assessment and Management*, John Wiley & Sons, Chichester, UK, pp357–363

IPCC (Intergovernmental Panel on Climate Change) (2007) *Climate Change 2007: Synthesis Report, Fourth Assessment Report*, Cambridge University Press, Cambridge

Janasoff, S. (ed) (2004) *States of Knowledge: The Co-production of Science and Social Order*, Routledge, London

Kabat, P., R. E. Schulze, R. E. Hellmuth and J. A. Veraart (eds) (2003) *Coping with Impacts of Climate Variability and Climate Change in Water Management: A Scoping Paper*, DWCSSO-01 International Secretariat of the Dialogue on Water and Climate, Wageningen, The Netherlands

Kabat, P., P. Vellinga, W. van Vierssen, J. A. Veraart and J. Aerts (2005) 'Climate proofing the Netherlands', *Nature*, vol 438, pp283–284

Matalas, N. C. (1997) 'Stochastic hydrology in the context of climate change', in K. D. Frederick, D. C. Major and E. Z. Stakhiv (ed) (1997) *Climate Change and Water Resources Planning Criteria*, Kluwer Academic Publishers, Dordrecht, The Netherlands

Meadows, D. H. (1992) *Beyond the Limits: Global Collapse or a Sustainable Future*, Earthscan Publications, London

Milly, P. C. D., J. Betancourt, M. Falkenmark, R. M. Hirsch, Z. W. Kundzewicz, D. P. Lettenmaier and R. J. Stouffer (2008) 'Climate change: Stationarity is dead – whither water management?' *Science*, vol 319, no 5863, 1 February, pp573–574

Pahl-Wostl, C. (2002) 'Participative and stakeholder-based policy design, evaluation and modeling processes', *Integrated Assessment*, vol 3, no 1, pp3–14

Pahl-Wostl, C. and M. P. Hare (2004) 'Processes of social learning in integrated resources management', *Journal of Community and Applied Social Psychology*, vol 14, pp1–14

Ruijgh-van der Ploeg, T. and A. Verhallen (2002) *Envisioning the Future of Transboundary River Basins with Case Studies from the Scheldt River Basin*, TU Delft, Wageningen University, The Netherlands

Scheffer, M., S. Carpenter, J. A. Foley, C. Folke and B. Walker (2001) 'Catastrophic shifts in ecosystems', *Nature*, vol 413, pp591–596

Smith, J. B. and S. S. Lenhart (1996) 'Climate change adaptation policy options', *Climate Research*, vol 6, pp193–201

Toth, F. (2000) 'Decision analysis frameworks', in R. K. Pachauri, T. Taniguchi and K. Tanaka (eds) *Guidance Papers on the Cross Cutting Issues of the Third Assessment Report of the IPCC*, IPCC, Geneva, pp53–68

Tuinstra, W. (2006) *Reducing Air Pollution in Europe: A Study of Boundaries between Science and Policy*, PhD thesis, Wageningen University, The Netherlands

Turton A., J. Hattingh, G. Maree, D. J. Roux, M. Claassen, W. F. Strydom (2007) *Governance as a Trialogue: Governance–Society–Science in Transition*, Springer Verlag, Berlin, Germany, pp1–28

Van der Heide, C. M., A. T. de Blaeij and W. J. M. Heijman (2008) *Economic Aspects in Landscape Decision Making: A Participatory Planning Tool Based on a Representative Approach*, Discussion Paper no 41, Manshold Graduate School of Social Sciences, Wageningen University, The Netherlands, pp1–18

Van der Sluijs, J. P. (2007) 'Uncertainty and precaution in environmental management: Insights from the UPEM conference', *Environmental Modelling and Software*, vol 22, pp590–598

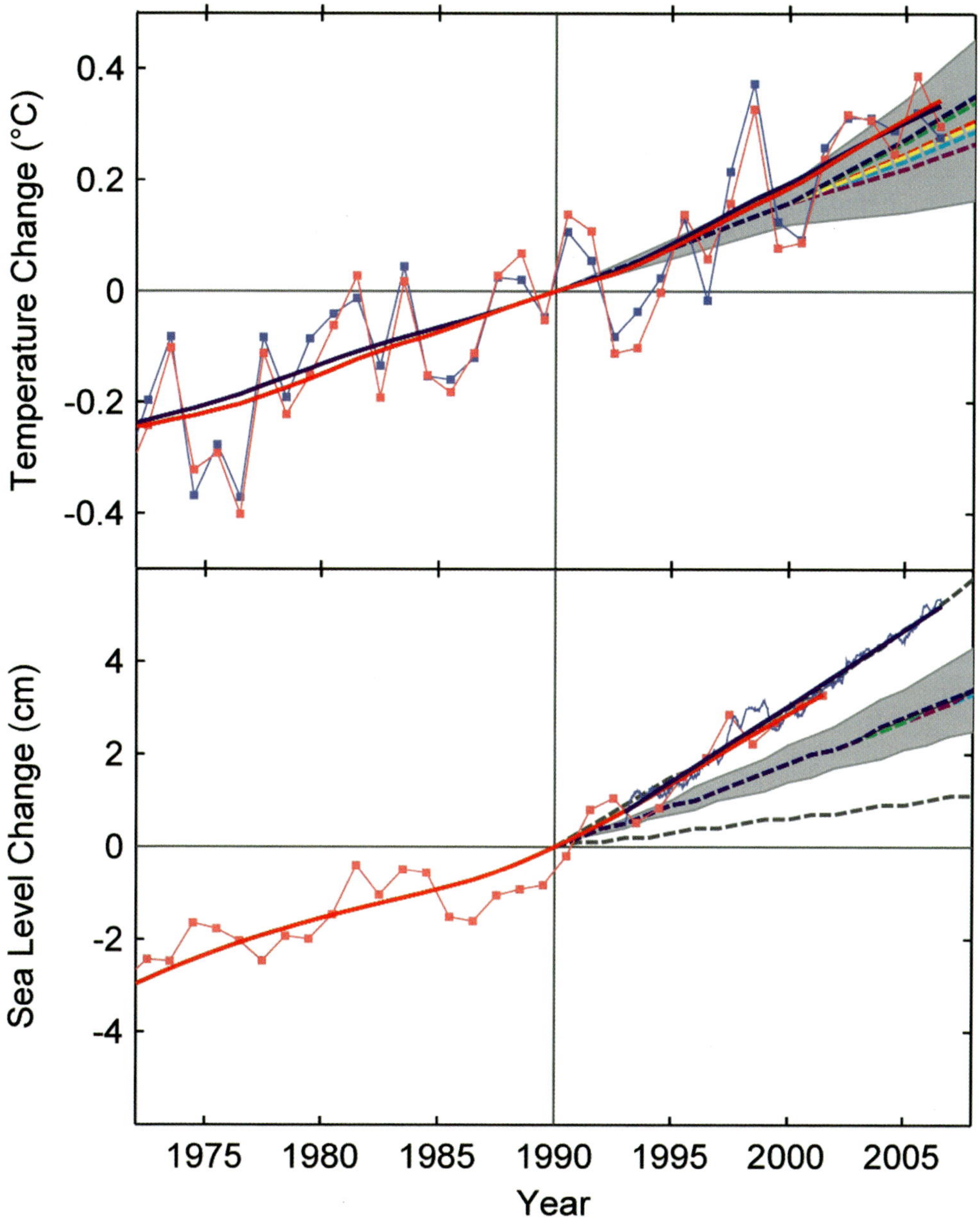

Plate 1 *Observed global mean temperature (top) and sea level (bottom), including projections published in the IPCC* Fourth Assessment Report *(see Chapter 2)*

Source: Rahmstorf, S., A. Cazenave, J. A. Church, J. E. Hansen, R. F. Keeling, D. E. Parker and R. C. J. Somerville (2007) 'Recent climate observations compared to projections', *Science*, vol 316, pp709–709

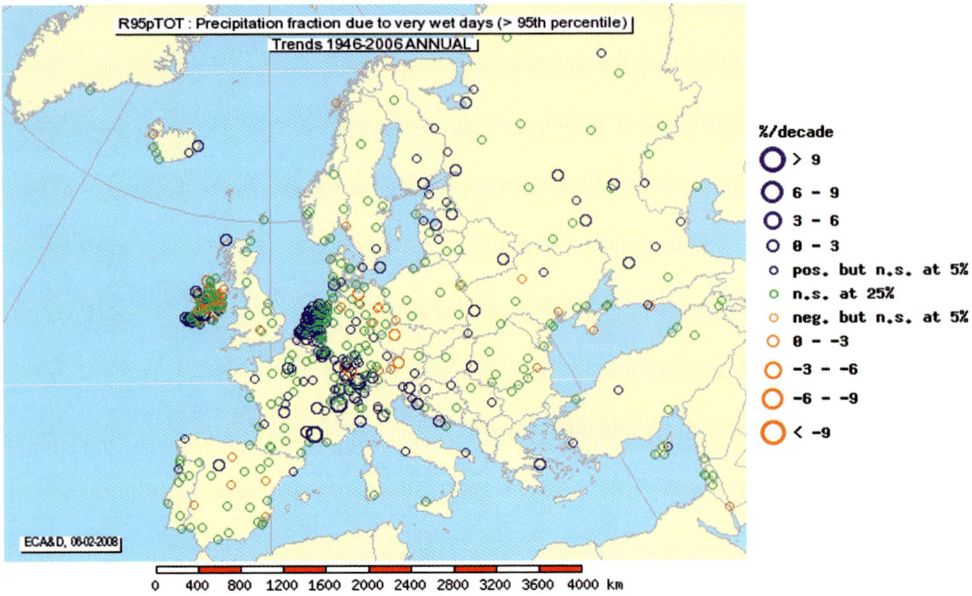

Plate 2 *Trend (1946–2006) of fraction of precipitation on very wet days (P>95 per cent) averaged over all seasons (see Chapter 2)*

Source: http://eca.knmi.nl

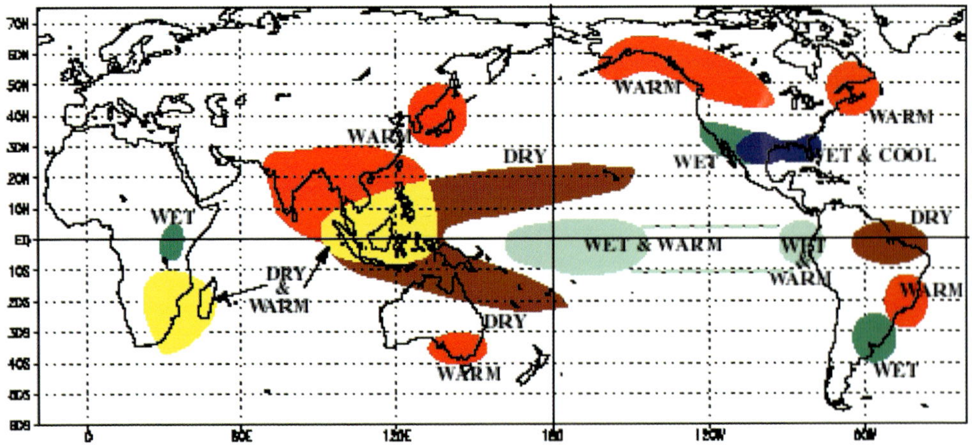

Plate 3 *Remote climate effects of ENSO during the SST peak period (December–February). Shown are areas where high Pacific SSTs correlate well with anomalously high or low seasonal mean precipitation or temperature values (see also Plate 4, opposite page) (see Chapter 2)*

Source: www.cpc.ncep.noaa.gov/products/analysis_monitoring/ensocycle/elninosfc.shtml

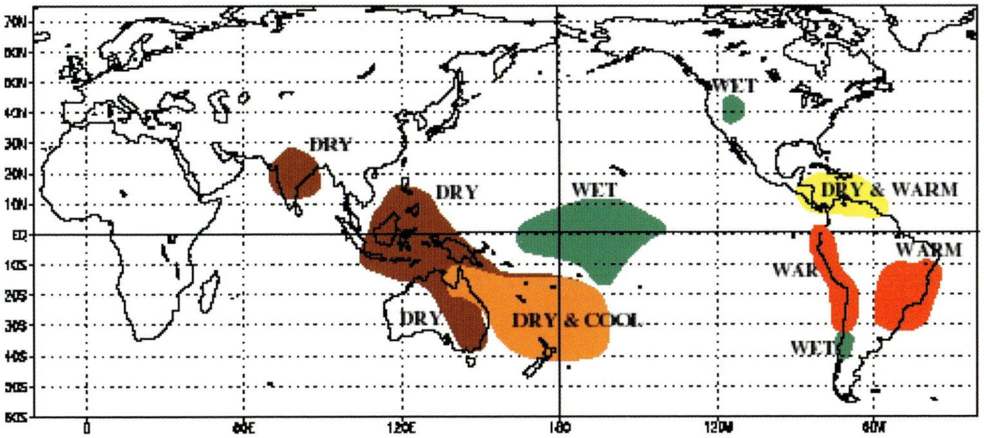

Plate 4 *Remote climate effects of ENSO during the boreal summer period (June–August) after the SST peak shown in Plate 3 (see Chapter 2)*

Source: http://www.cpc.ncep.noaa.gov/products/analysis_monitoring/ensocycle/elninosfc.shtml

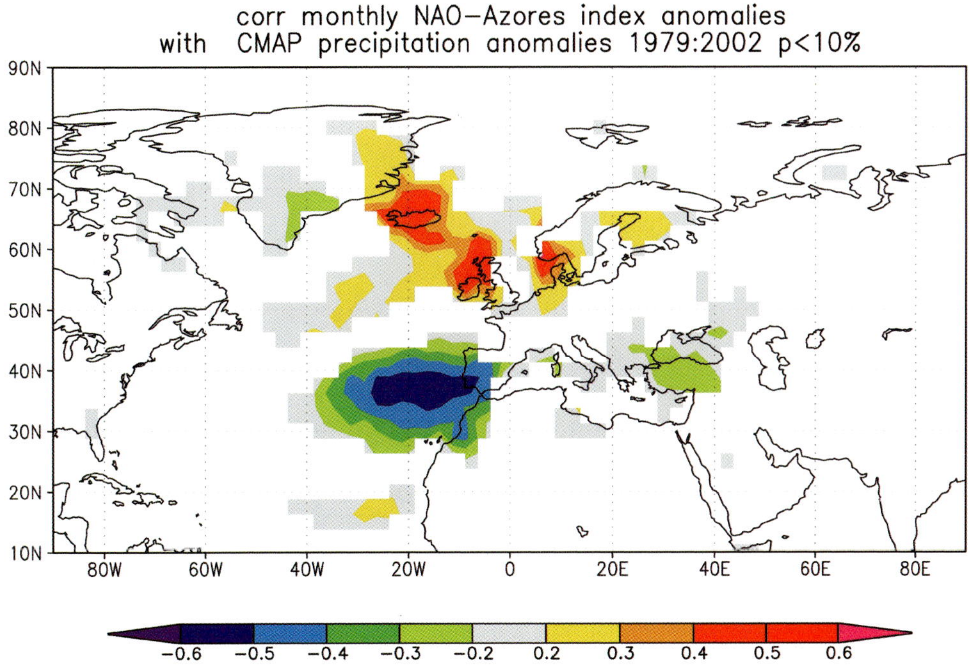

Plate 5 *Correlation between NAO index and annual mean precipitation from the CMAP database (see Chapter 2)*

Source: http://climexp.knmi.nl

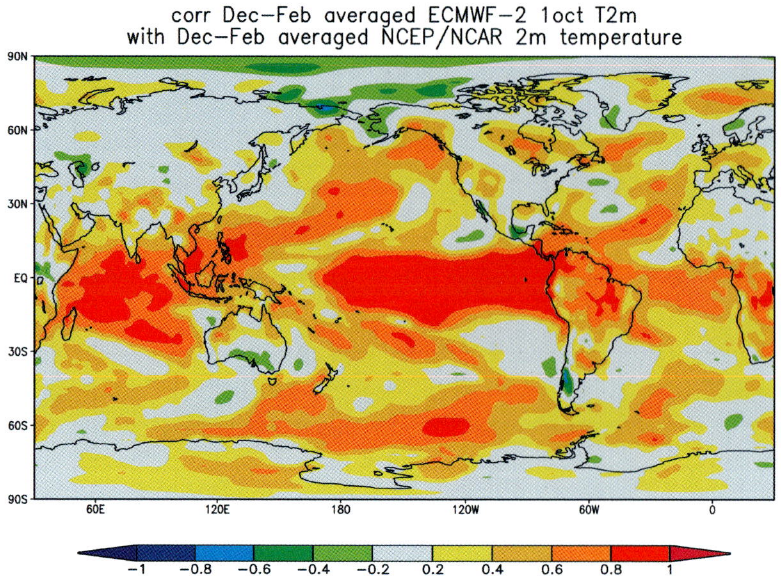

Plate 6 *Skill score of DJF temperature from the ECMWF coupled atmosphere–ocean general circulation model (see Chapter 2)*

Note: Red colours indicate good predictability; blue colours indicate predictions that are opposite to the realized weather; grey colours indicate poor predictability.
Source: Van Oldenborgh, G. J., M. A. Balmaseda, L. Ferranti, T. N. Stockdale and D. L. T. Anderson (2005) 'Evaluation of atmospheric fields from the ECMWF seasonal forecasts over a 15 year period', *Journal of Climate*, vol 18, pp3250–3269

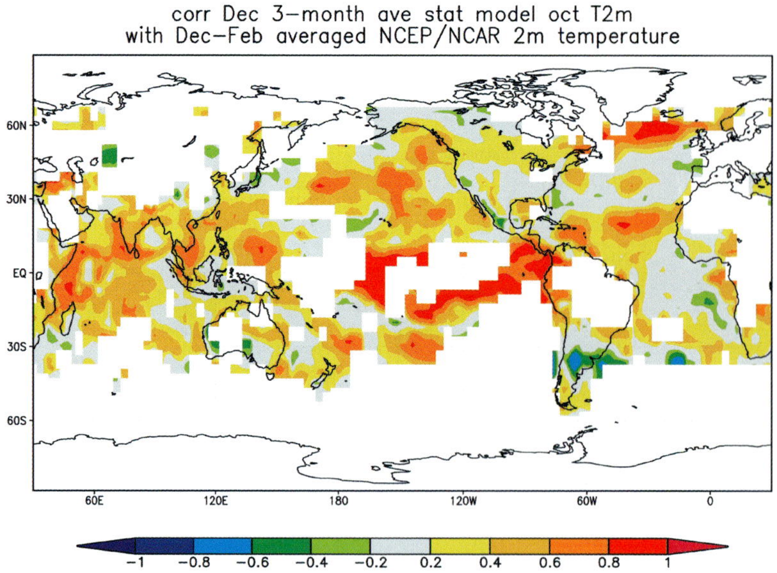

Plate 7 *Skill score of DJF temperature from a calibrated statistical forecast model (see Chapter 2)*

Note: Colour indications as in Plate 6.
Source: Van Oldenborgh, G. J., M. A. Balmaseda, L. Ferranti, T. N. Stockdale and D. L. T. Anderson (2005) 'Evaluation of atmospheric fields from the ECMWF seasonal forecasts over a 15 year period', *Journal of Climate*, vol 18, pp3250–3269

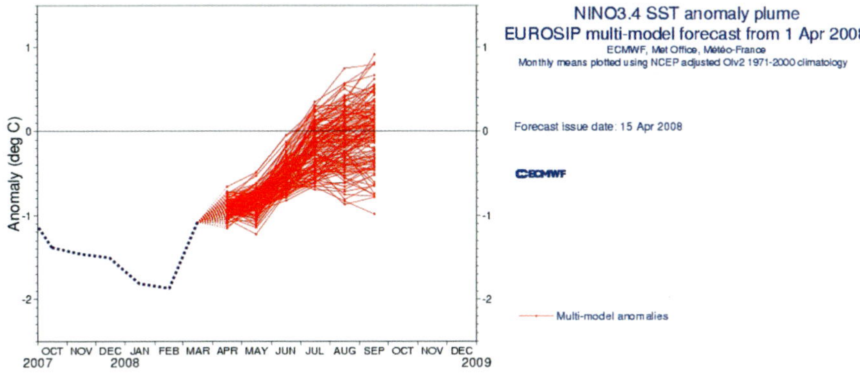

Plate 8 *Example of products from the seasonal prediction group at ECMWF: Forecast of the NINO3.4 index (see Chapter 2)*

Note: The blue dashed line shows the realized SST in the ENSO NINO3.4 area; the various red lines emerging from the observations denote the plume of predictions of the evolution of NINO3.4 in the coming months.
Source : www.ecmwf.int/products/forecasts/d/charts/seasonal/forecast/seasonal_range_forecast/nino_plumes_s3

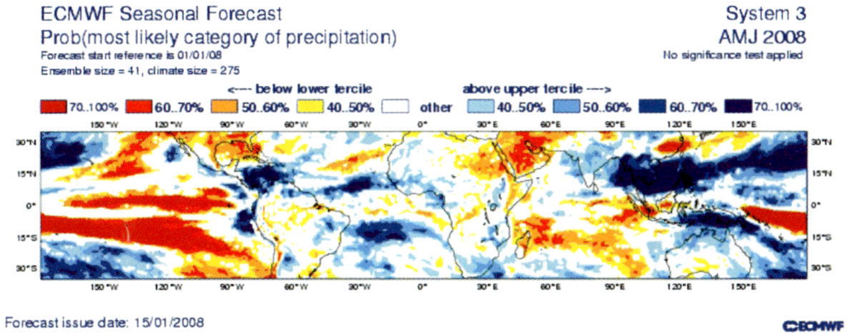

Plate 9 *Example of products from the seasonal prediction group at ECMWF: The probability for higher or lower than normal rainfall in the tropics in the coming months (see Chapter 2)*

Note: Red areas indicate a high likelihood of (much) lower than normal precipitation; blue areas indicate anomalously wet conditions.
Source: www.ecmwf.int/products/forecasts/d/charts/seasonal/forecast/seasonal_range_forecast/nino_plumes_s3

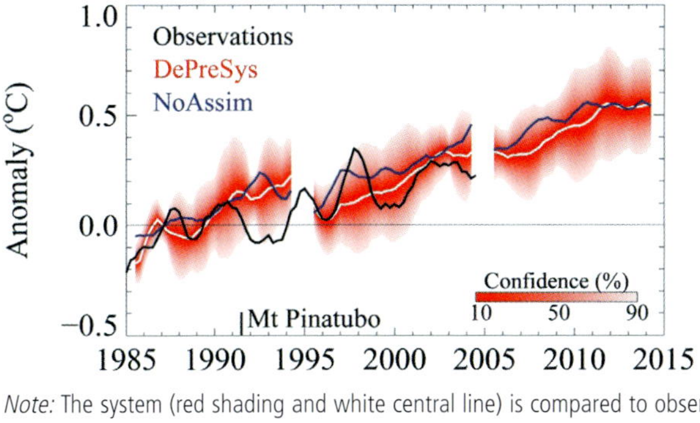

Plate 10 *Observed and projected global mean temperature change from a decadal forecasting system (DePreSys) presented by Smith and colleagues (see Chapter 2)*

Note: The system (red shading and white central line) is compared to observations and to a system that does not use updated initial ocean heat content (NoAssim, blue line). Projections started in 2005 to show a gradual increase of the global mean temperature; but in the DePreSys forecast this increase is compensated for by internal variability of the climate system (expressed as a reduced ocean heat content) in the years up to 2008. The confidence bar denotes the likelihood of the temperature falling inside the indicated shading.
Source: Smith, D. M., S. Cusack, A. W. Colman, C. K. Folland, G. R. Harris and J. M. Murphy (2007) 'Improved surface temperature prediction for the coming decade from a global climate model', *Science*, vol 317, pp796–799

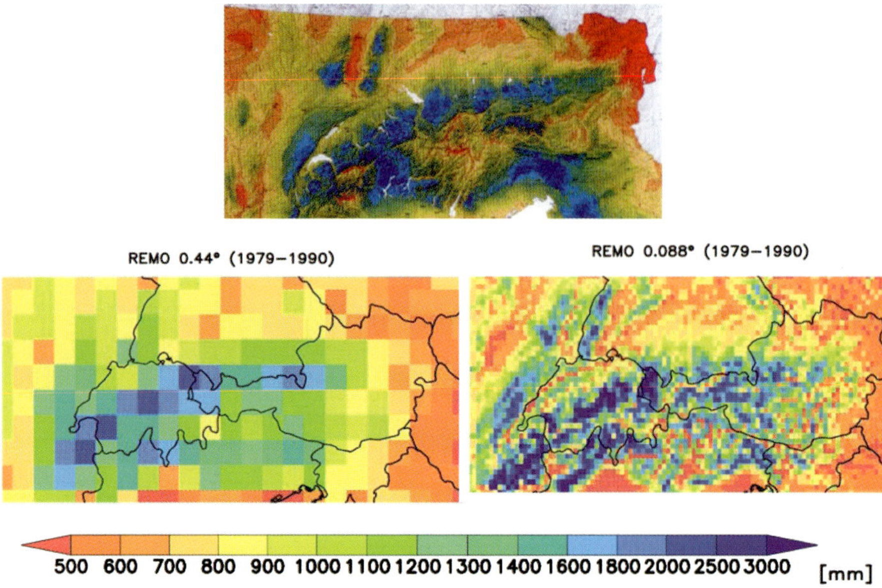

Plate 11 *Annual total precipitation (mm), observed (1971–1990, upper panel) and simulated with 50km grid lengths (bottom left) and 10km grid lengths (bottom right) (see Chapter 3)*

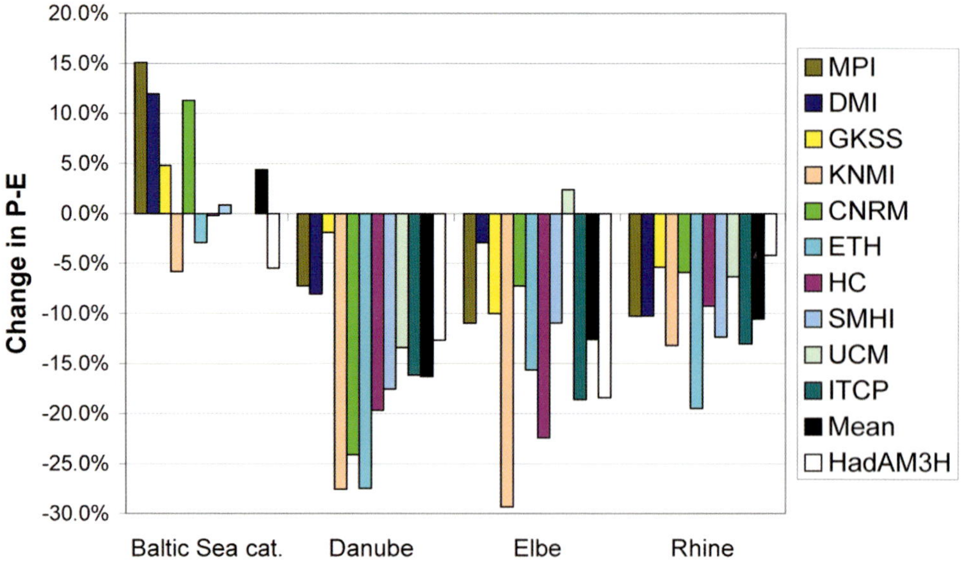

Plate 12 *Simulated and observed changes in river runoff for the period 2071–2100 compared to 1961–1990 for the Baltic Sea catchment, Danube, Elbe and Rhine (see Chapter 3)*

Note: y-axis: E = evaporation, P = precipitation. Calculated from a suite of RCMs driven by the same GCM data following the A2 emission scenario.

Source: Hagemann, S. and D. Jacob (2007) 'Gradient in the climate change signal of European discharge predicted by a multi-model ensemble', *Climatic Change*, PRUDENCE special issue, vol 81, supplement 1, pp309–327

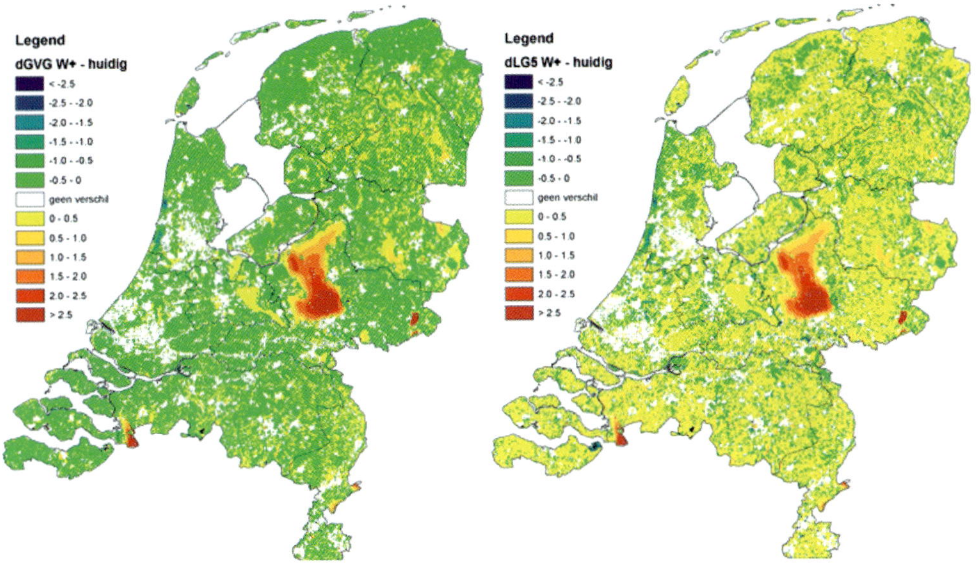

Plate 13 *Change in groundwater table as calculated for a so-called W+ climate change scenario (see Chapter 3)*

Note: Left: the mean groundwater table at the start of the growing season. Right: the seasonal lowest groundwater table, normally occurring at the end of the growing season. Positive numbers denote deeper groundwater tables.

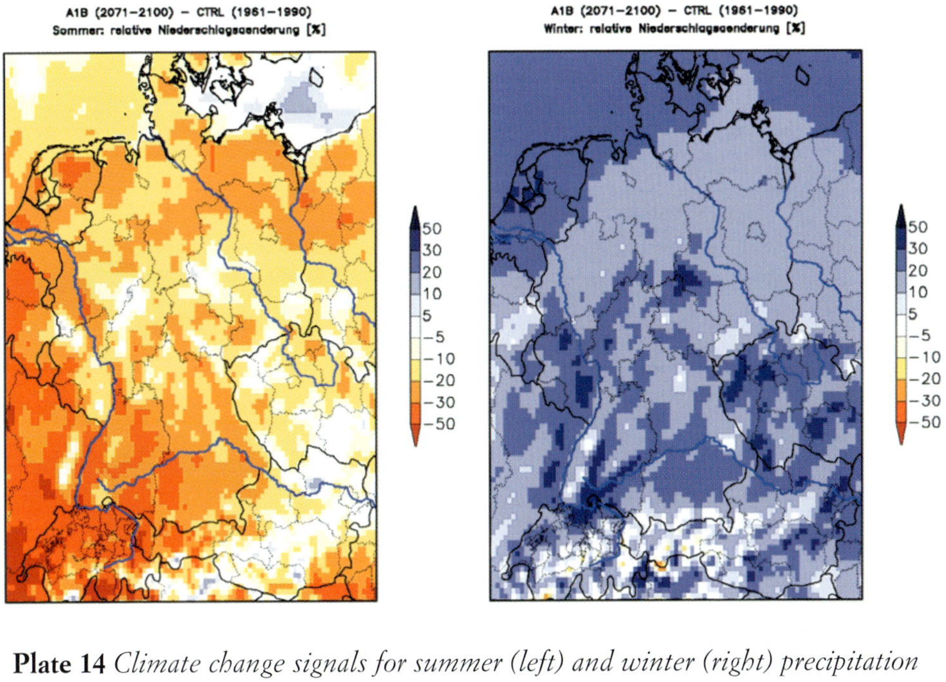

Plate 14 *Climate change signals for summer (left) and winter (right) precipitation (percentage) in A1B scenario for 2071 to 2100 compared to 1961 to 1990 (see Chapter 3)*

Source: Jacob et al (2008)

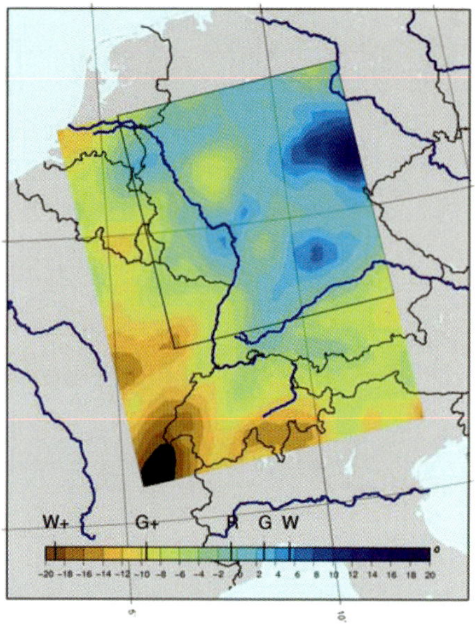

Plate 15 *Simulated relative change of the summer (left) and winter (right) mean precipitation (percentage) between 1950 and 2050 generated with the RACMO2 Regional Climate Model driven by the ECHAM5/OMI GCM following the A1B SRES scenario (see Chapter 3)*

Note: Also shown on the colour scale are the values corresponding to the KNMI'06 climate scenarios (G, G+, W, W+), and the domain mean value from the RCM (R).

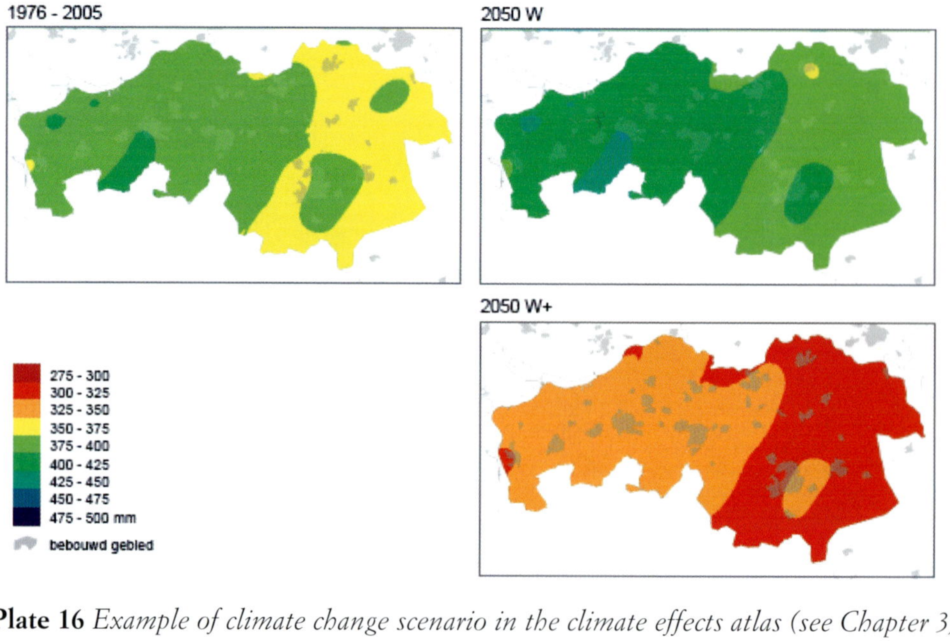

Plate 16 *Example of climate change scenario in the climate effects atlas (see Chapter 3)*

Note: Shown is the summer mean precipitation in Noord-Brabant (The Netherlands) for present-day conditions and according to the W and W+ scenarios.

Plate 17 *Map of estimated annual agricultural drought damage (€/year) in Noord Brabant (The Netherlands) according to the G scenario (see Chapter 3)*

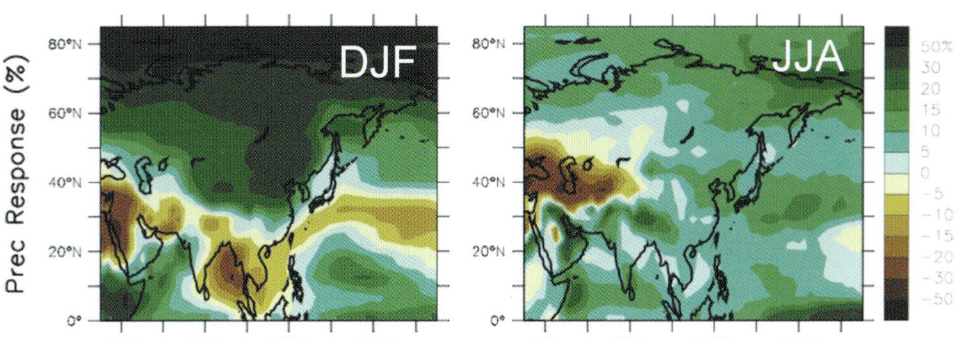

Plate 18 *Expected changes in rainfall patterns according to IPCC (2007) (see Chapter 9)*

Plate 19 *Bangkok floods, 11 October 2006 (see Chapter 9)*

Plate 20 *Bangkok floods, 11 October 2006 (see Chapter 9)*

Plate 21 *Preferred order of measures: Retain–store–discharge (see Chapter 10)*

Source: Christa Jesse

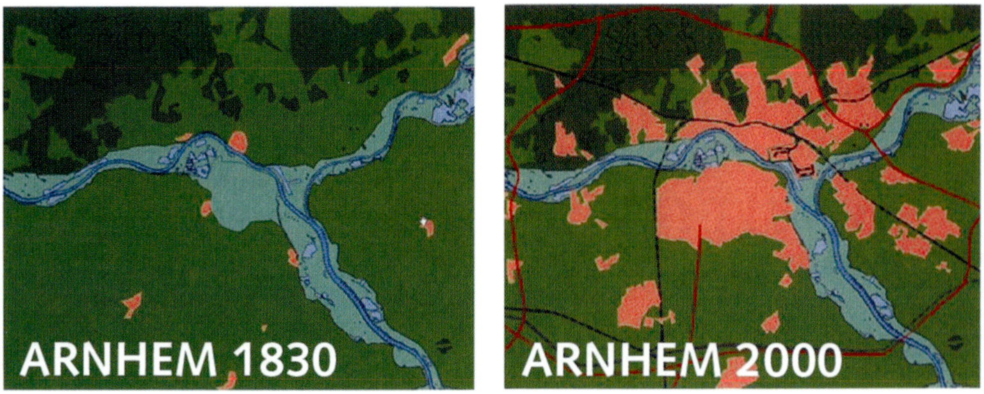

Plate 22 *Urbanization around Arnhem (see Chapter 10)*

Note: River beds are light green; urban areas are orange.

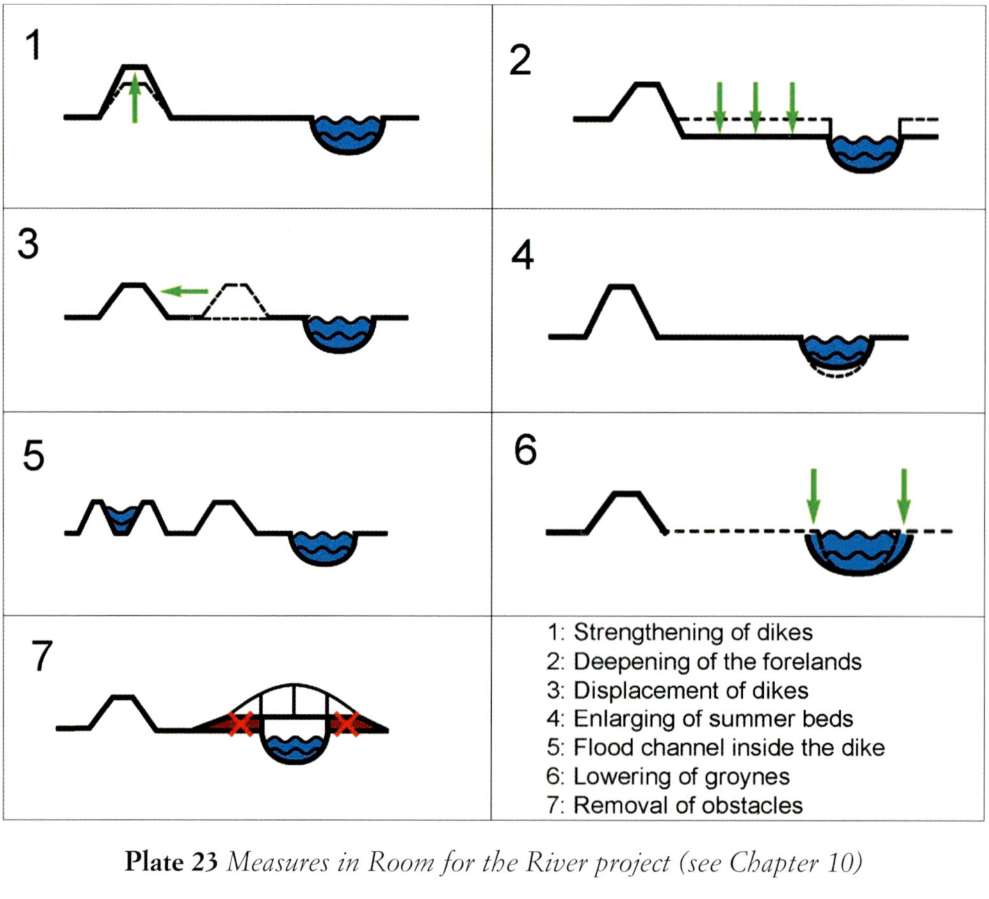

Plate 23 *Measures in Room for the River project (see Chapter 10)*

Plate 24 *Coastal erosion on the Isle of Ameland (see Chapter 10)*

Source: Johan Krol

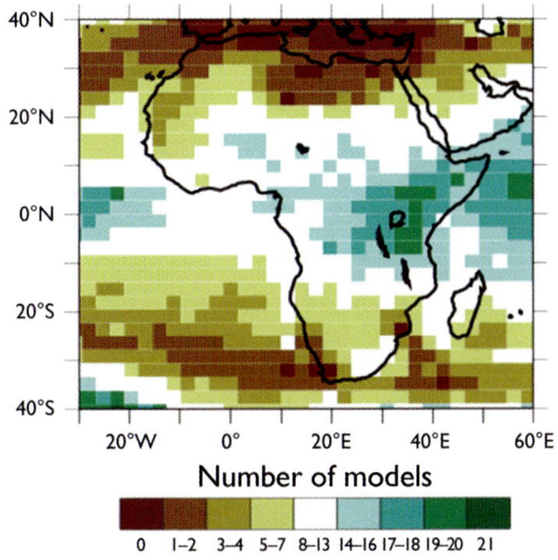

Plate 25 *Number of climate change prediction models, out of 21, that project increases in precipitation according to the IPCC (see Chapter 12)*

Source: Intergovernmental Panel on Climate Change (2007) *Climate Change 2007: Synthesis Report, Fourth Assessment Report*, Cambridge University Press, Cambridge, UK

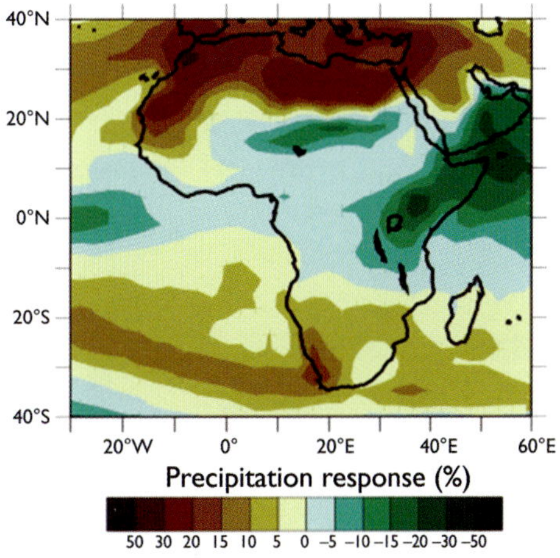

Plate 26 *Precipitation changes over Africa from the MMD-A1B simulations between 1980–1999 and 2080–2099 (see Chapter 12)*

Source: Intergovernmental Panel on Climate Change (2007) *Climate Change 2007: Synthesis Report, Fourth Assessment Report*, Cambridge University Press, Cambridge, UK

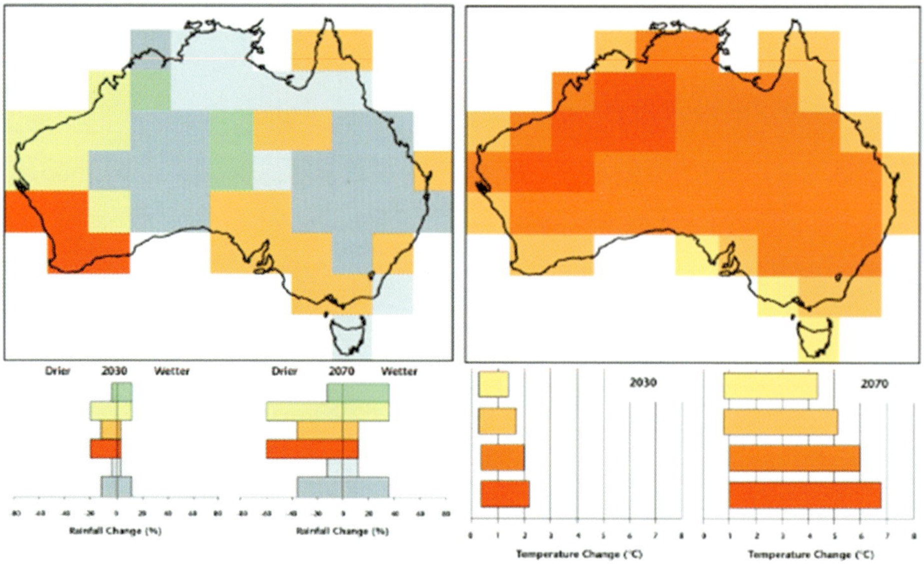

Plate 27 *Spatial distribution of projected changes in precipitation (left) and temperature (right) in 2030 and 2070 (see Chapter 13)*

Source: Preston, B. L. and R. N. Jones (2006) *Climate Change Impacts on Australia and the Benefits of Early Action to Reduce Global Greenhouse Gas Emissions*, Consultancy report for the Australian Business Roundtable on Climate Change, CSIRO Marine and Atmospheric Research, Melbourne, Australia

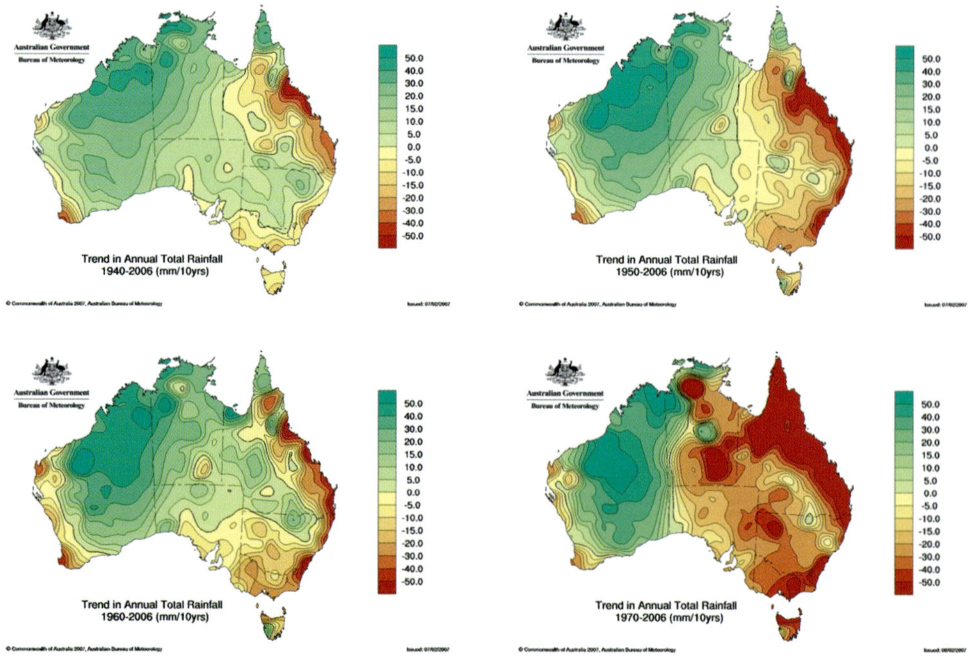

Plate 28 *Trend maps for Australian annual rainfall for four time slices (see Chapter 13)*

Note: Top left: 1940–2006; top right: 1950–2006; bottom left: 1960–2006; bottom right: 1970–2006. Units are millimetres per decade.

Source: Australian Bureau of Meteorology (www.bom.gov.au/cgi-bin/silo/reg/cli_chg/trendmaps.cgi)

Plate 29 *Elbe Basin (see Chapter 15)*

Source: Umweltbundesamt

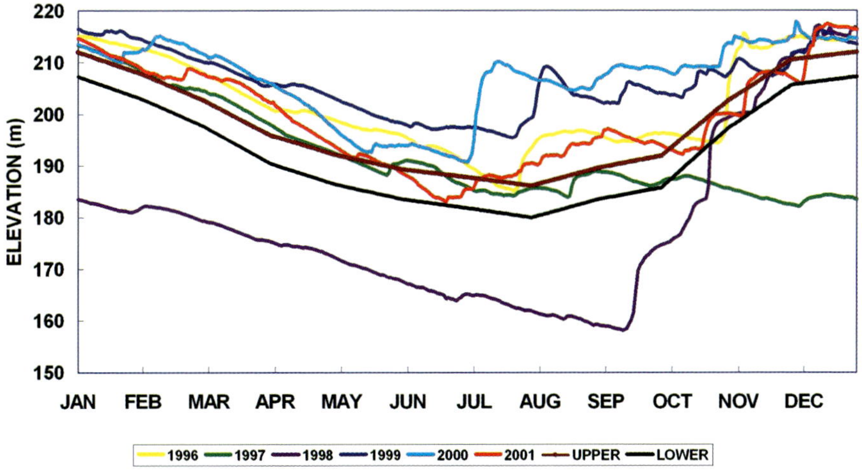

Plate 30 *Currently adopted rule curves (upper and lower) and reservoir storages from 1996–2001 in the Angat Dam (see Chapter 16)*

Note: Note the continued decrease in reservoir levels from October 1997 to September 1998.

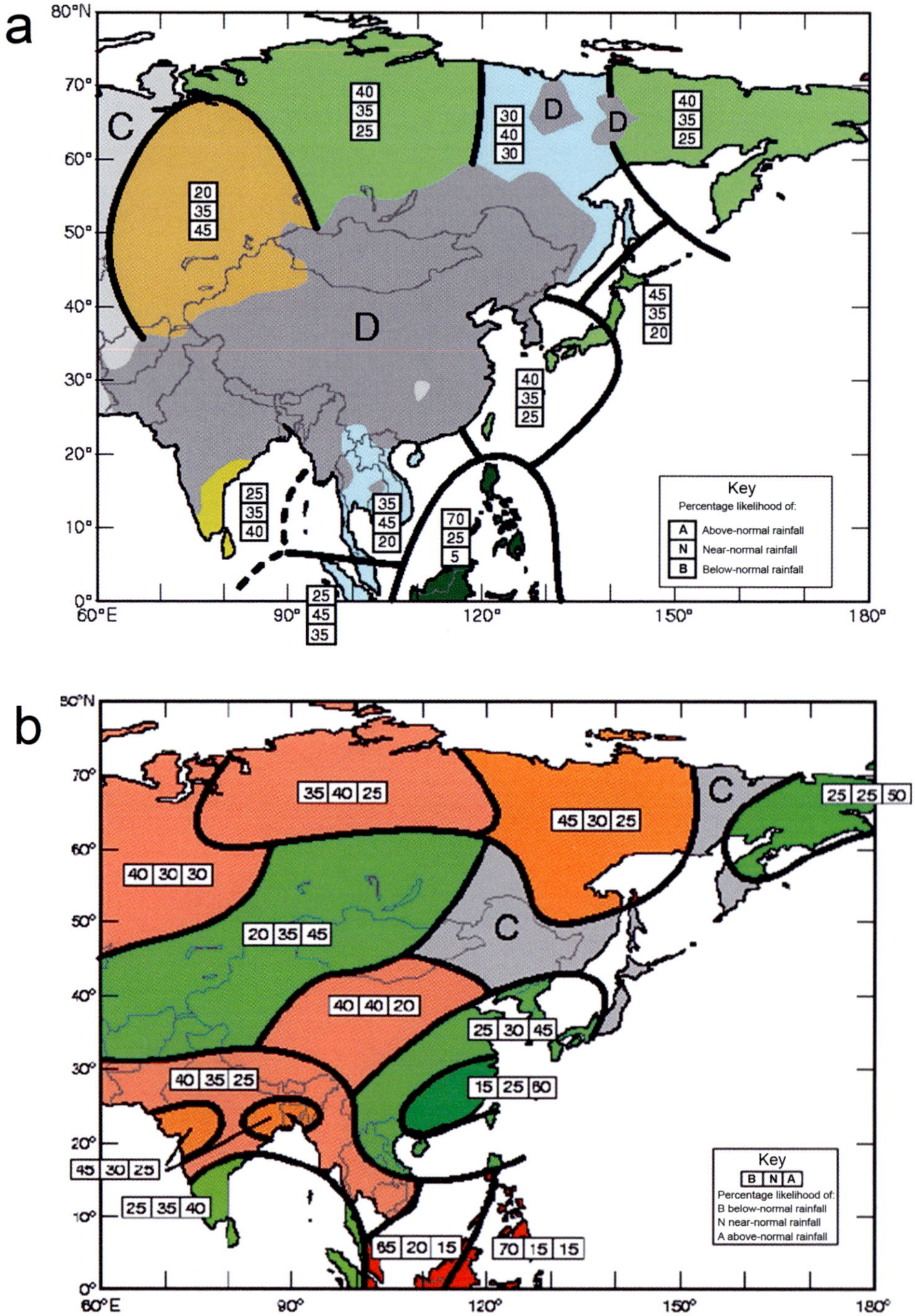

Plate 31 *Climate forecasts issued from the International Research Institute for Climate and Society (IRI) during the 1997–1998 ENSO events (see Chapter 16)*

Note: The charts show the OND-97 and OND-98 forecasts issued from the IRI in (a) October 1997 and (b) October 1998. Note the increased probability of below normal (45 per cent) and above normal (70 per cent) precipitation over the Luzon Island in the Philippines for OND-97 and OND-98.

Part II

Case Studies

Edited by Peter Droogers

9

Adaptation to Climate Change and Social Justice: Challenges for Flood and Disaster Management in Thailand

Louis Lebel, Tira Foran, Po Garden and Jesse B. Manuta

Introduction

Over the past 30 years, the number and impact of flood disasters has continued to increase across Asia (Dutta and Herath, 2004; ABI, 2005). This has occurred despite vastly improved abilities to monitor, warn and describe floods. In Thailand, this, in part, reflects growth in absolute numbers of people living in flood-prone areas and higher values of infrastructure at risk (Nicholls et al, 2007). Thus, around Bangkok, Chiang Mai and other urbanizing regions, new flood-sensitive settlements and land uses are expanding into low-lying wetlands and rice paddy landscapes (see Figure 9.1).

As elsewhere, flood waters are increasingly managed primarily to protect cities and related infrastructure (Takeuchi, 2001). Better early warning systems and improved emergency response capacities have helped to reduce losses of life. But infrastructure-based prevention measures are costly. Moreover, flood walls and diversions can also end up shifting, rather than reducing, some of the flood damage risks and costs onto others (Lebel and Sinh, 2007). Top-down policy-making and programme design on disasters can result in poor coordination among agencies, weak links among pre- and post-event actions and other institutional problems (Manuta et al, 2006). In the absence of effective insurance or transparent compensation schemes, managing flood disaster risks has emerged as an important social justice issue in Thailand.

The pursuit of social justice or fair access to resources and allocation of risks, benefits and burdens (Elster, 1992) in managing floods and disasters may be made more difficult by climate change in several ways (Thomalla et al, 2006; Lebel, 2007). First, the expected changes in burdens and risks are distributed very unevenly across peoples, places and generations (Adger, 2001; Thomas and Twyman, 2005). Second, international action and agreements on adaptation and mitigation are dominated by the interests of wealthy and powerful nations and therefore may not sufficiently take into account the interests, needs or capabilities of vulnerable groups (Paavola and

Figure 9.1 *Map of Thailand*

Note: Locations mentioned in this chapter can be seen on the map above: Chiang Mai in the north of the country, Bangkok in the centre and Hat Yai in the extreme south.

Source: based on a United Nations map

Adger, 2006). Third, the details of how climate change will affect seasonal precipitation and extreme rainfall events, and how this, in turn, will interact with other changes in land and water use to alter flood regimes, is filled with important uncertainties.

This case study focuses on issues of social justice in how floods and disasters are being managed in Thailand. Based on a critique of historical policies and practices, it draws inferences about the key challenges posed by altered flood regimes resulting from climate change and adaptation policies. These underline the importance of a politics of adaptation that emerges from contested and changing perceptions and experiences of risks. Our main conclusion is that persistent social injustices could be made worse by both inaction and misguided climate change adaptation policies.

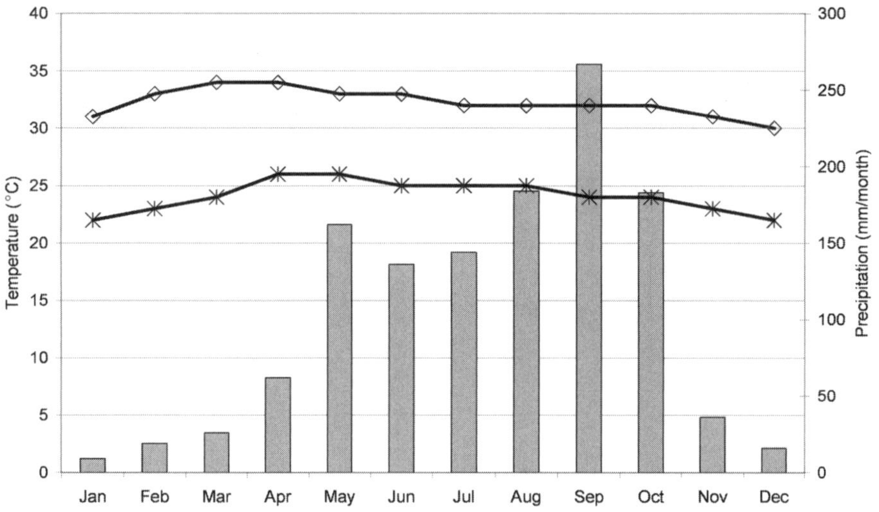

Figure 9.2 *Mean minimum and maximum temperature and precipitation for Bangkok, Thailand*

Source: http://www.wunderground.com

Climate change and flood regimes

Thailand has a monsoonal climate (see Figure 9.2). Floods are a normal part of the seasonal cycle and critical for agriculture. Thailand is the world's number one exporter of rice and also among the largest exporters of food products overall. Many rural households still recognize the benefits that floods bring to ecosystems and their livelihoods.

Floods are most likely to become disasters when they are unusual in timing or severity. Individual flood events pose risks and may contribute to disasters; but in the medium and long term, it is *changes* to flood regimes that redefine what is unusual. These changes pose important challenges to institutional development and adaptation. A flood regime is a historically experienced pattern of variability in onsets, durations, extents and frequencies. Here we highlight five types of flood (see Table 9.1).

Global warming is likely to cause additional changes to flood regimes and to affect different kinds of floods in different ways (see Table 9.1). Regional assessments in the latest Intergovernmental Panel on Climate Change report (IPCC, 2007) suggest likely increases in wet-season precipitation (June to August) and decreases in dry-season precipitation (December to February) (see Plate 18, centre pages). Where drying trends are being experienced or anticipated, reducing flood peaks or durations can be very important to wetlands, fisheries and agricultural ecosystems. More intense rainfall events (e.g. associated with more intense cyclones) increase flood peaks and durations, causing damage to property and posing risks to life in floodplains (see Table

Table 9.1 *Summary of how different types of flood may be affected by climate change and the consequences for vulnerability*

Types of floods	Anticipated impact of climate change on flood regime	Other factors affecting flood regime	Affected and vulnerable groups	How adaptation could exacerbate social justice issues
Flash	Higher frequency of intense rainfall events in urban areas	Increased runoff from impervious surfaces with urban development	Informal settlements near canals and drains	Eviction; no support for settlement
Landslides and floods	Higher frequency of intense rainfall events in mountain areas increases risks of landslides and flash floods	Altered hazard risks from land-use changes	Upland farmers and people living in rural towns near riverbanks	Relocation or restrictions on agricultural land use, which makes people more vulnerable to food shortages
Riverbank overflow	More prolonged rainfall episodes from more intense cyclones or depressions increasing bank overflow	Large-scale reductions upstream in tree cover for agriculture and urban development; irrigation schemes; structural failures (dams and embankments)	Human settlements, industry, infrastructure and agriculture	Diversion of water into farmers' fields to protect cities without compensation, claiming 'acts of nature'
Coastal floods	Increased risk of coastal flooding from sea-level rise	Land subsidence from groundwater pumping	Coastal farming and fisher communities	Embankments to protect hotels and valuable property that cause erosion and flood risks in surroundings
Seasonal floodplain inundation	Reduced flood heights and duration from decreased inland rainfall	Diversions, withdrawals and floodplain protection measures	Lowland farming communities; fishers and harvesters of products from wetland ecosystems	Draining and filling of wetlands as 'flood-prone areas'

9.1). They also increase risks of flash floods and landslides in mountain areas. Sea-level rise exacerbates flood risks in low-lying deltas. Finally, warmer temperatures may interact with flood patterns to alter exposure to water-borne diseases and thus alter risks of flood-related disasters.

In some basins, long-term trends in rainfall may have already altered flood regimes; but untangling the contributions of different factors is difficult. For example, total annual inflows from the Upper Ping River into the Bhumipol Dam has declined by about 0.47 per cent per year over the last 50 years. At the same time, irrigation areas have greatly expanded and forests have been converted to orchards, croplands and human settlements. Rainfall at the main Chiang Mai station upstream has declined about 0.28 per cent, or 3.3mm per year.

Altered flood regimes do not translate linearly into altered risks of flood and flood-related disasters. Modest changes in a flood regime may not have much impact until a threshold is reached, after which the impacts become large. Changes in flood regimes may interact with other processes of change (e.g. riparian land uses, styles of building construction or water withdrawals), which reduce or exacerbate the physical risks of disasters unfolding. Changes in flood regimes may interact with agricultural decision-making in complex ways as farmers try to adapt to changing risks of shortage or excess at different times of the year and, in doing so, alter runoff, groundwater recharge and return flows to rivers from their fields.

Implied rather than explicit in Table 9.1 is variation in how quickly waters rise and fall, how long they last, and what sediments, debris and other pollutants or disease risks they carry with them. Changes in water quality can be as important as quantity to the risks a flood poses to humans. Faster flows and flows with debris cause a lot more damage and loss of life. Contamination of drinking water supplies is often a critical factor in disease outbreaks after flooding.

Changes in flood regimes may also interact with other social factors affecting risk, such as access to resources, levels of convertible assets and wealth. A change to a more benign flood regime, for example, may reduce otherwise increasing social vulnerability. Conversely, a more adverse flood regime may not increase overall risks if it coincides with a decrease in social vulnerability – for example, arising from broad economic development or improved wetland ecosystem management.

The likely impacts of climate change on flood regimes in Thailand (see Table 9.1) are not known with much precision for specific locations. Regional differences across Thailand can be expected given current differences in climate: from dry and highly seasonal conditions in north-east and northern Thailand to the less strongly seasonal moist tropics of the southern peninsula (see Figure 9.1). Projections of future rainfall patterns are very uncertain and current modelling efforts do not yet provide much reliable information at the level of individual basins. This is an area of active research (e.g. Richey et al, 2007; Sharma et al, 2007). Table 9.1 outlines five types of impacts that need to be considered in adaptation. The large uncertainties have consequences for how challenges are articulated in particular places and basins and what make for appropriate and strategic responses.

Challenges

Government, business and civil society are beginning to respond to the challenges posed by climate change (Lebel, 2007). For flood and disaster management, five challenges stand out:

1 reducing the risks of exposure of vulnerable groups;
2 enhancing capacities to cope and respond;
3 securing the affected and vulnerable;
4 building and maintaining resilience; and
5 strengthening links between knowledge and practice (see Figure 9.3).

These challenges can be related to the conventional phases of the disaster management cycle: mitigation, preparedness, emergency and rehabilitation (see Figure 9.3). Adaptation to changes in flood regimes arising from climate change and other factors is subject to politics, particularly related to the allocation of burdens and risks. Fast-moving discourses, slower-changing institutions and practices, together with diverse interests and beliefs combine to shape these politics (see Figure 9.3).

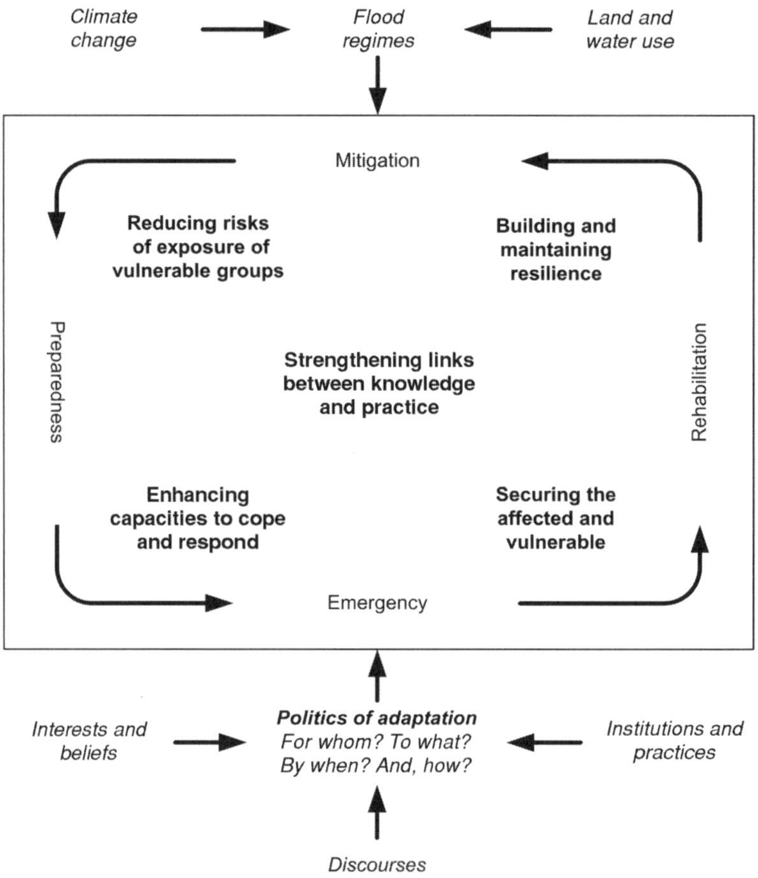

Figure 9.3 *Adaptation to changing flood regimes as a consequence of climate change and other factors poses multiple governance challenges for fair and effective flood and disaster management*

Source: Louis Lebel, Tira Foran, Po Garden and Jesse B. Manuta

The rest of this chapter will deal with each of these challenges in turn. Each section begins by describing the challenge, then proceeds to review how it is currently handled and how it is being (or could be) dealt with in a future climate. Each section ends with a brief reflection on the key elements of a strategic response.

Reducing the risks of exposure

The challenge of reducing the risks of exposure (see Figure 9.3) has traditionally been framed as a technical activity carried out by engineering, water and disaster management experts. Land-use planning and hazard mapping emphasize hazard characteristics rather than understanding of differences in risks of exposure. Broader public engagement is not sought.

The Department of Disaster Prevention and Mitigation (DDPM) was established in October 2002 as, quoting directly, 'the principal government agency to carry out the task and responsibility on disaster prevention and mitigation so as to remain in Thailand as the inhabitable and safe country' (DDPM, 2006). The concepts of disaster management are paternalistic and ambitious. The DDPM replaced the earlier Civil Defence Division and remains within the Ministry of Interior. It is organized through a hierarchy comprised of a national committee, a system of 12 regional centres and, finally, a local civil defence committee, as part of normal bureaucratic structure (province, district and local). The 1979 Civil Defence Act was replaced in August 2007 by the Disaster Prevention and Mitigation Act, which strengthened the legal status of DDPM and may thus help to improve inter-agency coordination. In its 2006 annual report, the DDPM (2006) ranked disaster risk from floods as the highest priority; the report makes no mention of climate change.

The Thai government more recently completed a first *Five-Year Strategy on Climate Change* (2008–2012). The strategy has six components:

1. building capacity to adapt and reduce vulnerability to climate change impacts;
2. promoting greenhouse gas mitigation activities based on sustainable development;
3. supporting research and development to better understand climate change, its impacts, and adaptation and mitigation options;
4. raising awareness and promoting public participation;
5. building the capacity of relevant personnel and institutions and establishing a framework of coordination and integration; and
6. supporting international cooperation to achieve the goal of climate change mitigation and sustainable development (ONEP, 2007).

What is noticeable so far in public forums is the linking of adaptation and mitigation issues and their placement in an environmental problems portfolio. Adaptation and other ways of reducing vulnerabilities to climate change are not seen as closely related to improving disaster management and only modestly related to the pursuit of sustainable development.

Thailand has more than 16 million people (26 per cent of its population) living in the low elevation coastal zone (< 10m), or ninth overall by population exposed (McGranahan et al, 2007). Low-lying areas of Bangkok, in particular, face flood challenges from both upstream and seaward. Land subsidence from groundwater withdrawals in Bangkok will magnify the impacts of modest sea-level rises due to global warming (see Table 9.1). In the unlikely event of large rises in sea levels, the challenges for coastal areas from flooding would be immense. Scenario-based analysis and long-term contingency planning are necessary to the strategic development of adaptation policies.

Massive infrastructure projects have been proposed to protect Bangkok from flooding. One recent example launched by disaster experts and politicians was a proposal to build a wall 80km long, 300m offshore and 3m higher than mean sea level to protect Bangkok and two surrounding provinces. Proponents claimed such a wall would allow mangroves to grow inside and slow down coastal erosion. Needless to say, the impacts on other people, ecosystems and the cost effectiveness of all such mega-projects need careful scrutiny as the side-effects may be larger than the risks they are proposed to address.

Alternatives should be considered, in both moderate and more extreme scenarios, including simply making more space for water as in, for example, restoring multi-functional seasonal wetlands and directing settlements further away from the low-lying coastal areas. Major shifts in land use are not going to be easy to achieve, but are probably essential (McGranahan et al, 2007; Nicholls et al, 2007). Such shifts will need high-quality public information, opportunities to debate and negotiate acceptable levels of risk, and forward-looking infrastructure investments to support the movement of built-up areas across the landscape.

Flood management in Chiang Mai in northern Thailand needs to take into account risks from overflow from the Ping River, which runs through the centre of town, as well as flash flooding from runoff from the adjacent Doi Suthep Mountain (Garden, 2007). Early warning systems in place are reasonably effective at providing adequate time for residents to prepare for riverbank overflow events caused by high rainfall further upstream. However, as with Bangkok, there is strong pressure to manage the main channel and various canals or flood barriers to reduce the risk of floods in the central business and tourism district (Manuta et al, 2006). Doing so increases water depths, velocities and inundation times in other adjacent areas. Conflicts often ensue among different quarters of the city as flood waters arrive (Garden, 2007) and different local agencies attempt to secure their areas. The redistribution of risks among rural and urban areas, as well as among poor and wealthy people in urban areas, is a central theme of flood politics in the Mekong region (Lebel and Sinh, 2007). Land- and water-use planning needs to take into account changes in climate and flood regimes; but doing so cannot be left to hidden processes in technical agencies.

Reducing the risks of exposure requires engagement with, and strong representation of, groups likely to be highly affected or especially vulnerable. Those at most risk should be given the opportunities to participate in reshaping and reducing the risks to which they are to be exposed. Scenario-based approaches can be helpful in handling uncertainties. Informed deliberation is critical to avoid inappropriate overreaction, for example, unnecessary relocations, as well as premature dismissals of risks. Equitably allocating scarce resources and, conversely, burdens and involuntary risks, will often be more of a governance rather than an engineering challenge.

Enhancing capacities to cope and respond

In the monsoonal climate, sound construction, reducing the speed of flood waters, and providing early warning so that people can move their belongings to higher ground may be more plausible and effective responses than overly ambitious attempts at preventing flood waters from ever entering areas of human settlement, or, worse, supporting calls for eviction from high-risk locations without providing alternative options for safe low-income housing. More powerful actors can be expected to mobilize private or public resources to protect themselves, for example, by raising the heights of the soil on which they construct their buildings in floodplains (e.g. Hara et al, 2007), often without regard to the side-effects on others.

Local adaptation to the changing flood regime arising from the urbanization of former irrigated agricultural areas is already under way. Lower-income households often maintain diversified livelihood and income sources (Rigg, 2006). This diversity reflects historical responsiveness to opportunities and demands that can be drawn on in the face of all kinds of challenges, including crop losses due to flooding.

Capacities to cope with and otherwise respond to floods and potential disasters (see Figure 9.3) can be enhanced or eroded by regulations and practices of state agencies (Manuta et al, 2006). Overall, it is not clear that the creation of dedicated centralized agencies for disasters has really helped to build response capacities (Manuta et al, 2005, 2006). One reason is the persistence of bureaucratic competition and fragmentation. With huge mandates, small resources and reluctant support from line agencies that still hold the real expertise (such as the Royal Irrigation Department, or RID), 'disaster' agencies are still finding their place in flood management. Local government, central agency branches and community initiatives often appear to be more important (Garden, 2007).

Central government agencies are beginning to recognize the importance of local government and community organizations in disaster preparedness and emergency response operations. Decentralization reforms in Thailand have helped to make local government more accountable for their constituencies on scales relevant to early warning flood and defence systems. But it also raises issues of capacity where responsibilities are transferred without matching financial resources, the equipment needed or building of relevant capacities. Capacities to cope and respond must remain multilevel because competencies and resources vary according to the level, as do the challenges posed by different disaster events.

Effective disaster preparedness and early warning systems require high-quality two-way communication between those at risk and those with expertise and resources that can help. Fieldwork suggests that redundancy of such channels is beneficial, for instance, when local radio stations are able to step in when the formal administrative system is on holiday or is itself adversely affected by flooding (Garden, 2007).

Flash floods in mountains (see Table 9.1) are particularly hard to address using centralized and remotely managed early warning systems. People at risk need to be able to recognize warning signs of high risk (e.g. intense prolonged rainfall, stream and river quality indicators) and take precautions during such periods. Radar can provide generic warnings about the likelihoods of intense rainfall events over broad areas; but such information alone will still often be inadequate for local disaster management. Tourists may be at higher risk because of unfamiliarity with the speed at which floods in mountain streams can rise. There have been several fatalities of visitors to national parks in Thailand in recent years after instances of intense rainfall.

Politicians increasingly view flood disasters as opportunities and have been instrumental in making the bureaucracy more responsive to public inputs. A series of floods that affected Chiang Mai in 2005 resulted in political responses at multiple levels, but with little real action (Garden, 2007). The mid 2007 municipal elections were notable for a campaign in which many posters pictured candidates standing waste deep in flood waters. In municipal Chiang Mai, traditional weir-based irrigation systems, known as *Muang-fai*, distribute flood waters through what were rice fields but are now, in part, built-up suburbs and commercial districts. Many citizens of Chiang Mai want these irrigation structures near the city removed and for the river to be allowed to adjust to natural levels; but there are also calls for high walls to block peak heights.

This tension between agricultural and urban interests is central to the politics of climate change adaptation in Thailand. For example, a recurrent annual challenge in operating water infrastructure upstream from Chiang Mai is to balance flood control (maintaining sufficient reserve to accommodate possible late wet-season rainfall) with the objective of maximizing storage for irrigation and domestic consumption during the dry season. Late depressions or cyclones can pose a major risk to urban flooding at a time when river levels are normally already high. Global climate change effects on regional cyclone activity or other features of the Asian monsoon could easily have major implications for capacities to cope and respond to flood disasters.

Securing the affected

Despite widespread improvements in well-being – decades of economic growth across Asia that have reduced poverty and improved the health and education of many people – the impacts of climate variability and extremes are distributed very unequally. Disaster policies, programmes and practices have not secured the most highly affected and vulnerable groups from flood disasters. Spatial coverage of relief operations can be limited, and in terms of rehabilitation programmes, even more so.

Difficulties often start from failing to acknowledge the large social differences in the impacts of floods and disasters – for example, between men and women. Complex procedures for compensation can multiply vulnerabilities that begin as social differences. Communities of ethnic minority upland farmers or coastal fishers struggle to get equivalent levels of support in recovering from disasters and often need to appeal through non-state channels to mobilize resources.

In 2004 in Ban Mapota, Om Koi District, in Chiang Mai Province, high flood waters and associated debris from landslips destroyed fields and houses (Manuta et al, 2006). Emergency relief by helicopter was provided; but very little assistance was offered in the recovery stage to a very vulnerable group. Language differences, unfamiliarity and unrealistic bureaucratic procedures that insist reports of damage to dwellings must be made within three days for compensation to be considered discriminate strongly against ethnic minorities living in remote areas. The insistence that compensation only be given to households possessing citizenship cards is particularly unfair, given that the state's failure to adequately provide citizenship cards and services to remote areas created such vulnerabilities in the first place (Manuta et al, 2006).

Flood protection measures to protect central business districts may redistribute risks and burdens to neighbouring urban areas or surrounding rural locations. In October 2006, for example, the Thai government diverted flood waters to agricultural fields upstream to protect key parts of Bangkok after His Majesty the King allowed the Royal Irrigation Department to flood some of his own land to protect Bangkok. Many other areas were subsequently flooded (see Plates 19 and 20) with promises of compensation provided that farmers followed RID planting and harvesting instructions. The RID also argued the need for a law to give it authority to flood areas during high-water periods. Despite a long history of similar events (see Manuta et al, 2006), inadequate prior consultation with farmers and absence of proper institutional mechanisms for compensation meant that serious conflict ensued, causing substantial hardship to farming communities.

Although the 2004 tsunami has nothing to do with climate change risks *per se*, it was the largest disaster to strike Thailand. The challenges that it posed to the social and ecological resilience of coastal communities and the institutional responses that it triggered are insightful for understanding disaster governance and its current limitations. The rush to control coastal land uses (and to deter coastal resettlement) made small fisher households, whose livelihoods were already highly vulnerable from depleted coastal fisheries and competition with larger trawlers, at greater risk in the post-tsunami recovery process (Manuta et al, 2005; Lebel et al, 2006b).

These examples underline that securing the most affected and vulnerable people has not been a priority of governments, and there is little reason to expect that emerging policies to adapt to climate change will change this.

Building and maintaining resilience

Reducing risks from floods does not just depend upon proximate actions before and after floods. Ultimately, some of the more profound impacts could come from building and maintaining the resilience of highly vulnerable groups (see Figure 9.3). Resilience is a measure of the amount of change that a system can undergo and still retain the same controls on structure and function or remain in the same domain of attraction (Walker et al, 2002; Lebel et al, 2006a). In situations of uncertainty and change in ecological services or social organization, building resilience could involve, for example, expanding livelihood opportunities or restoring ecosystems upon which people depend for food, shelter and income.

In Thailand, there is a history of assuming large-scale water infrastructure, such as dams to store and conveyances to transfer water among basins, as the best solution to problems of both excess and shortage. Changing climate and flood regimes imply a careful rethinking of this approach. Infrastructure that is proposed needs to be assessed with respect to future environmental conditions and resource demands, not just short-term political opportunities. Large dams that displace people but are never more than half-full do not help; nor do walls and diversions that just temporarily displace risks in space or time. It may be possible to greatly reduce vulnerabilities by paying much more attention to those factors which enhance social and ecological resilience.

The wetlands in the delta of the Chao Phraya River have been transformed by five decades of interventions of water infrastructure, including embankments, dams, irrigation and drainage canals (Haruyama, 1993). Local rainfall in the lower central plains is prevented from draining naturally by roads and irrigation infrastructure, also increasing risks of deeper and longer flooding in flood-prone areas further inland in the Suphanburi and Angthong provinces (Haruyama, 1993). Within the city, flood waters tend to collect in low-lying eastern parts of the city for prolonged periods. Bangkok's canal system, which had supported a lifestyle that fitted the monsoonal pulse, has been partially converted to allow for road expansion (Ross et al, 2000). The loss of resilience has been compounded by groundwater extraction. Land subsidence in the Bangkok metropolitan area has increased the risk of floods in urban areas (Babel et al, 2006). Deep-well groundwater extraction has resulted in the compaction of sand and clay layers. Observed subsidence of 0.5m to 1m has been measured in some areas. More recent studies suggest that land subsidence still occurs, but at reduced rates in the heavily built critical zones of the city where rates are now around 1cm per year. Higher subsidence rates are now observed in more coastal areas, increasing the risk of seawater flooding. The resilience of Bangkok to climate change has been reduced by patterns of urban development, making future adaptation more difficult and costly.

In the Songkram River wetlands, in contrast, the majority of residents still view flood events as positive: there are times when they catch more fish, have higher incomes and more food to eat (Friend, 2007). More extreme floods can, however, damage paddy rice crops and affect drinking water supplies. Overall, the expectation is that droughts would have more adverse impacts than floods, but that there is substantial resilience to changing climate in the society, especially with ongoing expansion of livelihood options on and off site (MWBP, 2005).

Thailand's intermediate level of resources and technical capacities means that adaptation in some areas will continue to rely, at least in part, on 'living with changes in flood regimes'. Fortunately, there is already substantial local expertise within the region on how to do this well, which in some locations, with proper support, may be a better outcome for local livelihoods than interventions that prevent all flooding. A useful response to challenges posed by climate change, therefore, is to restore and protect seasonal floodplains, mangroves and wetland ecosystems, allowing for compatible uses. In making more space for water and nature, the risks posed to less flood-tolerant infrastructure elsewhere can also be lessened.

Strengthening links between knowledge and practice

It may not be easy to tell if flood regimes are changing because of climate change until well after it has happened. Flood regimes and the risks that they pose to particular social groups, in any case, are affected by multiple factors (see Table 9.1). Planning will invariably take place in the presence of significant knowledge uncertainties and interest-based barriers to new information. A better understanding of how floods affect different groups, and how flood and disaster management systems are performed through a disaster cycle in the public domain, would be valuable to efforts aimed at reducing the risks of disaster more fairly (see Figure 9.3).

Understanding the causes of floods and flood regime changes is also important in taking appropriate remedial actions. Unfortunately, much policy on floods appears to have been made with little reference to evidence-based reasoning. It is common practice in Thailand, for example, to blame all floods on land-use changes in the mountains (Forsyth, 1998; Walker, 2003), without paying any attention to the amount or intensity of rainfall. An example from southern Thailand illustrates some of the pitfalls of quick attribution. Major floods occurred in Hat Yai, southern Thailand, in 1988 and 2000. The floods of 21–24 November 2000 killed 30 people and caused more than US$220 million in damage. In 1998 about 5.6 per cent, and in 2000 about 22 per cent of the municipal area was inundated to depths of more than 2m. Some communities were submerged more than others. Tanavud et al (2004) assert that the vulnerability of Hat Yai to floods increased as a result of the reduction in forest cover from 20 per cent to 11 per cent in the upstream of the Khlong U-Taphao Basin between 1982 and 2002. Most of the land, the authors note, was converted to rubber plantations. They argue that the conversion 'disturbed the finely tuned equilibrium of the natural ecosystems to such a degree that environmental stability was compromised'.

The impacts of changes in land use on watershed hydrology in mountains with complex landscapes, when studied carefully, are complex (Bruijnzeel, 2004; Sidle et al, 2006; Richey et al, 2007). This contrasts with the quick and simplistic attribution of floods in urban areas to events upstream in the watershed, which is a common practice in Thailand (Manuta et al, 2006; Lebel and Sinh, 2007). Careful scrutiny of changes within urbanizing regions is much rarer.

Thailand's *Initial Communication* (OEPP, 2000) under the United Nations Framework Convention on Climate Change (UNFCCC) makes few specific references to how it proposes to adapt to climate change impacts upon floods. The standard line has been that 'reforestation, afforestation, protection of conservation forests, land and water conservation also support the adaptation process' (OEPP, 2000, p69). In effect, climate change has simply been incorporated within existing environmental debates and policy in which reforestation is prescribed as a solution to almost all environmental ills (Forsyth, 2003). The problem is that, in practice, this usually means focusing on restricting land use in vulnerable mountain areas, relocation and other drastic policies that would make marginal people even more vulnerable, while major factors affecting risks within floodplains are completely ignored. Moreover, misleading attributions of recent individual flood events or disasters to climate change may also prevent learning because they also become an argument for blaming others – global warming – rather than examining those contributions to vulnerability and differences in risk that can be addressed locally. Poorly reasoned and unfair adaptation policies could increase risks for vulnerable groups (see Table 9.1). This is the 'dark side' of adaptation.

Overall, capacities for assessing vulnerability and adaptation options in Thailand are fairly limited – indeed, so much so that the *Initial Communication* to the UNFCCC published in 2000 states: 'the lack of comprehensive research in this area seriously limits the ability to make appropriate policy recommendations' (OEPP, 2000). A review carried out in 2007 suggests only minor improvements (Jesdapipat, 2007). Public awareness, however, seems to be growing substantially with much more media coverage of climate change events and international issues in 2007 then during earlier years.

Discussion and conclusions

The government has struggled to cater for the interests and needs of the poor and other disadvantaged groups under current climate and flood regimes. Issues of social justice have been ignored when they should have been made central to the pursuit of reducing risks of disasters.

Flood regimes and risks are already changing in Thailand as a result of human activities. The prospect of additional risks from the impacts of climate change on flood regimes makes the need for forward-looking action greater than ever before. But interventions in the name of adaptation to climate change can create winners and losers (see Table 9.1). Interventions can shift the distribution of benefits or involuntary risks from one group to another. Adaptation may even exacerbate injustice, as when actions in the logic of protecting national assets and interests make some disadvantaged groups even more vulnerable than they were before.

There are four main reasons why climate change could significantly exacerbate existing unfairness and inequities corresponding to each of the major management challenges (see Figure 9.3). First, risks of exposure vary hugely across different social groups despite profound improvements in average measures of well-being, health and economic development. Second, capacities to influence decision-making on behalf of, or by, vulnerable groups remain limited. Opportunities for building capacities to cope with, and respond to, floods are limited. Third, ethnic minorities, migrants, women and other second-class citizens continue to be at a disadvantage when accessing key relief and rehabilitation services and resources. This is the result of discriminatory policies and practices. Fourth, some of the most vulnerable groups are dependent upon seasonal floods for the maintenance of wetland and farming systems. Excessive river regulation in the name of flood protection and for other objectives has made these social–ecological systems less resilient. High risks of exposure, weak political influence, limited access and neglected dependencies, together, spell disaster.

New approaches are needed to more fairly and equitably address current and future challenges posed by changing flood regimes, including the anticipated impacts of a changing climate. The political and institutional dimensions of disaster management need to be acknowledged and become a foundation for improved disaster governance (Lebel et al, 2006c). Important guiding principles might include putting the most vulnerable groups first (Paavola and Adger, 2006); building social and ecological resilience (O'Brien et al, 2006; Berkes, 2007); and what might be described as 'democratization' of disaster management (Lebel and Sinh, 2007).

From mitigation through to rehabilitation phases of the disaster cycle, there is a need to empower and enable the affected and disadvantaged (see Figure 9.3). They require places to articulate their needs and aspirations, and space to build and develop their capacities. Access to, and control over, resources is necessary, not just top-down allocations to agencies acting on 'their behalf'. Informed, deliberative and collaborative approaches to major decisions about flood and disaster management hold promise. The links between knowledge and practice should be strengthened in both directions, recognizing the valuable contributions which local experience and understanding of conditions can make to reducing risks, while also making best use of science and technology to serve groups in greatest need. A focus on building and maintaining resilience of affected and vulnerable groups rather than managing floods as generic hazards could also help to address current injustices.

Reducing the risk of disasters should be central to climate adaptation (Thomalla et al, 2006; Bouwer et al, 2007). Incorporating climate change adaptations within flood and disaster management should be seen as an opportunity to address inequities, insecurities and unfairness that have created large disparities in well-being, vulnerability and opportunity. But we should not wait for more catastrophic confirmations of climate change: there are many actions that would benefit disadvantaged and vulnerable groups now which do not need climate change or any other excuse as a justification.

Acknowledgements

The Asia-Pacific Network for Global Environmental Change Research and START (global change SysTem for Analysis, Research and Training) supported initial case study work and regional meetings. Follow-up work has been supported by International Fund for Agricultural Development and Echel Eau through the Challenge Program on Water and Food (CPWF) for the M-POWER programme (www.mpowernet.org) under project grant PN50.

References

ABI (Association of British Insurers) (2005) *Financial Risks of Climate Change*, Summary report, June, ABI, London

Adger, N. W. (2001) 'Scales of governance and environmental justice for adaptation and mitigation of climate change', *Journal of International Development*, vol 13, pp921–931

Babel, M. S., A. D. Gupta and N. D. S. Domingo (2006) 'Land subsidence: A consequence of groundwater over-exploitation in Bangkok, Thailand', *International Review for*

Environmental Strategies, vol 6, pp307–327

Berkes, F. (2007) 'Understanding uncertainty and reducing vulnerability: Lessons from resilience thinking', *Natural Hazards*, vol 41, pp283–295

Bouwer, L., R. Crompton, E. Faust, P. Hoppe and R. A. Pielke Jr. (2007) 'Confronting disaster losses', *Science*, vol 318, p753

Bruijnzeel, L. A. (2004) 'Hydrological functions of tropical forests: Not seeing the soil for the trees', *Agriculture, Ecosystems and Environment*, vol 104, pp185–228

DDPM (Department of Disaster Prevention and Mitigation) (2006) *Thailand Country Report*, DDPM, Ministry of Interior, Thailand

Dutta, D. and S. Herath (2004) 'Trend of floods in Asia and flood risk management with integrated river basin approach', paper presented at Second Asian Pacific Association of Hydrology and Water Resources Conference, Singapore

Elster, J. (1992) *Local Justice: How Institutions Allocate Scarce Goods and Necessary Burdens*, Russell Sage Foundation, New York, NY

Forsyth, T. (1998) 'Mountain myths revisited: Integrating natural and social environmental science', *Mountain Research and Development*, vol 18, pp126–139

Forsyth, T. (2003) *Critical Political Ecology: The Politics of Environmental Science*, Routledge, London

Friend, R. (2007) *Securing Sustainable Livelihoods through Wise Use of Wetland Resources: Reflections on the Experience of the Mekong Wetlands Biodiversity Conservation and Sustainable Use Programme (MWBP)*, Mekong Wetlands Biodiversity Conservation and Sustainable Use Programme, Vientianne, Lao PDR

Garden, P. (2007) *The Chiang Mai Floods of 2005*, USER Working Paper WP-2007-19, Unit for Social and Environmental Research, Chiang Mai University, Chiang Mai

Hara, Y., D. Thaitakoo and K. Takeuchi (2007) 'Landform transformation on the urban fringe of Bangkok: The need to review land-use planning processes with consideration of the flow of fill materials to developing areas', *Landscape and Urban Planning*, vol 84, no 1, pp74–91

Haruyama, S. (1993) 'Geomorphology of the central plain of Thailand and its relationship with recent flood conditions', *GeoJournal*, vol 31, pp327–334

IPCC (Intergovernmental Panel on Climate Change) (2007) *Climate Change 2007: Synthesis Report, Fourth Assessment Report: The Physical Science Basis*, Cambridge University Press, Cambridge

Jesdapipat, S. (2007) *Report on Capacity Building Survey to Address Thailand Vulnerabilities, Adaptation and Resilience to Climate Risks*, SEA START RC, Chulalongkorn University, Bangkok, Thailand

Lebel, L. (2007) 'Adapting to climate change', *Global Asia*, vol 2, pp15–21

Lebel, L. and B. T. Sinh (2007) 'Politics of floods and disasters', in L. Lebel, J. Dore, R. Daniel, and Y. S. Koma (eds) *Democratizing Water Governance in the Mekong Region*, Mekong Press, Chiang Mai, pp37–54

Lebel, L., J. M. Anderies, B. Campbell, C. Folke, S. Hatfield-Dodds, T. Hughes and J. Wilson (2006a) 'Governance and the capacity to manage resilience in regional social–ecological systems', *Ecology and Society*, vol 11, no 1, pp11, 19, www.ecologyandsociety.org/vol11/iss11/art19/

Lebel, L., S. Khrutmuang and J. Manuta (2006b) 'Tales from the margins: Small fishers in post-tsunami Thailand', *Disaster Prevention and Management*, vol 15, pp124–134

Lebel, L., E. Nikitina, V. Kotov and J. Manuta (2006c) 'Assessing institutionalized capacities and practices to reduce the risks of flood disasters', in J. Birkmann (ed) *Measuring Vulnerability to Natural Hazards: Towards Disaster Resilient Societies*, United Nations University Press, Tokyo, pp359–379

Manuta, J., S. Khrutmuang and L. Lebel (2005) 'The politics of recovery: Post-Asian tsunami reconstruction in southern Thailand', *Tropical Coasts*, July, pp30–39

Manuta, J., S. Khrutmuang, D. Huaisai and L. Lebel (2006) 'Institutionalized incapacities and practice in flood disaster management in Thailand', *Science and Culture*, vol 72, pp10–22

McGranahan, G., D. Balk and B. Anderson (2007) 'The rising tide: Assessing the risks of climate change and human settlements in low elevation coastal zones', *Environment and Urbanization*, vol 19, pp17–37

MWBP (Mekong Wetlands Biodiversity Conservation and Sustainable Use Programme) (2005) *Vulnerability Assessment of Climate Risks in the Lower Songkhram River Basin, Thailand*, Vientiane, Lao PDR

Nicholls, R. J., S. Hanson, C. Herweijer, N. Patmore, S. Hallegatte, J. Corfee-Morlot, J. Chateau and R. Muir-Wood (2007) *Ranking of the World's Cities Most Exposed to Coastal Flooding Today and in the Future*, Organisation for Economic Co-operation and Development (OECD), Paris

O'Brien, G., P. O'Keefe, J. Rose and B. Wisner (2006) 'Climate change and disaster management', *Disasters*, vol 30, pp64–80

OEPP (Office of Environmental Policy and Planning) (2000) *Thailand's Initial National Communication under the United Nations Framework Convention on Climate Change*, OEPP, Ministry of Science, Technology and Environment, Bangkok, Thailand, p100

ONEP (Office of Natural Resources and Environmental Policy and Planning) (2007) *Five-year Strategy on Climate Change (2008–2012)*, ONEP, Ministry of Natural Resources and the Environment, Bangkok, Thailand

Paavola, J. and N. W. Adger (2006) 'Fair adaptation to climate change', *Ecological Economics*, vol 56, pp594–609

Richey, P. T., D. Thomas, S. Rodda, B. Campbell and M. Logsdon (2007) 'Effects of landuse change on the hydrologic regime of the Mae Chaem River Basin, NW Thailand', *Journal of Hydrology*, vol 334, pp215–230

Rigg, J. (2006) 'Land, farming, livelihoods, and poverty: Rethinking the links in the rural South', *World Development*, vol 34, pp180–202

Ross, H., A. Poungsomlee, S. Punpuing and K. Archavanitkul (2000) 'Integrative analysis of city systems: Bangkok 'Man and the Biosphere' programme study', *Environment and Urbanization*, vol 12, pp151–161

Sharma, D., A. D. Gupta and M. S. Babel (2007) 'Spatial disaggregation of bias-corrected GCM precipitation for improved hydrologic simulation: Ping River Basin, Thailand', *Hydrology and Earth System Sciences Discussions*, vol 4, pp35–74

Sidle, R., A. Ziegler, J. Negishi, A. R. Nik, R. Siew and F. Turkelboom (2006) 'Erosion processes in steep terrain – truth, myths and uncertainties related to forest management in Southeast Asia', *Forest Ecology and Management*, vol 224, pp199–225

Takeuchi, K. (2001) 'Increasing vulnerability to extreme floods and societal needs of hydrological forecasting', *Hydrological Sciences Journal*, vol 46, pp869–881

Tanavud, C., C. Yongchalermchai, A. Bennui and O. Densreeserekul (2004) 'Assessment of flood risk in Hat Yai Municipality, southern Thailand, using GIS', *Journal of Natural Disaster Science*, vol 26, pp1–14

Thomalla, F., T. Downing, E. Spanger-Siegried, G. Han and J. Rockström (2006) 'Reducing hazard vulnerability: Towards a common approach between disaster risk reduction and climate adaptation', *Disasters*, vol 30, pp39–48

Thomas, D. S. G. and C. Twyman (2005) 'Equity and justice in climate change adaptation amongst natural resource-dependent societies', *Global Environmental Change*, vol 15, pp115–124

Walker, A. (2003) 'Agricultural transformation and the politics of hydrology in northern Thailand', *Development and Change*, vol 34, pp941–964

Walker, B., S. R. Carpenter, J. Anderies, N. Abel, G. S. Cumming, M. A. Janssen, L. Lebel, J. Norberg, G. D. Peterson and L. Pritchard (2002) 'Resilience management in social–ecological systems: A working hypothesis for a participatory approach', *Conservation Ecology*, vol 6, article 14, www.ecologyandsociety.org/vol6/iss1/art14/print.pdf

10

Water and Spatial Planning in The Netherlands: Living with Water in the Context of Climate Change

Michelle J. A. Hendriks and Joost J. Buntsma

Introduction

One quarter of The Netherlands is located below sea level and for centuries its inhabitants have been accustomed to dealing with an abundance of surface water and the threat of flooding. The Dutch have thus developed a lifestyle which they term 'living with water'. Over 50 per cent of the total population of 16 million lives under sea level in the heavily populated areas of the western part of the country. This 'living with water' is evident throughout the landscape. Flood defences, such as dikes and dunes bordering rivers, lakes and the sea, prevent 65 per cent of the country from being flooded on a regular basis (see Figure 10.1). One sixth of the country's surface area is covered by open water. Water resources are managed at a square metre scale in order to create optimal conditions for agriculture, buildings, infrastructure and natural ecosystems. This is a complex task in a country with a population density of 483 people per square kilometre.

This 'living with water', combined with high population pressure, has led to a high degree of competition for space between water, man and ecosystems. Current and future situations in The Netherlands thus require an integrated approach to managing the spatial distribution of water with a long-term view and in close collaboration with all stakeholders.

This chapter describes the way in which The Netherlands is taking up the challenge of climate change. We start with a brief overview of the physical, historical and organizational context of water resources in The Netherlands. This is followed by a description of the Dutch way of dealing with water in the past and new ways of dealing with it in the future. Some of the main projects that are currently undertaken to manage water throughout the country (urban and rural, coastal and river zones) will be discussed. We finally remark on how The Netherlands is adopting a more forward-looking approach to adapt the Dutch way of 'living with water' to the impacts of climate change.

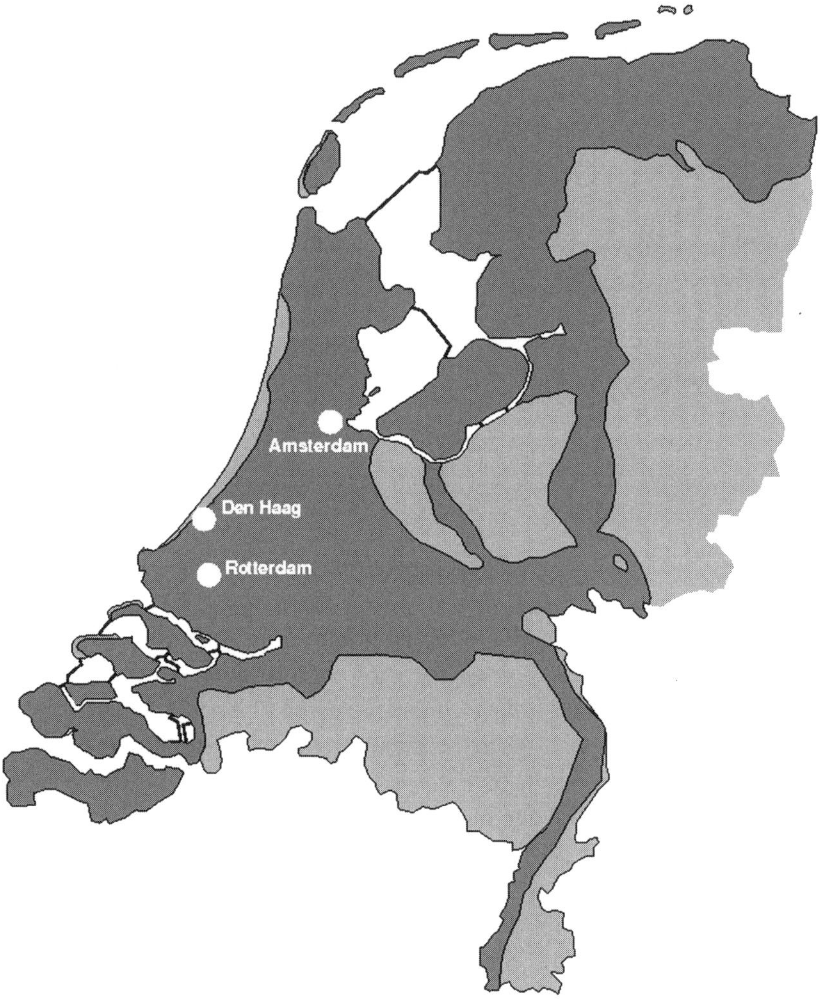

Figure 10.1 *Floodable areas (darker grey) of The Netherlands if there were no flood defences*

Source: http://web.inter.nl.net/hcc/Gbm.Delahaye/marais.htm

The Dutch living in the lowlands

It is well known that The Netherlands has fought a long and successful battle against water. The country is a delta area fed by three main rivers: the Rhine, the Meuse and the Scheldt (see Figure 10.2). It is famous for its land reclamations from lakes and from the sea. All of this has created a situation where The Netherlands continuously has to live with water. Centuries ago, the Dutch more or less accepted the power of water. To prevent damage, they built their houses on artificially raised hills (mounds or '*terps*') and had their luxury rooms with expensive furniture in a higher part of the house. This was followed by centuries in which flooding was prevented by constructing dikes and draining water with pumps. More space for living and working was created by making 'polders' (a wet area where the water has been pumped and drained).

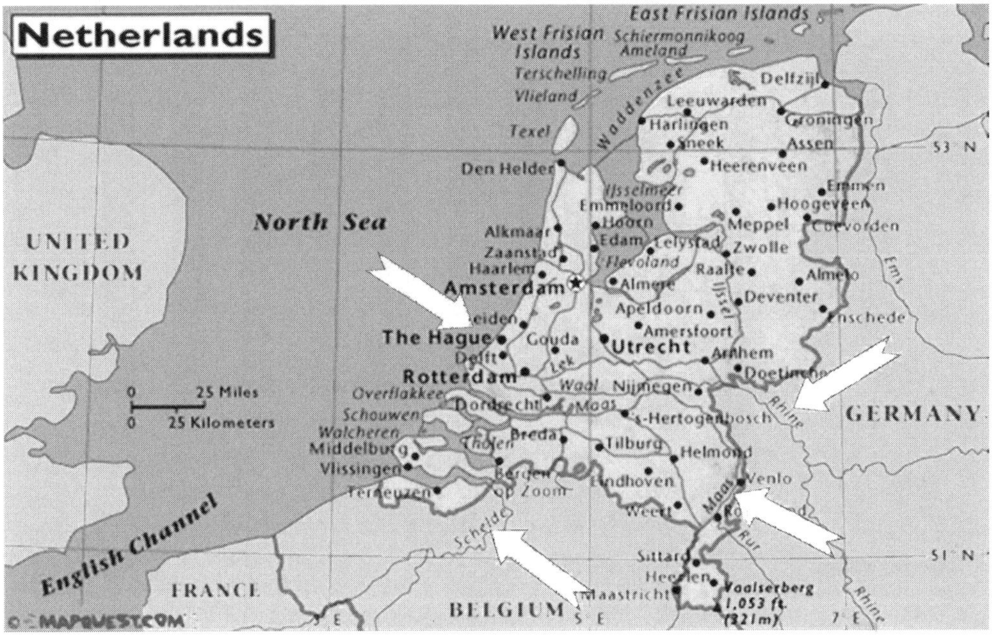

Figure 10.2 *Map of The Netherlands*

Note: Rivers mentioned in the text are indicated with arrows. On this map, the Scheldt is labelled Schelde and the Meuse is labelled Maas.

Source: http://www.mapquest.com

The lowest point of The Netherlands is 6.74m below mean sea level. More than half of the total population of 16 million live in areas below sea level and about 70 per cent of gross domestic product (GDP) is produced in these areas.

Water enters The Netherlands from the south and east through several rivers. The catchment areas of the main rivers are 185,000km^2, 32,000km^2 and 22,000 km^2 in extent for the Rhine, Meuse and Scheldt rivers, respectively, whereas The Netherlands itself only covers 34,000km^2 (excluding water). Annual rainfall (800mm) in The Netherlands is greater than evaporation (560mm) (see Figure 10.3). However, rivers are the main contributors of inflow of water into the country, with mean annual discharges of 69 billion m^3 and 8 billion m^3 for the Rhine and Meuse, respectively. Converting this into millimetre equivalent water depth over the entire country yields 1700mm and 200mm, respectively.

Dutch history is scattered with water-related disasters of various degrees of impact. This has always affected the way in which water in the country is dealt with. One of the most significant events in recent history was the flooding of the province

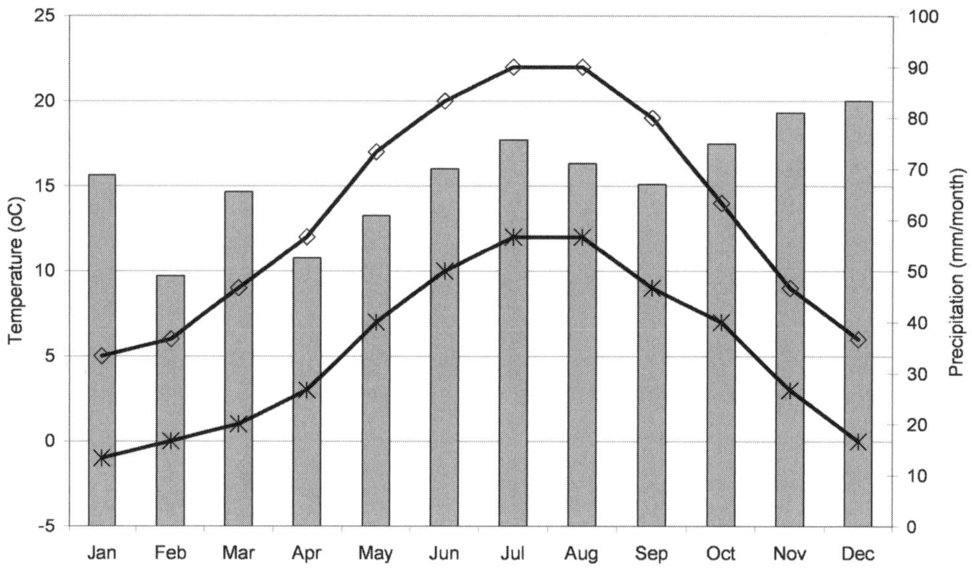

Figure 10.3 *Mean minimum and maximum temperature and precipitation for De Bilt, located in the middle of The Netherlands*

Source: Royal Netherlands Dutch Meteorological Institute (KNMI)

of Zeeland in February 1953. The dikes were breached in many places, large areas became inundated and many people and livestock lost their lives. This event triggered the initiation of the so-called Delta Works. Many waterworks were built in order to protect the south-west area of the country against storm surges and to prevent salinization. The Delta Works are designed for the 1 in 10,000-year storm occurrence in the densely populated coastal provinces. The storm surge barrier, the Maeslantkering, in the Rotterdam Waterway was completed in 1997 and was the last component of the Delta Works project.

Two other major events that influenced the Dutch water policies were the occurrences of extremely high water levels in the main rivers. In 1993 and, in particular, in 1995 the water levels were so high that people feared a breach of the dikes and many were evacuated from their premises. Although this turned out to be unnecessary as the dikes stayed in place, these events speeded up plans for dike reinforcement through the *Delta Plan for the Major Rivers*. This made it possible to maintain a normative river discharge of 15,000m^3/s (1 in 1250 years) for the Rhine River, which was in line with the 1997 Flood Protection Act.

The events also created the awareness among politicians and the general public that physical protection in the form of dikes and draining of land could not work forever. This awareness was strengthened by the increasing frequency of excessive rainfall that The Netherlands experienced at the end of the 20th century. More water was coming from all directions. On top of this, the country faced some severe droughts in more recent years. Surprisingly, one of these events led to the collapse of a small secondary dike due to instability caused by shrinking of the dike body. This resulted in damage to an entire neighbourhood and required the evacuation of people.

These types of events might be attributed to climate change. To be able to predict climate change and its consequences, the Royal Netherlands Meteorological Institute (KNMI) develops sets of climate scenarios on a regular basis. Based on Intergovernmental Panel on Climate Change (IPCC) studies, observations and local climate models, KNMI provides climate change scenarios in an easy and understandable format to be used by relevant stakeholders (see Table 10.1). According to the most recent projections (Van den Hurk et al, 2006) temperature will have risen by 1°C to 6°C by the year 2100 (base year 1990). Rainfall will increase as well, with less frequent but more extreme events in summer and longer periods of rain in winter. Projected sea-level rise will be between 35cm and 85cm. However, soil subsidence in The Netherlands caused by the continuous drainage of marshy land consisting of peat and clay is common. This aggravates the problem of sea-level rise. Dutch water policies are based on an assumed rise of 60cm during this century.

Future challenges related to climate change can be classified as follows:

- *Sea-level rise.* The projections for the year 2050 are between 15cm and 35cm. The main threat is flooding from the sea during extreme storm events. Draining water surpluses will require more regular pumping.
- *Erratic river flow from other countries.* More frequent and larger floods will, for example, influence river transportation. More frequent extreme low flows will influence irrigation.
- *Inundation of downstream areas.* This issue is mainly covered under the *Nationaal Bestuursakkoord Water* (to be discussed later).
- *Rainfall surplus and drought.*
- *Local extreme rainfall events.* Figure 10.4 shows a typical example of expected local water surpluses – potentially not life-threatening, but causing economic damage.

The classical way of responding to these issues has been to construct higher dikes, more dams and increased pumping capacity. As indicated earlier, this is not considered sustainable any more. New approaches based on (innovative) adaptation strategies are currently implemented in the country.

Institutional context

Water management in The Netherlands is administered at national (1), provincial (12) and municipal (483) levels. In addition, water boards (27) have a specific mandate for managing water. These democratic institutions all have distinct tasks and responsibilities regarding water management.

Table 10.1 *Royal Netherlands Meteorological Institute (KNMI'06) climate change scenarios for 2050 relative to 1990*

	G	G+	W	W+
Summertime values				
Mean precipitation (%)	2.8	−9.5	5.5	−19.0
Wet day frequency (%)	−1.6	−9.6	−3.3	−19.3
Precipitation on wet days (%)	4.6	0.1	9.1	0.3
10-year return level daily precipitation sum (%)	13.0	5.0	27.0	10.0
Potential evaporation (%)	3.4	7.6	6.8	15.2
Wintertime values				
Mean precipitation (%)	3.6	7.0	7.3	14.2
Wet day frequency (%)	0.1	0.9	0.2	1.9
Precipitation on wet days (%)	3.6	6.0	7.1	12.1
10-year return level, 10-day precipitation sum (%)	4.0	6.0	8.0	12.0
Annual values				
Sea-level low estimate (cm)	15	15	25	25
Sea-level high estimate (cm)	20	20	35	35

Note: Only water-related aspects are included. G indicates global temperature increase by 1°C; W indicates increase by 2°C; + scenarios indicate changes in dominant wind directions.

The coordinating party at national level is the Ministry of Transport, Public Works and Water Management. It is responsible for national policies on flood protection and water management, and for overseeing implementation. *Rijkswaterstaat*, the implementing directorate of the ministry since 1798, is responsible for operational management of certain waters only. The major rivers and waterworks of national interest, such as the primary dikes and dunes, are their responsibility. The other ministries, who have partial water management responsibilities, are the Ministry of Housing, Spatial Planning and Environment and the Ministry of Agriculture, Nature Management and Food Safety.

Twelve provinces make up the intermediate level. Their role is to ensure that regional and local parties implement the national and provincial policies on water and spatial planning. An important feature of this role is the coordination of the different policy sectors, such as water, environment, housing and economics. They execute their role through provincial spatial plans, water extraction permits, etc.

At the local level, the municipalities have responsibility for spatial planning and, more specifically, for managing sewerage systems. There is a tendency of municipalities to hand over some of their responsibilities to the water boards.

The water boards play a big role in water management and flood protection at regional level. Water boards are typically Dutch. They originated from local communities electing community members whose role was to take responsibility for dikes, ditches and flumes in their area. They are the oldest democratic organizations in The Netherlands. Being governmental bodies, the water boards function according to the

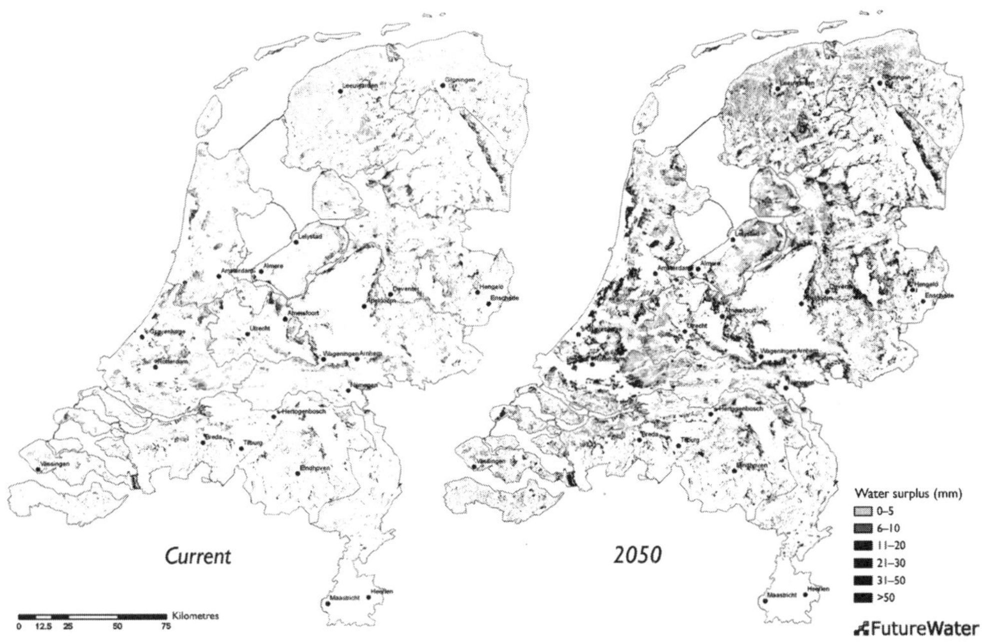

Figure 10.4 *Local water surplus under climate change and land-cover changes*

Source: Immerzeel and Droogers (2008)

basic principle of 'interest–payment–say'. Landowners, residents and wastewater dischargers (firms and households), who all have an interest in water (quantity and quality), elect members for the water boards. They also pay for the services delivered by the water boards. During the last 60 years, the number of water boards has decreased from 2500 to 27. Their main tasks are flood control, management of regional water resources (quantity and quality) and treatment of urban wastewater. They implement this through maintaining dikes and waterworks, licensing discharges, etc. Water boards are entitled to raise tax and are financially self-supporting.

Legislation and institutional arrangements are part of the organizational context as well. An important act with regard to rising sea level and river discharges is the 2005 Flood Protection Act. This act divides The Netherlands into 95 areas surrounded by flood defence structures: so-called levee rings. For each area, the act defines a safety standard. For example, for the most densely populated coastal provinces, the dikes, dunes and waterworks have to be able to resist the 1 in 10,000-year storm event and they are tested every five years. If necessary, the waterworks are adapted to ensure that the set standard will be met.

A recent institutional rearrangement in water management is the introduction of a river basin district approach that stems from the requirements of the European

Water Framework Directive. The boundaries of river basins do not fully coincide with those of the existing administrative bodies. This has led to new networks of cooperation between parties in the (sub-)river basins. This approach enhances cooperation and integration, as well as synergetic solution, for water quality and quantity problems.

A more collective approach

In June 2003, the four different governmental bodies dealing with water entered into an agreement on a national level for the first time. In the so-called National Governmental Water Agreement (*Nationaal Bestuursakkoord Water*, or NBW) the three ministries, the Interprovincial Platform (IPO), The Netherlands Association of Municipalities (VNG) and the Union of Water Boards (UvW) expressed their shared responsibility for a more sustainable Dutch water system by the year 2015 and for maintaining such a system by anticipating expected future changes. They agreed on goals and measures to be taken, where possible, through an integrated approach.

The NBW has influenced Dutch water management processes positively. Common goals are clearer and there is more practical cooperation and communication. It has also led to better insight into the potential measures with regard to water quantity problems at regional level.

The agreement comprises some relatively new instruments and approaches. One of them is the once-only offer of €100 million for subsidies from the national government to support projects that address regional water quantity problems. This led to the initiation of a list of projects with a total budget of €400 million. Through this subsidy, these projects are timely, are safely set on the political agenda and execution is ensured. The partners also chose a new approach to communicate water issues to the broader public. They agreed to take up communication collectively under one flag: 'The Netherlands lives with water.' This has turned out to have a positive effect on public awareness of water problems and measures.

The organizational approach and the instruments chosen in the NBW were a logical transition to the paradigm shift in thinking about water management that was triggered by the near flooding events in 1993 and 1995. The paradigm shift found substance in a new policy, *Water Management in the 21st Century*, better known as WB21. This was the conclusion of a commission advising on the necessary changes in the way in which water is managed, keeping in mind the consequences of climate change, sea-level rise, land subsidence and spatial developments. The basic principles of WB21 have become a connecting thread in Dutch water management. A key premise of WB21 is that water must be provided space before it takes it itself. For centuries, rivers have been narrowed and straightened, ditches have been filled and areas with impermeable surfaces have increased rapidly. In current situations where water flows (rainfall and river discharges) are increasing in volume and intensity, water tends to break out from tight systems, finding its own way and taking its own space.

WB21 is based on three basic principles. The first principle is to anticipate (instead of reacting to) changes in the water system. Although this seems obvious, it has far-reaching consequences. Water will have a much more prominent role in spatial planning. The second principle is to combine technical solutions with the smart use of space – for example, multipurpose uses of space such as houses on water. The third principle of WB21 is considered the most important one and deals with excess water. It is referred to as 'retain–store–discharge'. The aim is to retain excess rain where it falls by, for example, making it easier to drain into the soil. The local storage of water or storage in another area (e.g. by maximizing capacity of existing canals) is the second best option. Discharging water to another water system has the lowest preference as it is a way of shifting the burden to another water system (see Plate 21, centre pages).

Putting more emphasis on water in spatial planning is also an important measure with respect to flood safety as it was doubtful whether raising and strengthening the dikes is a durable solution. In addition to this, the government is considering new safety standards based on other potential risks rather than on the risk of flooding only. At the same time, the government is looking for other safety measures, such as evacuation plans. It was decided to swiftly act on the advice of WB21 by setting out concrete projects (the pragmatic approach) and by developing a new, more forward-looking approach at the same time.

A pragmatic approach

The advocated pragmatic approach in WB21 was further refined for the three areas under threat in The Netherlands:

1 urban and rural areas;
2 areas close to rivers; and
3 coastal areas.

In each of these three domains, specific actions were defined and translated into concrete projects.

Urban and rural development

The basic principles of WB21 find their best application in dealing with local low-risk water problems (where 'risk' is considered as the combination of chance and consequence). A concrete example of the more prominent role of water in spatial planning is the introduction of a mandatory Water Impact Assessment (*Watertoets*) in spatial plans. By including an obligatory paragraph on water management in spatial plans, the consequences of changes in spatial planning on water management are made clear in advance.

Another example of linking water management and land-use in urban and rural areas is the iterative process of setting standards of acceptable levels of local inundation. First, a working standard is set, based on the type of land use that was established for all areas. For example, inundation of pasture is acceptable once every 10 years and once every 100 years in urban areas. To enhance transparency, a democratic approach is used where all regional parties work together on an acceptable standard using the necessary technical and spatial methods. Specific areas may thus diverge from the general standard. A typical example is lowering the criterion for an urban or agriculture area in combination with financial compensation of inhabitants for these higher chances of potential damage. In this more dynamic approach, parties balance costs and benefits, looking for area-specific solutions and combinations of water and other interests.

River basin regions: Meuseworks and Room for the River projects

In areas close to rivers (river basin regions), the principles of WB21 play a major role as well. Urban development along the main rivers has resulted in higher safety requirements (see Plate 22, centre pages). After the life-threatening flood events in 1993 and 1995, the national government started with accelerated dike improvements of primary flood defences. In the area of the Meuse River, the *Meuseworks* project was initiated. This is one of the first projects to use an integrated approach to river flood prevention (chance of flooding 1 in 250 years), nature development, gravel exploitation and shipping. National and provincial governments and regional partners jointly decided on goals and measures. All of this makes the process rather complex and time consuming, but creates a higher acceptance of chosen measures and leads to synergy in solving problems. Nevertheless, under political pressure to ensure desired safety, it was initially decided to build more or higher dikes anyway. In tandem with this action and in light of new insights, other measures were further developed. Typical examples of these rather innovative measures are broadening of the river bed; allocation of areas for water storage; adjustment of bridges and weirs; and nature conservation development.

In conjunction with the implementation of these measures, the Integral Exploration of the Meuse project (*Integrale Verkenning Maas*) was initiated in 2003. The project was designed to explore measures in greater spatial range, such as preventing areas from being occupied by buildings in order to preserve space for future higher discharges. A set of potential measures is shown in Plate 23 (centre pages), where options to select from are location specific.

Whereas the areas in the Meuse region are subject to the risk of shallow inundation, flooding in the Rhine region can have more serious impacts. Therefore, safety standards were set at levels between 1 in 1250-year and 1 in 4000-year flood return periods. Converting these standards to expected Rhine discharges in 2015 results in 16,000m^3/s. Considering dikes as the only protection measure would require a rise of 30cm of all primary river dikes. This was undesirable as higher dikes also mean higher risks of larger impacts. In 2000, the government decided on a new approach of flood protection where open space areas in the vicinity of the river will be used as retention zones. In this project, referred to as Room for the River, the main goals are to reach the desired safety level for the Rhine branches by 2015 and to simultaneously improve environmental quality in the area.

Various parties, including ministries, water boards, provinces, municipalities, non-governmental organizations (NGOs) and inhabitants participated in this Room for the River project, which led to over 600 potential measures. 39 of these measures have now been selected for further studies in a Spatial Planning Key Decision (SPKD) process. Selected measures consist of displacement of dikes, deepening of forelands, lowering of groins, etc. (see Plate 23, centre pages). Reinforcement of dikes is included only if other measures are too expensive, inadequate or impossible. In conjunction, the SPKD demarcated ten potential spatial reservations. These are areas that might be needed for creating space for the river, looking at a possible discharge of 18,000m^3/s in 2100. At the moment, the chosen package of measures is being worked out in more detail in a participatory approach with stakeholders.

The broader public showed a significant interest throughout the process, mainly due to the greater awareness of the need for measures after the evacuations in 1995. A specially developed planning kit has proven to be a good participatory tool. It has helped to create insight into possible effects because it demonstrates in a simple way how several measures could add up to the overall impact and, as a result, has created public support.

Coastal area: Weak links

In the coastal area, the approach to flood safety has also changed (see Plate 24, centre pages). Flood safety as an objective is no longer to be treated in isolation to other objectives related to a specific area. The national government is now collaborating with other parties for solutions. Through the Weak Links project, these changes were introduced into the coastal policy. The project started in 2003 after research indicated that waves were having more impact upon the coastal flood defence structures (dunes, dikes and waterworks) than considered so far. It turned out that ten sites would need additional strengthening within 20 years in order to maintain the desired safety level. For eight of those locations, called priority weak links, the government decided to link flood safety to spatial quality. For example, parties looked at options of flood safety measures that would also enhance internationally attractive bathing resorts. The coastal provinces were appointed to study the possibilities and to choose the desired solution, in consultation with a broad range of regional participants. The national government, which is responsible for the safety level of coastal flood defences, now only sets conditions for the process and solutions.

There was an innovative shift from the so-called 'hard measures' approach to a 'soft measures' approach (infrastructural to non-infrastructural). This means that preference is given to flexible measures, such as strengthening with extra sand instead of building immovable structures (e.g. dikes). The reason behind this is that flexible measures would increase adaptability in future scenarios. This approach stimulated innovation in coastal flood defence and forced parties to have a more integrated and long-term view on safety and spatial development. Beginning in 2007, the chosen solutions are now being implemented. The safety aspect is financed by the national government and the spatial quality measures by regional and local parties (public and private).

Forward-looking approach

The adaptation measures described in the previous section are essential to maintain standard safety levels. In addition to these so-called pragmatic approaches, The Netherlands is taking up a more forward-looking approach to water management in order to deal with climate change. Changes to be dealt with are not only the shifting dynamics of the natural system, but the addition of increased intensity of land use and the economic value of the delta area – all of this make flood-risk planning and management particularly complex. The type of land use has direct bearing on hydrology. In paved areas, for example, rainwater runs off more rapidly. Land use also influences other hydrological components, such as water loss through evapotranspiration and replenishing of groundwater through drainage. The type of land use and its economic value also determine the potential risks that water poses. The possibility of flooding a town simply leads to a higher risk than the flooding of pasture.

This forward-looking approach consists of a combination of research and policy. Some research projects, such as the earlier mentioned KNMI climate scenario study, consider the consequences of hydrology on living, working and recreation in The Netherlands. All of the new insights stemming from research or real-life experience continuously feed into the policy processes. Keeping a broader scope – in time, in space and in possible solutions – is becoming a standard policy. For example, the new vision on development of the coastal area looks forward to 2050, keeping in mind climate change and spatial developments. The new safety policy will not only focus on keeping the water out, as this is untenable in future situations, but also on minimizing the consequences of flooding. Three typical examples of large research initiatives are described in the following sections.

Research: Working together on knowledge of water and space

The fact that a substantial part of national research grants is spent on water-related issues indicates the government's degree of concern for adequate water management. One of the large national research funds, the so-called Besluit Subsidies Investeringen Kennisinfrastructuur (BSIK) fund, focuses on creating more knowledge and research capacity for five themes that are important for Dutch society. The programmes funded under the BSIK began in 2004 and are all carried out in networks of various parties (private parties, knowledge institutes, universities, governmental bodies, NGOs, etc.). The programmes provide opportunities for fundamental and applied scientific research, as well as for innovative implementation of research results. Knowledge-sharing, international cooperation and the execution of pilot projects are common aspects of the programmes.

The two programmes within BSIK that most prominently focus on water and climate are Climate changes Spatial Planning (€40 million for five years) and Living with Water (€20 million). The goal of the first programme is to provide the government, the private sector and the scientific world with a shared knowledge infrastructure on the relation between climate change and spatial use (www.klimaatvoorruimte.nl). It includes projects on climate scenarios, mitigation, adaptation, climate dialogue and integration (to generate consistency and to combine results).

The mission of Living with Water is to achieve changes in water management that improve on traditional methods which have reached their limits. Technical measures alone to improve on water management are also insufficient (www.levenmetwater.nl). The innovative projects in the programme look at hydrology, management, communication and/or appreciation of water in specific spatial areas (e.g. urban areas and low-lying areas). Another part of the programme is to create a network where water managers and knowledge creators meet each other, develop knowledge and coordinate their activities.

Knowledge and experience from both programmes will feed into the new Knowledge for Climate research programme, for which €50 million will be available from 2008 to 2013.

The results of all of these research projects are continuously feeding into current water policies. They help to adjust spatial planning in The Netherlands in order to make the effects of climate change 'acceptable'.

Adaptation strategy on space and climate

In 2007, a new research initiative was launched: the Adaptation Programme for Spatial Planning and Climate (ARK). ARK is a cooperative programme between various ministries, provinces, municipalities, water boards, research communities and the private sector. The aim of the ARK programme is to formulate the means of making The Netherlands 'climate-proof'. It also defines different responsibilities in this respect. The point of departure of the ARK programme is the notion that climate change is inevitable. However, it is not exactly clear to what extent temperatures will rise, rainfall will increase or sea level will rise. Therefore, ARK is looking for a flexible and robust spatial plan that makes it easy to react to both foreseen and unforeseen changes. This also requires a new governmental approach. Guiding principles in this approach are as follows: adaptation to climate change is paramount in spatial development; natural processes such as sand dynamics in coastal areas should be utilized; and risk prevention includes the minimization of possible impacts.

To reach its desired flexibility, ARK has a four-pronged approach. The first is to increase awareness and willingness to take action with all kinds of parties, including the public. The second is to adjust instruments of implementation (rules and laws, plans, financial instruments) in order to embed changes in behaviour. This requires adaptation of existing plans and development of new instruments. The third is to stimulate innovation and development of knowledge. Innovative concepts for design and for organizational structures, such as public–private alliances, are necessary for the new approach. The fourth and last approach is to encourage the government to be more future oriented. All initiatives must work together with a long-term view; this calls for decision-makers with a clear vision.

To achieve a 'climate-proof' Netherlands, the complete package must have a documented vision of operations between the parties involved. In tandem with current big projects, such as Room for the River and Weak Links, action is also undertaken on a smaller scale. In current planning processes, parties already keep in mind the notions of flexibility. The package has attempted to make use of current projects, such as the renewal of a sewerage system in an urban area in such a way as to create more space for water. Parties directly make use of the outcomes of the research projects executed in the knowledge facility of the ARK programme.

Two interesting components of ARK are the so-called 'experimental gardens' and 'hot spots'. A typical example of such a hot spot is the Zuidplaspolder in the coastal province of Zuid-Holland (6m below sea level). The area has been designated for urbanization (including greenhouse farming). Parts of the area are sensitive to land subsidence. A river that connects to the large rivers and to the sea borders the polder. The latter, combined with potential impacts of climate change, poses a challenge to the future planning of land use. The province has therefore taken the initiative to constitute a consortium of governmental bodies, stakeholders and knowledge institutes that will study the long-term effects of climate change on the planned development of the area. This must lead to integrating solutions within current plans in order to make them more 'climate-proof'. Possible solutions are innovative buildings or smart measures to reduce the effects of disasters.

Conclusions

The various projects as described above all have a common denominator which is 'the Dutch way of coping with climate change'. Of course, this approach is adapted to the specific Dutch physical and organizational context, and simply replicating it is not always possible. However, some features might be worth highlighting.

The notion that water needs space makes it necessary to have an integrated approach to manage surface water. Even in countries where land is less intensively used than in The Netherlands, such an approach could open the way to more flexible and robust water management. It makes it possible to look for solutions that benefit not only the water sector, but also other sectors such natural ecosystems, tourism and housing. Integrated goal-setting can create synergy and lead to cost savings.

An integrated approach asks for collective actions to water problems by all of the different parties. Working together may cause slower decision-making in the beginning, but it will lead to more accepted and, thus, durable solutions. Experiences have shown that participation of regional parties and communication with interested parties is an important factor for a successful collective approach. It has shown that spreading the same message on water by all public actors contributes to public awareness.

Dutch politicians and the public show a relatively high awareness of water-related issues. Adaptation to changing circumstances is part of the system (e.g. through the five-yearly tests of dikes that are enforced by law). The Dutch approach is characterized by a combination of a relatively long-term view and taking 'no-regret' measures at the same time. This action-oriented attitude is demonstrated by the fact that each project, even if it is long term, has to have a component to be initiated immediately. The potential benefits are generally recognized: speeding up processes, raising awareness, managing the initial problems and, last but not least, stimulating innovation. Another way to encourage innovative improvements on existing approaches and measures is to combine research and policy-making or implementation. Researchers and practitioners give each other new insights.

Notwithstanding that innovative and big solutions are attractive, they must not eliminate the need to look into smaller solutions as well. For example, in The Netherlands, it turned out that most low-risk water problems foreseen until 2015 can be solved with small-scale solutions. In many cases, these types of solutions are easier to implement. Above all, they give the opportunity of delivering tailor-made solutions, a prerequisite for flexibility and robustness in many situations.

The Netherlands is one of world's safest delta areas. This is a result of constantly looking forward and designing measures bearing in mind sea-level rise. Climate change as a cause of changing conditions has been given prominence over the last decade. Politicians and citizens are more aware of the need to look for different coping mechanisms. This leads to a willingness to have a long-term view, to give more space to water, to choose more flexible solutions and to consider potential impacts rather than only considering the chance of events occurring. This is a remarkable transition in a country that has fought the dangers of water for ages. However, this transition cannot be completed in the short term. It is a gradual and iterative process in which all parties must reset their visions on dealing with problems and solutions. Political choices have to be made on acceptable uncertainty and on balancing different values. All of this is to reach the common goal: to have and to keep a durable water system against acceptable social costs.

References

Immerzeel, W. W. and P. Droogers (2008) *Climate Change and Local Precipitation Surplus, FutureWater Report 73*, The Netherlands (in Dutch)

Van den Hurk, B., A. Klein Tank, G. Lenderink, A. van Ulden, G. J. van Oldenborgh, C. Katsman, H. van den Brink, F. Keller, J. Bessembinder, G. Burgers, G. Komen, W. Hazeleger and S. Drijfhout (2006) *KNMI Climate Change Scenarios 2006 for the Netherlands*, KNMI Scientific Report WR 2006-01, KNMI, De Bilt, The Netherlands

11

Climate Change and Alluvial Aquifers in Arid Regions: Examples from Yemen

Jac A. M. van der Gun

Introduction

It is well known that on a global scale, at least 100 times more water is stored in aquifers than in rivers and lakes combined (Shiklomanov and Rodda, 2003). This makes groundwater systems comparatively resilient to short-term, seasonal and even longer-term shortage of rainfall. Climate change is, nevertheless, likely to impact upon the quantity and quality of groundwater resources, and alluvial aquifers in arid regions rank as the most vulnerable to climate change. Groundwater is often, especially in arid regions, the most reliable source of water – if not the only one – for domestic water supply and irrigation of crops. This case study in Yemen explores to what extent groundwater in alluvial aquifers in arid regions may be affected by climate change during the 21st century.

Yemen, with its capital Sana'a (see Figure 11.1), is currently home to about 23 million people. The long-term mean rate of renewal of surface water and groundwater combined is less than 200m^3 per capita per year. The country thus has a severe water shortage. Only about 3 per cent of the total area of about 528,000km^2 is arable land – making Yemen a large importer of food. Although agriculture contributes a modest share of the gross domestic product (GDP), it provides employment and family income to more than half of the country's labour force and consumes 95 per cent of the available water resources of the country. A significant part of the country's agriculture is dependent upon alluvial aquifers, which underlines the need to slow groundwater depletion down in these aquifers and to implement adequate adaptation measures.

Alluvial aquifers and their importance in arid regions

Arid regions have low annual rainfall characterized by irregular and infrequent events. As a consequence, perennial streams are rare. Occasional surface water runoff tends to be discharged by natural drainage channels called *wadis*. Wadis have intermittent or seasonal flow and have no significant base flow during the long dry periods.[1] Wadi

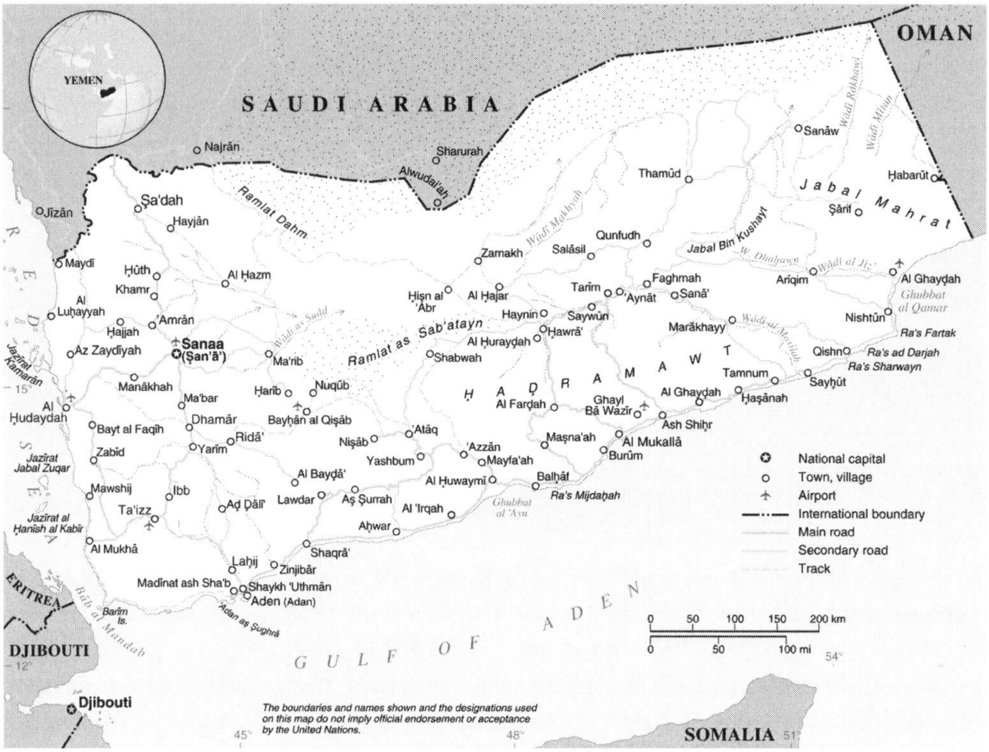

Figure 11.1 *Map of Yemen*

Source: based on a United Nations map

channels are usually accompanied by shallow, narrow and elongated alluvial aquifers ('strip aquifers'). They consist of sediments carried and deposited by the wadis. The groundwater that they contain is recharged mainly by wadi flows. Wadi channels and alluvial aquifers tend to be so inseparably linked that the word 'wadi' is often used to refer to both the wadi channel and the related alluvial aquifer.

Alluvial aquifers are particularly important components of the hydrological system in arid zones: they are more actively recharged than any other aquifer. They contain and carry water permanently. Being strongly linked to wadis, alluvial aquifers in arid regions are often called wadi aquifers.

Historically, wadis are preferential zones for human settlement and for economic and cultural development in arid regions, very much like perennial rivers in humid climates. Many wadis are well known for their archaeological or historical interest. Examples in Yemen are Wadi Hadramawt, with its historic towns Sayun, Tarim and Shibam; Wadi Adhana, where the famous Marib dam was built more than 2500 years ago; and Wadi Zabid, for which an old document – dating from 1704 – describes the principles and rules for the allocation of spates (DHV, 1988). Wadi Dayqah in Oman, Wadi Natrun in Egypt and Wadi Dahr in Yemen are renowned for their natural beauty. Wadis of historical interest in Jordan are Wadi Rum, with its graffiti in the rocks dating back to various eras, and Wadi Musa, where the famous rock temples of Petra are located. Common denominators of wadis are relatively dense population and economic – in particular, agricultural – activity, facilitated by the presence of water. With water as the uniting factor, a wadi often provides marked identity and social cohesion among the population which inhabits its catchment area.

Types of wadi aquifers and examples from Yemen

Descriptions

Wadi aquifers are Quaternary sedimentary aquifers with strong links to wadis: wadi flows transport and deposit the aquifer sediments and are the origin of at least a significant part of the aquifer groundwater recharge. The characteristics of wadi aquifers are well documented. The most important reports, also relevant to this review, are listed in the references. Emphasis was placed on those that illustrate concepts and methods with practical examples (Falkenmark and Chapman, 1989; MWR, 1991; TNO, 1992; van der Gun and Ahmed, 1995; Euroconsult et al, 1996; Khater and Al-Weshah, 2002; Vasak, 2002).

The two main types of wadi aquifers are *alluvial strip aquifers* and *wadi plain aquifers*. Both can be subdivided further into several subtypes, depending upon their setting. Figure 11.2 shows five elementary subtypes. This subdivision is not exhaustive: more subtypes may be defined by combining and modifying the ones shown. A brief description of the wadi aquifer types illustrated in Figure 11.2 is given below.

Alluvial strip aquifers (wadi aquifer type I)

The typical alluvial strip aquifer can be found anywhere in mountainous areas; it is composed of recent unconsolidated sediments that form a wadi bed and adjoining alluvial terraces. Both the wadi bed and the alluvial terraces are relatively narrow and shallow, and their total thickness usually varies between a few metres to a few tens of metres. The typical elongated shape of the aquifer explains the term 'strip aquifer'. The surface area of strip aquifers is only a small fraction of the total area of the wadi catchment; but runoff from the entire catchment area flows towards them, which produces an apparent 'multiplier effect'. For instance, an average runoff depth of 20mm over the entire catchment area may result into more than 200mm of recharge depth averaged over the alluvial strip aquifer's area. This explains why the seasonal variation of groundwater levels in alluvial strip aquifers is often in the order of metres, in spite of high effective porosity of the aquifers and low rainfall rates. In alluvial strip aquifers, the groundwater recharge derived from runoff (indirect recharge) is normally much more important than the recharge produced by local infiltration of rainfall on its surface area (direct runoff). Depending upon climate and upon aquifer dimensions, the mean residence time of groundwater in alluvial strip aquifers tends to be in the range of a few years to a few tens of years.

The *alluvial strip aquifer of type Ia* (embedded in impervious rocks; see Figure 11.2) is a very common wadi aquifer type. After recharge caused by rain, the limited

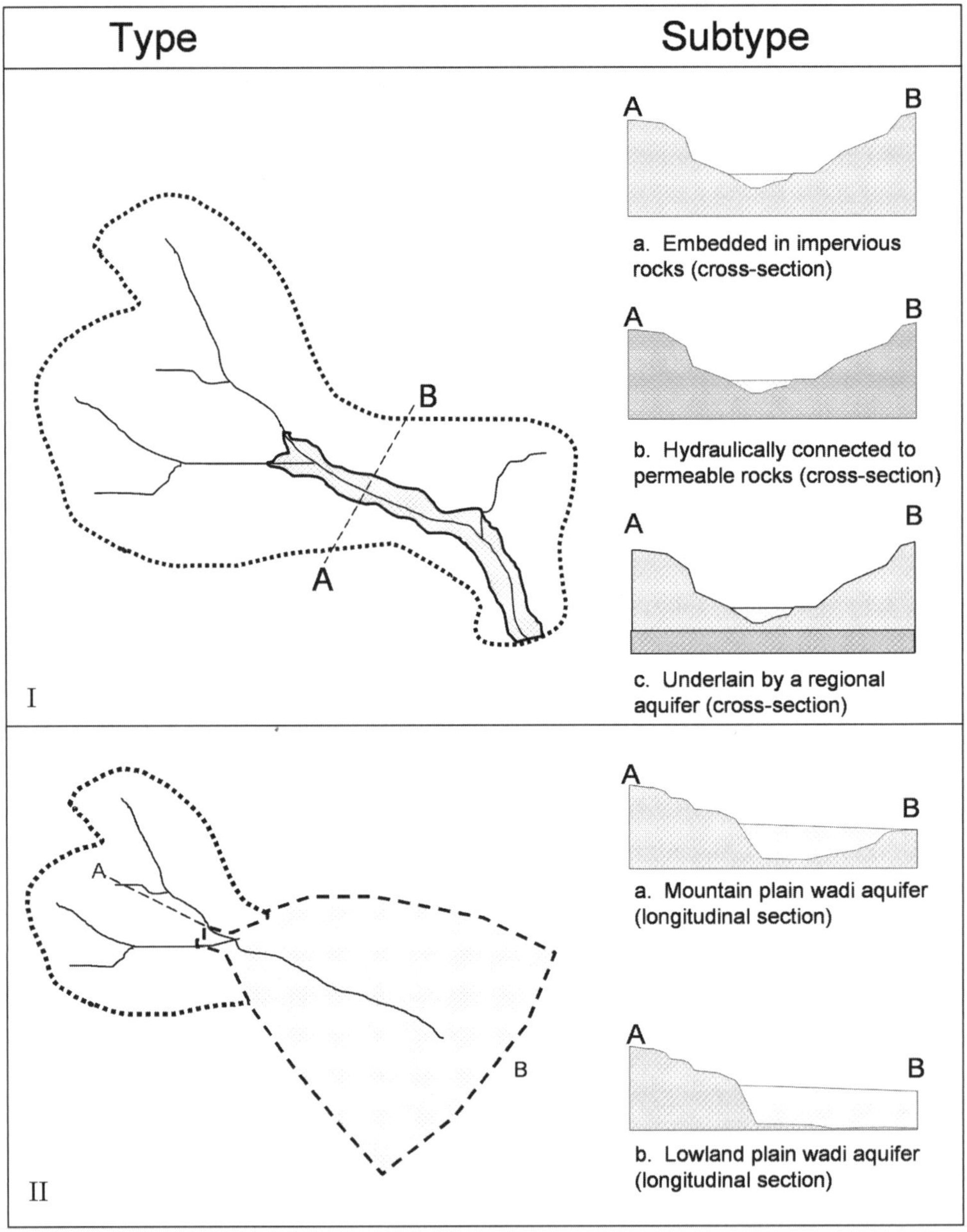

Figure 11.2 *Different types of alluvial aquifers in arid regions ('wadi aquifers')*

Source: Jac A. M. van der Gun

volume of water stored in the alluvial material is quickly lost to downstream flows, to abstractions and sometimes to evaporation. There is little buffer over periods longer than one year. Consequently, the aquifers are sensitive to variations in annual rainfall and may become nearly depleted during dry years.

The *alluvial strip aquifer of type Ib* (alluvial fill hydraulically connected to a regional aquifer; see Figure 11.2) is another common wadi aquifer type. Wadis cut into water-bearing carbonate rocks, belong to this type (e.g. Wadi Dayqah in Oman; MWR, 1991; TNO, 1992). After recharge during rainfall events, the alluvial aquifer volume of limited amount may continue to be replenished through the connected regional aquifer for a long period after the rain event. In such cases, the groundwater levels remain more stable during the year than in the case of aquifer type Ia. In fact, the alluvial strip aquifer and the regional aquifer may be considered to form a single complex groundwater system, with the alluvial aquifer component forming a preferential zone for recharge and flow.

The better the water-bearing properties of the connected regional aquifer, the more stable the wadi flow regime will be, up to a point where the typical ephemeral or seasonal regime is lost. It should be noted that it is also possible for the alluvial aquifer to lose water to the regional aquifer. In this case, the strip aquifer loses water even more rapidly than aquifer type Ia under comparable climatic and geological conditions.

Less common is the *aquifer type Ic* representing an alluvial strip aquifer embedded in poorly permeable rocks, but sufficiently close to an underlying regional aquifer to allow an artificial hydraulic connection by means of wells penetrating the separating lithological unit. Wadi As Sirr alluvial aquifer in Yemen is an example of this aquifer type (SAWAS, 1996). Under virgin conditions, this aquifer type is similar to type Ia; but after wells have produced significant hydraulic connection with the underlying regional aquifer, its behaviour may become more like that of aquifer type Ib.

Wadi plain aquifers (wadi aquifer type II)

Where topographic or tectonic conditions force the wadi gradients to be suddenly reduced, alluvial plains tend to develop by deposition of the sediments carried by the wadis during floods. Steep alluvial fans of coarse and permeable material are formed in the piedmont zone near the foothills, while finer sediments are laid down in nearly horizontal layers further downstream. The wadi plain aquifers expand laterally in directions perpendicular to the main direction of the wadi and therefore tend to have much greater surface areas than alluvial strip aquifers. Due to their tectonic setting, most of these wadi plain aquifers are also much thicker than alluvial strip aquifers: a thickness of several hundreds of metres is not uncommon. For an equal size of catchments recharging them, wadi plain aquifers have much larger storage capacity than strip aquifers. As a consequence, the average residence time of water in them is much longer, often in the order of hundreds or thousands of years.

Wadi plain aquifers belong to the category of extensive or regional aquifers; but they can be considered as 'wadi aquifers' as long as the wadi is a major source of their recharge. Although wadi flows are still an important source of groundwater recharge (indirect recharge), the 'multiplier effect', as described above, for the alluvial strip aquifers it is much less pronounced. Direct recharge may be important as well, provided that there is enough rainfall on the plain.

Two subtypes are distinguished: the *mountain plain wadi aquifer* (IIa) and the *lowland plain wadi aquifer* (IIb). The mountain plain wadi aquifer is limited downstream by impermeable rock and usually fills a tectonic 'graben', drained by a wadi that has cut a narrow outlet through the rocks. The Amran Plain aquifer in Yemen (van der Gun and Ahmed, 1995) and the Valle Alto in Bolivia (SERGEOMIN and TNO, 1998) are good examples. The lowland plain wadi aquifer, however, slopes towards the sea and gradually transforms into a coastal plain where natural groundwater discharge is either through evaporation in *sebkhas* or through groundwater outflow into the sea. Examples are the Tihama Plain in Yemen (van der Gun and Ahmed, 1995) and the Batinah Plain in Oman (MWR, 1991).

Illustrative examples of alluvial aquifers in Yemen

Yemeni alluvial aquifers selected as illustrative examples for analysis of climate change impacts are listed in Table 11.1; their locations are shown in Figure 11.3. A brief description of these aquifers will follow. Special emphasis will be given to the particular sensitivity of each aquifer and the most suitable strategy to overcome the potential negative aspects of climate change.

Table 11.1 *Selected examples of different types of wadi aquifers in Yemen*

Type and subtype	Aquifer name	Remarks
1 Alluvial strip aquifers		
a Embedded in impervious rocks	Wadi Fallah alluvial aquifer	Near Sa'dah Plain
b Hydraulically connected to other permeable rock units	Wadi Hadramawt alluvial aquifer	Shibam-Sayun-Tarim zone
c Underlain by a regional aquifer	Wadi As Sirr alluvial aquifer	North-east of Sana'a Plain
2 Wadi plain aquifers		
a Mountain plain aquifer	Sana'a Plain alluvial aquifer	In the central highlands of Yemen
b Lowland plain aquifer	Tihama Plain aquifer	Bordering the Red Sea

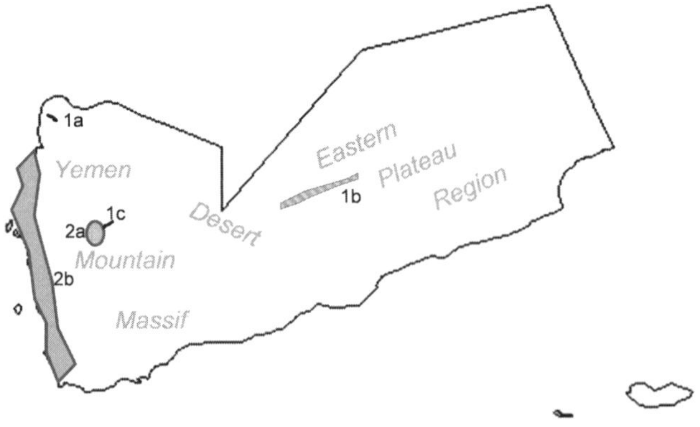

Figure 11.3 *Location of the aquifers mentioned in Table 11.1*

Source: Jac A. M. van der Gun

Wadi Fallah alluvial aquifer

Wadi Fallah is located in the north-western part of Yemen, 20km to 30km north-west of the town of Sa'dah. It is typical of the water resources conditions of many small alluvial aquifers scattered over the territory of Yemen and other Middle Eastern countries. The wadi is steep: it runs from an elevation of more than 2500m to about 1900m – the local surface level of the Sa'dah Plain – over a horizontal distance of only 15km to 20km. The wadi and its tributaries are cut into uplifted Precambrian bedrock composed of impermeable gneisses and granites. Below approximately 2100m above mean sea level, the wadi beds are filled with alluvial deposits forming a shallow aquifer that becomes gradually thicker (10m to 20m) and wider (20m to 150m) in the downstream direction.

Runoff in the Wadi Fallah is in the form of small flash floods. An estimated average annual rainfall of 150mm on its approximately 100km^2 catchment area causes only a few runoff events annually, all of them of short duration. The runoff that infiltrates on its way down to the Sa'dah Plain is the main source of recharge to the alluvial wadi aquifer.

This aquifer, although small, is the only permanent source of water in its immediate vicinity and is tapped by several tens of wells that provide the inhabitants of the wadi zone with drinking water and water for small-scale irrigation. Because the stored volume of groundwater (between 1 million and 2 million cubic metres) is not more than three times the mean annual groundwater recharge, the aquifer is sensitive to temporal variations of recharge and this sensitivity is aggravated by intensive pumping. As a result, groundwater levels decline significantly throughout the dry season to the extent that wells have reduced yields or even become completely dry at the end of the dry season, particularly during years of below-average rainfall.

Prior to its exploitation, this aquifer may have overflowed and sustained permanent base flow in the wadi. The presence of pumped wells – with an estimated aggregated abstraction rate of 0.5 million cubic metres per year – has reduced groundwater storage and, consequently, weakened the aquifer's resilience to climatic variations.

Wadi Hadramawt alluvial aquifer

Wadi Hadramawt is located in a west-south-west–east-north-east running canyon incised into sequences of Cretaceous Mukalla sandstones, covered by marls, and topped by Tertiary carbonate rocks (see Figure 11.4). The nearly vertical sides of the canyon rise to 300m, on average, above the top of the Quaternary deposits, which form the wadi bed inside the canyon. These Quaternary deposits constitute an alluvial aquifer approximately 90km long, 20km (west) to 1.5km (east) wide, and more than 100m thick. Not only does this aquifer support significant groundwater flow and storage, it is also very favourably located with respect to sources of recharge. Although rainfall is low (less than 100mm per annum), the runoff volumes are relatively high because the catchment area yielding water to the canyon is large. Various substantial tributaries rapidly bring flood water from around 22,500km^2 of the bare limestone-covered Jawl plateaus to the Wadi Hadramawt canyon, where it mostly infiltrates into the Quaternary aquifer.

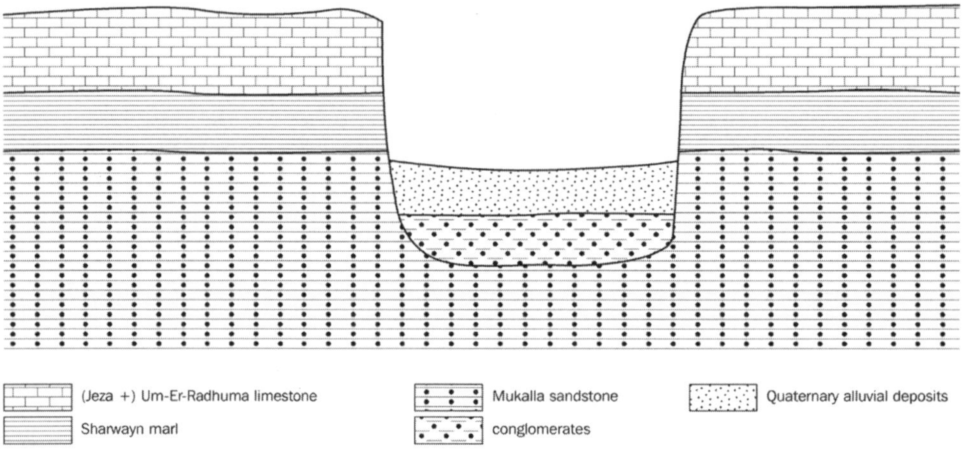

Figure 11.4 *Schematic geological cross-section across Wadi Hadramawt*

Source: Van der Gun and Ahmed, 1995

Groundwater levels in the Quaternary deposits are relatively shallow, generally 20m to 30m below surface. But according to McDonald (1988), they declined some 20m to 25m on average during the period of 1952 to 1984. The presumed originally very shallow water tables are consistent with field observations made by Van der Meulen and Von Wissmann (1932), who observed *wadi* base flow, small pools of stagnant water and white salt crusts on the soil surface, especially in the eastern part of Wadi Hadramawt.

Figure 11.4 shows that the wadi locally has removed the limestone rocks of the Um-Er Radhuma formation and the Sharwayn marls completely, as well as part of the underlying Mukalla sandstones. Relatively fresh groundwater of the Mukalla sandstones flows into the alluvial aquifer as lateral flows at the borders of the canyon. Water also moves upward through the conglomerates that form an aquitard between sandstones and alluvium inside the canyon. Groundwater in the alluvial aquifer has a high mineral content, with salt concentration rising to more than 10,000mg/litre along the longitudinal axis of the wadi. Relatively fresh groundwater (salt content less than 2000mg/litre) is only encountered immediately along the flanks of the steep-sided canyon and in the mouths of the wadi tributaries (Van der Gun and Ahmed, 1995).

During 2001, there were more than 3000 wells in Wadi Hadramawt, together abstracting about 248 million cubic metres of groundwater (Vasak, 2002). The greater part of the total abstracted volume is from the sandstone aquifer, while about 10 per cent is drawn from the alluvial aquifer. Approximately 95 per cent of all groundwater pumped is used for irrigation of date palms, citrus, sorghum, wheat, alfalfa and other crops, occupying about 22,500ha of land.

The volume of groundwater stored in the alluvial aquifer is estimated to be between 5 billion and 10 billion cubic metres, which exceeds the mean annual rate of renewal by more than two orders of magnitude. In addition, the alluvial aquifer and the sandstone aquifer are hydraulically united into one large regional system – the Extended Mukalla Complex – with a huge combined storage equivalent to more than 10,000 times the present-day mean annual recharge.

Wadi As Sirr alluvial aquifer

At first glance, conditions in Wadi As Sirr – a wadi debouching into the Sana'a Plain – look rather similar to those of Wadi Fallah. However, it supports a much more intensive groundwater use as a result of its different hydro-geological setting. The catchment area of Wadi As Sirr, located between 2600m and 2200m + mean sea level is approximately 200km^2, while the mean annual rainfall is between 100mm and 150mm. A detailed well inventory carried out in 1993 (SAWAS, 1996) revealed the existence of 1257 wells in Wadi As Sirr; about half of them are dug wells. A more recent inventory (WEC, 2004) mentions 1359 wells, with a total annual abstraction of 41 million cubic metres of groundwater. This abstracted groundwater is mainly used for the irrigation of about 3600ha of agricultural lands, with *qat* and grapes as predominant crops.

In spite of receiving similar rainfall and having a catchment size only twice as large, Wadi As Sirr apparently sustains an aggregated groundwater abstraction that is approximately two orders of magnitude higher than the abstraction from Wadi Fallah's aquifer. This is partly because all drilled wells tap the regional Cretaceous Tawilah sandstone aquifer lying underneath the alluvial aquifer and extending westward over most of the Sana'a Plain and neighbouring zones. Curiously, and unexpectedly, groundwater levels in the alluvial aquifer of Wadi As Sirr remain relatively stable, although the aquifer is being pumped by dug wells at a rate of approximately 6 million cubic metres per year, which is far beyond the estimated mean natural recharge. Induced recharge by excess irrigation water pumped from the Tawilah aquifer explains this behaviour.

Groundwater levels in the alluvial aquifer are higher than those in the underlying sandstone, suggesting that there is a poorly permeable formation between the two aquifers and that their local hydraulic interconnection is mainly through wells. Induced recharge has greatly enhanced the exploitation potential of the alluvial aquifer, while at the same time reducing the average residence time to only a few years. The short residence time is comparable to that in Wadi Fallah, but Wadi As Sirr's alluvial groundwater system's resilience to intensive exploitation and climate change is much higher. This is because a second groundwater resource is present in the form of the Tawilah sandstone aquifer.

Sana'a Plain alluvial aquifer

The Sana'a Plain, an intermountain plain located in the Yemen Highlands, is underlain by a sequence of sedimentary rocks (sandstones and limestone), dipping southward under Tertiary volcanic rocks and covered, unconformably, by a blanket of unconsolidated Quaternary deposits. These Quaternary deposits form an alluvial aquifer that varies in thickness from less than 25m near the borders of the plain to almost 200m under the international airport. The plain has an average elevation of around 2200m above mean sea level and is approximately 300km^2 in extent. *Wadis* surrounding the plain in a radial pattern bring their occasional flood waters into the plain to recharge the alluvial aquifer and underlying aquifers. Their catchment areas combined with the Sana'a Plain constitute the Sana'a Basin of approximately 3200km^2. Mean annual rainfall is about 230mm. Other contributions to groundwater recharge in the plain are direct recharge resulting from heavy rainfall, recharge by percolation of excess irrigation water, and recharge by infiltrated wastewater. Natural outflow of water from the Sana'a Plain is minimal. Wadi Kharid is the only outlet of surface water and releases water only during rare storm events. Natural outflow of groundwater today is exclusively through subsurface flows, while previously also through springs.

Groundwater in the Sana'a Basin was traditionally abstracted from the alluvial aquifer and from springs emerging from the volcanoes in the southern part of the basin. Numerous dug wells have been used for centuries to draw water from the alluvial aquifer to supply the Sana'a city and the surrounding rural area. Since the early 1970s, however, after it was discovered that a regional sandstone aquifer (Tawilah sandstone) underlies the alluvial aquifer, deep wells pumping from the sandstone rapidly increased groundwater abstraction in the plain. As a result, part of the alluvial aquifer has been depleted, while in many other areas groundwater levels are continuously declining (as in the sandstones). Only small parts of the alluvial aquifer have stable groundwater storage volumes due to return flows. The total annual groundwater abstraction in the Sana'a Basin increased from 16 million cubic metres in 1972 to about 182 million cubic metres in 1993. It will have increased further since then. Approximately 10 per cent of the 1993 abstraction was drawn from the alluvial aquifer (SAWAS, 1996). Annual recharge of about 40 million cubic metres around 1972 may have doubled by now due to return flows.

The volume of groundwater under the Sana'a Plain is more than 1 billion cubic metres, of which today perhaps 50 per cent is still stored in the alluvial aquifer. Although this volume is substantially larger than the difference between annual groundwater discharge (abstraction plus natural outflow) and recharge, the imbalance between discharge and recharge already presents significant problems. In a steadily increasing number of zones, the alluvial aquifer is being depleted, while groundwater levels in the underlying sandstones are falling by a rate of several metres a year.

Tihama Plain alluvial aquifer

The Tihama is an almost 500km long and 25km to 60km wide lowland plain, bordered in the west by the Red Sea and in the east by the western foothills of the Yemen Mountain massif. Its origin is related to the Red Sea graben system, which started to develop at the end of the Cretaceous and has trapped thick sequences of Tertiary and Quaternary sediments. The Quaternary sediments are 200m to 300m thick and form an important regional alluvial aquifer, with highly permeable zones where the larger wadis enter the plain, and somewhat less permeable zones in the 'inter-*wadi* zones'. The Tihama enjoys a hot and dry climate, with mean annual rainfall declining east–west from about 300mm to about 50mm. The alluvial aquifer is recharged by ephemeral wadi floods, by direct infiltration of surplus rainfall at the eastern edge, and by percolating excess irrigation water. The mean annual groundwater recharge is about 550 million cubic metres (van der Gun and Ahmed, 1995). Groundwater discharge is through groundwater outflow into the Red Sea, through evaporation on the near-shore sebkahs and through abstraction from numerous wells.

The Tihama Plain is an important agricultural area, irrigated by wadi spates and by groundwater pumped from many thousands of wells. Total annual groundwater abstraction by 1994 was estimated to be 810 million cubic metres, which exceeds the mean annual recharge. Part of the alluvial aquifer's groundwater is saline or brackish, owing mainly to seawater intrusion and connate saline groundwater moving upward from the Tertiary deposits. The quantity of fresh groundwater in the Tihama alluvial aquifer is at least 250,000 million cubic metres, which exceeds mean recharge by two to three orders of magnitude. Although this huge stored volume contributes to the aquifer's resilience to stress, it does not prevent declining groundwater levels – especially near the foothills – and steadily increasing groundwater salinity problems in the coastal area.

Climate change impacts

Observed and predicted climate change

In February 2007, the Intergovernmental Panel on Climate Change (IPCC) released summary information from its *Fourth Assessment Report* on climate change (IPCC, 2007). This information confirms earlier hypotheses on global warming and supports with very high confidence the opinion that human activities play a significant role in global warming, particularly through the emission of greenhouse gases. Observed changes include an increase of global average surface temperature, a rise of the global average sea level, and reductions in snow cover, mountain glaciers, polar ice sheets and permafrost areas. Changes in atmospheric precipitation are observed as well, varying from decreasing amounts in some regions to increasing amounts in others. In general, heavy precipitation events have become more frequent over most land areas.

The IPCC's predictions on climate change during the 21st century follow a number of greenhouse gas emission scenarios that are based on different assumptions about population, economic development and environmental protection. Predictions vary, depending upon the varying scenarios; but generally a mean global warming of 0.6°C to 4.0°C is predicted for the 21st century. Sea level is predicted to rise between 0.18m and 0.59m. Despite relatively large spatial differences, both temperature and sea levels are expected to rise globally.

Spatial variation in precipitation is more complex: projected patterns of precipitation changes over the 21st century show likely decreases in annual totals in some regions (in particular, most subtropical land regions) and very likely increases in precipitation in other regions (especially in high-latitude regions). It is considered very probable that hot extremes, heat waves and heavy precipitation events will continue to become more frequent across the Earth.

This chapter focuses on Yemen and the only meaningful global change predictions (according to the report) relevant to Yemen are those on surface temperature. Projections on local surface temperature change in Yemen during the 21st century show a rise of about 2.5°C (B1 scenario) to 4.0°C (A2 scenario), which is 25 to 40 per cent higher than the corresponding predicted global average changes. World maps on projected precipitation change differ, depending upon the models used to construct them. There is thus no conclusive evidence of either an increase or decrease of precipitation in Yemen. Of the 21 general circulation models (GCMs) included in the latest IPCC *Fourth Assessment Report* (IPCC, 2007), about half projected an increase in rainfall.

Impacts upon alluvial aquifers

Different mechanisms and impacts

Climate change may have direct or indirect impacts upon alluvial aquifers. Under the direct impacts category, we may group those changes in the groundwater associated with natural processes and without dominating human interference. The most important corresponding mechanisms are groundwater recharge, groundwater discharge, and interactions between sea water and groundwater. The indirect impacts are caused by humans responding to climate change in the form of changes in groundwater abstraction patterns and land use. The different mechanisms and impacts are described and assessed in a preliminary way below.

Changes in 'natural' groundwater recharge

'Natural' groundwater recharge in arid regions is not only generally low, but also tends to be very sensitive to changes in rainfall and temperature regimes. A minor change in rainfall in such regions may cause a disproportionally large change in groundwater recharge due to strong non-linearity of the rainfall–recharge relationship (threshold effect). Similarly, an expected increase in temperature will lead to higher evaporation/evapotranspiration rates, which may result in lower groundwater recharge rates. On the other hand, a steadily higher frequency of heavy precipitation events – which is generally expected – will favour groundwater recharge.

Since reliable predictions on the most significant variable (i.e rainfall) are missing for the region concerned, it is not yet possible to make a reliable estimate of changes in 'natural' groundwater recharge during the 21st century. It is not even clear whether such changes would be positive or negative. Among the five selected Yemeni aquifers, Wadi Hadramawt alluvial aquifer may most likely experience more prominent changes in 'natural' groundwater recharge than any of the other four aquifers due to the extremely low rainfall in the zone.

Changes in 'natural' groundwater discharge

Natural groundwater discharge may occur through different routes: through springs, through seepage to streams (base flow), through evaporation and/or evapotranspiration in shallow water table zones, and through subsurface outflow.

The latter two mechanisms are active in the Tihama alluvial aquifer. There is a significant fresh groundwater outflow into the sea: due to the groundwater system's inertia, it is still almost equal to the mean annual natural recharge to the aquifer. A sea-level rise in the order of decimetres – as expected to occur during the 21st century – will reduce the subsurface seaward hydraulic gradients only slightly because the fresh groundwater level decreases a few hundred metres from east to west. Sea-level rise is thus expected to have a minor influence on the submarine fresh groundwater outflow (around 0.1 per cent reduction).

However, the near-shore zone of the alluvial aquifer will be significantly modified. Zones of submarine fresh groundwater outflow will gradually shift eastward and the same will happen to the sebkhas. The latter are characteristic near-shore windows of diffuse shallow groundwater discharge through evaporation. Similarly, the near-shore saline and brackish groundwater domains will start expanding further east. All of these shifts will be over a horizontal distance in the order of 100m only.

Sea-level rise does not influence groundwater discharge from the four other selected Yemeni aquifers. The natural discharge from these aquifers is predominantly through subsurface outflow, which is expected to adjust to changing groundwater storage and hydraulic gradients over the long term.

Changes in the interaction between sea water and groundwater

As mentioned before, only the Tihama aquifer is subject to interaction between sea water and groundwater. An expected sea-level rise will lead to further saline water encroachment, which will be accompanied by a slight reduction of fresh groundwater storage (around 0.1 per cent reduction). Other changes in groundwater regime – as described above and below – may lead to further modification of the interaction between sea water and fresh groundwater.

Changes in groundwater abstraction

The climate in Yemen will become warmer. Mean temperature is likely to increase by 3°C towards the end of the 21st century. Ignoring the effects of possible, but unconfirmed, changes in precipitation, it is assumed that unit water demands are also likely to increase. The expected temperature change will, for example, increase the potential evapotranspiration of crops by about 75mm per year. This means that irrigated lands will, by the end of the century, require about 10 per cent more irrigation water than at present. Eventually, a proportion of the currently rain-fed agricultural lands will have to be converted to irrigation lands as well. A rise in temperature will similarly increase unit water demands in the domestic water use sector. Altogether, one may expect that the impact of climate change alone will lead to a 5 to 10 per cent higher water abstraction (i.e. 200 million to 400 million cubic metres a year), which will mainly be derived from groundwater. This corresponds to an 8 to 15 per cent increase of the current groundwater abstraction rate that is already unsustainable.

Changes in land use

Climate change may lead to adaptive measures related to land use. Farmers may respond to higher potential evapotranspiration rates by intensifying irrigation water applications. But under certain conditions (if water is a really limiting factor), they may decide to reduce their irrigation areas. The latter is likely to occur in aquifer systems such as those of Wadi Fallah. Furthermore, changes in land use – and even changes in natural vegetation – may influence the recharge process in the sense that reduction of evapotranspiration by crops and natural vegetation will enhance groundwater recharge.

Changes in stored volume of fresh groundwater

The impacts described above add up to changes in fresh groundwater storage that are reflected in the groundwater level. For practical purposes, the change in fresh groundwater storage or groundwater level is the most significant single parameter to use as an indicator of the extent to which the local population may be affected.

Based on the properties and setting of the alluvial aquifer described in the section on 'Alluvial aquifers and their importance in arid regions', the following tentative assessment can be made for aquifer storage conditions by the end of the 21st century:

- *Wadi Fallah alluvial aquifer.* Groundwater storage will be very significantly reduced, particularly during the dry season. It is to be expected that this aquifer will diminish significantly in value.
- *Wadi Hadramawt alluvial aquifer.* Groundwater storage in the alluvial aquifer will be reduced significantly in the axial zones of the wadi. Near the canyon walls, however, the depletion process may be slowed down considerably by recharging groundwater flows from the Mukalla sandstone into the alluvial aquifer beds. The groundwater regime in the Mukalla sandstone is expected to react extremely slowly to climate change because it is a very large groundwater system with a huge groundwater volume in storage.
- *Wadi As Sirr alluvial aquifer.* Groundwater storage in this alluvial aquifer may remain relatively stable as long as conditions in the Tawilah sandstone aquifer allow local groundwater-based irrigation to continue at its present intensity. This is because irrigation water return flows constitute the main share of groundwater recharge in the alluvial aquifer. However, it is questionable whether the Tawilah sandstone aquifer will remain in good shape for a long time in the future because it is not only exposed to climate change, but, in particular, to intensive exploitation in the Sana'a Plain and immediate surroundings.
- *Sana'a Plain alluvial aquifer.* Parts of this alluvial aquifer are dry already and other parts will gradually become dry as well, mainly as a consequence of very

intensive groundwater pumping, but accelerated by climate change. By the end of the 21st century, the saturated part of the Sana'a Plain alluvial aquifer will probably have shrunk to only a few small zones. One zone is near the international airport, where the alluvial deposits reach 200m of thickness, and another one encompasses part of the urban zone where used water is recycled to the alluvial aquifer.

- *Tihama Plain alluvial aquifer.* Groundwater storage in the Tihama Plain alluvial aquifer is already being depleted by abstraction. It is most pronounced in the eastern zone of Tihama, near the foothills. The current rate of storage depletion per annum is not more than about 0.1 per cent of total fresh groundwater storage. It is thus not the complete exhaustion of the resource in itself that is of main concern, but rather the problems triggered by declining groundwater levels (e.g. increasing groundwater salinity and higher pumping cost).

Implications for the local population

Seriously threatened livelihoods for those depending upon small isolated alluvial aquifers

Wadi Fallah is a typical example of alluvial aquifers where livelihoods will be seriously threatened by expected climate change. Climate change will lead to higher water demands and most probably to lower recharge rates as well, which will result in higher stress on the groundwater system. Groundwater storage will diminish, thus reducing the aquifer's buffer capacity during dry seasons. This will progressively hamper irrigated agriculture and domestic water supply. Not only are the periods of water stress for agriculture lengthening, but there are also no other water resources at hand to substitute for the dwindling capacity of the alluvial aquifer. People will have to adapt to the gradually deteriorating conditions, probably by reducing their irrigated areas and perhaps even by importing drinking water at high cost from outside the area. This is a gloomy prospect for the entire *wadi* zone. Its carrying capacity for supporting an economically sustainable rural population is eroding, which in the longer term may result in complete abandonment of the area.

Need for relocating groundwater abstractions from alluvial aquifers to other sources of water, followed on in the longer term by reducing abstractions

The alluvial aquifers of Wadi Hadramawt, Wadi As Sirr and the Sana'a Plain are located in zones where other aquifers, more resilient to climate change, are present as well. This means that water users may relocate their abstractions (or part of them) to another aquifer if the alluvial aquifer becomes deficient. In the Sana'a Plain, for instance, many wells in the alluvial aquifer have become dry and have had to be abandoned. Replacing them by wells sunk into the Tawilah sandstone aquifer is possible and effective in most zones, but requires additional investment and often higher running costs as well. Furthermore, the Tawilah sandstone aquifer is being depleted rapidly, resulting in additional costs and an uncertain future. The alluvial aquifer of

Wadi As Sirr, located near the edge of the Sana'a Basin and in a rather similar hydrogeological setting, is still quite productive. In both Sana'a Plain and Wadi As Sirr, the situation will probably have changed radically in 10 or 20 years from now as a result of the rapid depletion of groundwater resources in the Tawilah sandstone. When the axial zone of Wadi Hadramawt's alluvial aquifer has been depleted, alluvial groundwater storage near the canyon wall is likely to change, but only very slowly. This is because of the rather stable state of the huge Mukalla sandstone aquifer under the limestone plateau.

In summary: climate change contributing to depleting groundwater in the alluvial aquifers of Wadi Hadramawt, Wadi As Sirr and the Sana'a Plain will not, in the short to medium term, deprive the local population from all-season access to water. However, except for the marginal zones of Wadi Hadramawt's alluvial aquifer, a substantial part of groundwater abstractions will need to be relocated from the alluvial aquifers to the more sustainable sandstone aquifers. Even after such relocations, the finiteness of groundwater resources in the Tawilah sandstone will oblige abstractions in Wadi As Sirr and the Sana'a Plain to be soon reduced to sustainable rates.

Deteriorating groundwater economy

All impacts described above reduce the benefits that groundwater provides to the inhabitants of the areas concerned. This is due to steadily higher costs of groundwater abstraction (e.g. relocating/deepening wells, more powerful pumps and higher pumping lifts) and smaller available quantities of groundwater. This is true not only for the aquifers of Wadi Fallah, Wadi As Sirr, Sana'a Plain and Wadi Hadramawt, but also for the Quaternary Tihama aquifer. The latter will not be exhausted soon; but steadily falling water levels will have consequences that will make the economic use of groundwater more difficult.

Increasing groundwater salinity in the near-shore zones

Increased salinity is mainly caused by sea-level rise and results in a loss of freshwater resources. This is a relatively minor impact and occurs only in a very narrow belt in the aquifers bordering the coast (e.g. Tihama).

Uncertainties

There are many uncertainties that may affect the future conditions of the aquifers and consequences for their stakeholders. In the first place, the estimates of the IPCC are for future periods that depend upon specific scenarios, and no conclusive predictions were possible for some parameters in some regions (e.g. rainfall in Yemen). Second, the IPCC's (2007) report is not undisputed: several scientists have expressed their criticism of the information and/or the methodologies used. A third important source of uncertainty is the behaviour of people: will they find appropriate ways to adapt to changing conditions or will they just maintain current practices until their local economies collapse?

Conclusions

In spite of all the uncertainties, there is no doubt that alluvial aquifers in Yemen are seriously at risk. The intensive interaction of people with these aquifers – especially by ever-increasing water abstractions and by polluting activities – has put these aquifers in a state of stress. Their functions cannot, consequently, be fulfilled to desired levels. It is clear that the expected change in climate in the foreseeable future will significantly aggravate the stress on groundwater resources.

It is in arid zones, more than anywhere else on Earth, that climate change will contribute to stressed groundwater systems. And within these arid zones, alluvial aquifers tend to be more sensitive to climate change than other aquifers. The paradox is that they are relatively well endowed with water resources within their arid environment, but at the same time are very vulnerable because they are at the heart of the changing hydrological cycle. Small alluvial aquifers embedded in poorly permeable hard rock (isolated alluvial strip aquifers), characterized by low average residence times of their groundwater, are the most vulnerable and many of them may cease to support local communities adequately within a few decades.

The threats imposed by the impacts of climate change on groundwater should obviously be assessed against other threats. The most important threat to sustainability of groundwater in the alluvial aquifers of Yemen is the direct interaction of people with groundwater – through abstraction, through land use and through polluting activities. The impacts of climate change in Yemen may be small, but will accelerate water resource problems, which will complicate adequate adaptation.

Coping with the consequences of degenerating groundwater resources in alluvial aquifers in Yemen is difficult. Technical measures such as artificial recharge and improved water-use efficiencies will not be sufficient to counter the negative impacts. Rigorous water resource planning and management are necessary. Unconventional and innovative measures will have to be developed, including control of demographic pressure and transition to a less water-dependent economy. The government and water resources managers face the difficult task of developing timely adaptive strategies in order to avoid a collapse of rural economies. They also need to develop institutional capabilities for implementing measures without overlooking public awareness and public support as indispensable factors for success.

Although the analysis presented in this case study is based on examples of aquifers in Yemen, the overall conclusions may be extrapolated to alluvial aquifers elsewhere in arid zones. The mechanisms are similar: ever-increasing human pressure on scarce and dwindling groundwater resources with its related set of complex problems, escalated by climate change. As demonstrated above, the aquifers and their stakeholders become more vulnerable to climate change if the ratio of stored groundwater volume to groundwater recharge is smaller, and if, locally, there are no other water resources available. Isolated alluvial aquifers of the 'strip aquifer' type – and the people depending upon them – are likely to be among the most seriously and directly affected victims of climate change. It will change the livelihoods of many rural communities scattered over the entire arid zone.

Note

1 The word *wadi* or *oed* is used in Western Asia and Northern Africa (Arab region); but the same type of system is known as *arroyo* or *quebrada* in Latin America and as *nullah* in India and other South- and East-Asian countries.

References

DHV (1988) *Tihama Basin Water Resources Study, Technical Report no 7: Wadi Irrigation Monitoring*, Report prepared for the Tihama Development Authority, Ministry of Agriculture and Fisheries Resources, Yemen

Euroconsult, TNO and IHE (1996) *Future of Recharge and Similar Schemes in Oman*, Report prepared for the Ministry of Water Resources, Oman

Falkenmark, M. and T. Chapman (1989) *Comparative Hydrology: An Ecological Approach to Land and Water Resources*, UNESCO, Paris

IPCC (Intergovernmental Panel on Climate Change) (2007) *Climate Change 2007: Synthesis Report, Fourth Assessment Report: The Physical Science Basis*, Cambridge University Press, Cambridge, www.ipcc.ch/SPM2feb07.pdf

Khater, A. and R. El-Weshah (eds) (2002) *Status of Groundwater Protection in the Arab Region*, UNESCO, Cairo Office, IHP no 13

McDonald, Sir M. and Partners (1988) *Wadi Hadramawt Agricultural Development Project, Phase II: Borehole Construction Supervision and Groundwater Studies*, Final Report, Prepared for Ministry of Agriculture and Agrarian Reform, People's Democratic Republic of Yemen

MWR (Ministry of Water Resources, Oman) (1991) *Water Resources of the Sultanate of Oman: An Introductory Guide*, Yemen

SAWAS (1996) *Final Report SAWAS Project*, Prepared by TNO and NWSA, Yemen

SERGEOMIN and TNO (1998) *Estudio para el Control y la Protección de las Aguas Subterráneas en el Valle Alto*, Final report of the CPAS Project, Cochabamba/Delft

Shiklomanv, I. A. and J. Rodda (eds) (2003) *World Water Resources at the Beginning of the Twenty-first Century*, UNESCO/University Press Cambridge, Cambridge

TNO Institute of Applied Geoscience (1992) *Proper Development of the Water Resources of the Wadi Dayqah*, Report prepared for the Ministry of Water Resources, Oman

Van der Gun, J. A. M. and A. A. Ahmed (1995) *The Water Resources of Yemen: A Summary and Digest of Available Information*, Report WRAY-35, Sana'a/Delft

Van der Meulen, D. and H. von Wissman (1932) *Hadramaut – Some of Its Mysteries Unveiled*, E. J. Brill Ltd, Leiden, The Netherlands

Vasak, S. (2002) *Water Resources Management Plan for the Hadramawt Region*, Report prepared for the National Water Resources Authority at Sana'a (Yemen) and UNDESA/Government of the Netherlands, The Netherlands

WEC (2004) *Well Inventory Wadi As Sirr*, report by the Water and Environmental Centre, University of Sana'a, Yemen

12

A Water Utility's Approach to Addressing the Potential Impacts of Climate Change

Steve W. Gillham and Mark J. Summerton

Introduction

The United Nations Millennium Development Goal (MDG) 7 ('Ensure environmental sustainability') includes a clear statement on drinking water: 'Reduce by half the proportion of people without sustainable access to safe drinking water'. According to the *MDG Monitor* (2008), South Africa is well on track with MDG 7, stating that 'significant achievements have already been made in areas, such as access to basic water supply'. The South African government has also committed to the provision of safe potable water to all by 2008. According to the *MDG Monitor*, 88 per cent of the population had access to improved drinking water sources in 2004.

However, climate change might put even more stress on achieving this MDG by the year 2015. The Intergovernmental Panel on Climate Change (IPCC) stated in its *Fourth Assessment Report* that for Africa: 'By 2020, between 75 and 250 million people are projected to be exposed to increased water stress due to climate change' (IPCC, 2007).

This case study focuses on a water utility in South Africa and the steps taken by this water utility to include climate change in its general policies and procedures. The focus will be on Umgeni Water, located in the eastern part of the country (see Figure 12.1) and currently serving about 5 million people in Durban, Pietermaritzburg and their surroundings with a total of 340 million cubic metres of potable water annually. The area can be described as having a warm subtropical climate. Summer (December to March) is generally hot and humid, averaging 28°C, and experiences the majority of the annual rainfall, while winter, with an average temperature of 23°C, is warm and dry. Mean annual precipitation (MAP) ranges from more than 1200mm in the main water source catchments of the Drakensberg in the west to around 700mm towards the coast in the east.

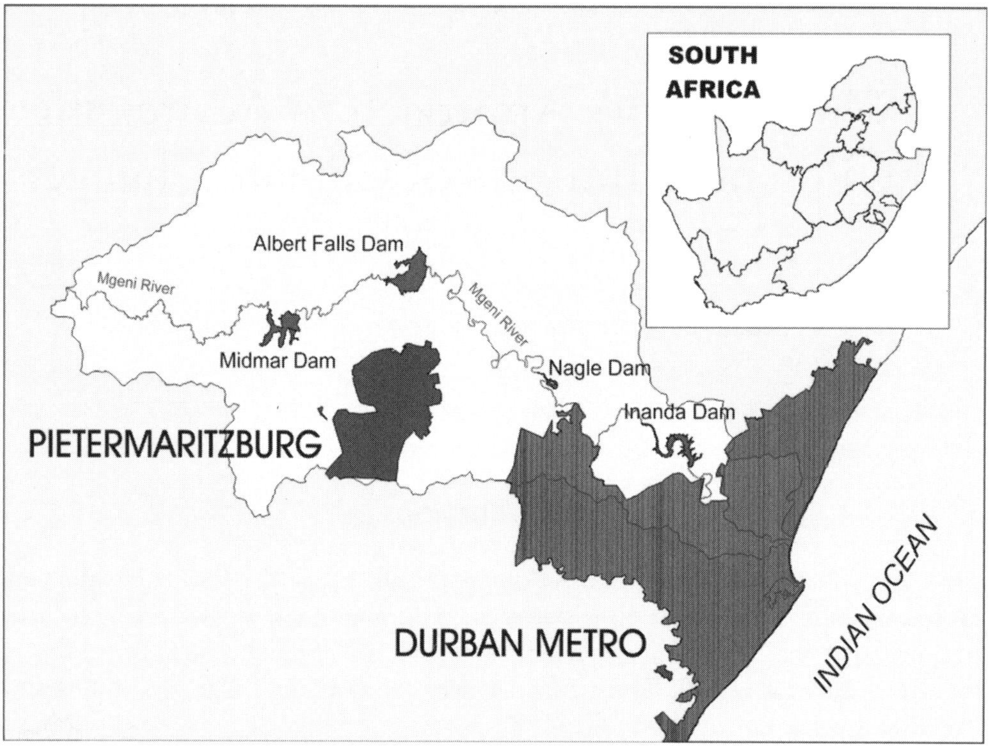

Figure 12.1 *Locality map of the Mgeni catchment and the major demand centres*

Source: Steve W. Gillham and Mark J. Summerton

Umgeni Water

The institutional framework of the water sector in South Africa consists of a number of role players: the Department of Water Affairs and Forestry (national government), which is the custodian of the country's water resources and the regulator within the water sector; water service authorities (WSAs), which are the municipalities (local government) that have the executive authority to provide water services within their areas of jurisdiction; and water service providers (WSPs), which are organizations that have contracts with WSAs or other WSPs to sell water to and/or accept wastewater from that authority or provider for the purposes of treatment.

Umgeni Water is one of three water utilities operating within the province of KwaZulu-Natal in South Africa. It is classified as a regional bulk WSP, whose primary function is the treatment and distribution of bulk potable water on a regional scale to a number of WSAs, who in turn are responsible for reticulation to the end users. Umgeni Water also owns and operates a few wastewater treatment plants and is involved in various water resource management activities to ensure the quality and sustainability of the water resources upon which it depends.

The primary water resource for Umgeni Water's operations is the Mgeni River, which is fully utilized with four large storage dams on the river. This water resource is augmented (when required) through an inter-basin transfer scheme from the adjacent Mooi River. Water is supplied from this system at a 99 per cent level of assurance (i.e. 1 in 100-year risk of failure) to the Greater Durban-Pietermaritzburg area, which is one of the most important economic hubs of the country.

Gillham (2003) details the water resource assessment techniques used by the utility. Effectively, two categories of assessment exist:

1 the current situation, which is based on short timeframe analyses where current demands are balanced against current supply availability, leading to possible changes to the system operation rules; and
2 the future situation, which is based on long timeframe analyses where future demands are balanced against future supply availability. These results guide the choice of options that forward-looking planning investigations are based upon.

Climate change would impact most significantly upon the latter type of assessment. The need was therefore recognized to incorporate climate change within the utility's business environment as these potential impacts could dramatically affect the utility's future business endeavours and, hence, its customers. In order to do this, two issues needed to be addressed: how to quantify the potential impacts of climate change for the region and, consequently, the utility; and how to get the utility to take cognizance of the potential impacts in order to initiate a proactive approach.

Incorporating climate change

In the past, few articles pertaining to climate change were published or broadcast in the media, resulting in an irregular and inconsistent message that did little to alert the 'man in the street' to the potential impacts that were pending in the years ahead. Unless one had a specific interest in this topic and was prepared to delve a little deeper, there were few indications of the potential risks at a local scale.

The water resource planners within Umgeni Water are required to continuously scan the external environment for factors that may possibly influence the utility's future ability to provide an adequate supply of water at the required levels of assurance, and to incorporate these potential impacts into future planning scenarios. Climate change is one such factor.

In order to ensure acceptance of results and recommendations relating to climate change once they were incorporated within analyses, it was important that top management were educated on the issues at the beginning of the process, and then kept abreast of developments during the analyses. In order to achieve this, an internal workshop was convened with the utility's board, chief executive, senior managers, all planning staff and other relevant key staff members within the utility. The objective of this workshop was to:

- create an awareness of what climate change is all about;
- improve the level of understanding of the potential local impacts; and
- identify actions required by the utility to mitigate against these impacts.

A range of external experts were invited to speak on various aspects relating to climate change in the KwaZulu-Natal region of South Africa: climate-related concerns and considerations; impacts upon water resources; impacts upon biodiversity; vulnerability; the legal framework; global warming scores for urban water systems; and the implications of climate change for strategic planning in the city of Durban. Furthermore, it was equally important to present a balanced view on climate change so that the delegates could draw their own conclusions. A well-known academic who regularly voices strong views against the existence of climate change was thus invited to present his point of view.

This workshop achieved its objectives and proved to be a milestone for the utility. Discussions on climate change within the utility were elevated to a higher level based on a better understanding of the topic. A set of recommendations on further actions that the utility should take was compiled (Umgeni Water, 2006a) based on the output from the workshop. This document enjoyed a smooth passage through the approval process within the utility as there was common understanding and agreement on what was required. The main recommendations emanating from the workshop included investigations into:

- hydrological impacts;
- energy management;
- water demand impacts; and
- water quality impacts.

Umgeni Water has an integrated risk management process (Umgeni Water, 2006b) where risks that affect, or may affect, its strategic objectives are identified, assessed, regularly reviewed and rated to determine significance. It also requires the planning, arranging and controlling of activities and resources to minimize the impacts of all risks to levels that can be tolerated. A corporate risk register that contains a consolidated risk record (currently for 33 risks) is maintained.

Climate change is currently ranked third of the risks associated with the natural environment. The organization's current highest risk relates to the availability of water resources to meet development demands, which also has a direct linkage to climate change risks. The second highest risk deals with the mechanical/physical failure of strategic infrastructure. The inclusion of climate change in the risk register and its ranking is based mainly on the heightened awareness and the improved understanding (albeit at a superficial level) by management of the potential impacts of climate change on water resources.

The future actions to minimize this particular risk are twofold. First, the aim is to obtain a better understanding of the potential impacts associated with climate change in terms of magnitude, severity, vulnerability and timeframes. This is aligned to the further investigations approved after the climate change workshop. Second, based on this improved information, a plan of action can be established towards actually minimizing this risk and, hence, reducing its ranking on the register.

Local climate change implications

The possible impacts of climate change have been widely debated; however, it is only recently that improved prediction modelling has resulted in credible future climate change and impact scenarios. If these scenarios are to be believed, they will have far-reaching global consequences, especially in Africa, which the IPCC (2001) predicts will be the continent most adversely affected by climate change in terms of its impacts and capacity to respond.

Regional climate change scenarios for South Africa (IPCC, 2001) support an increased occurrence of extreme weather events such as droughts and floods. A synopsis of potential impacts on rainfall from the Climate Systems Analysis Group (CSAG) at the University of Cape Town, South Africa, for 2070 to 2099 shows a decrease in average rainfall in the west and an increase in the east of South Africa (including Umgeni Water's operational area); however, the intensity of these events in the east as indicated by fewer rainy days will be greater.

Schulze et al (2005) used the Conformal Cubic Atmospheric (CCAM) regional climate model to predict that most of South Africa's future mean annual precipitation is set to decrease slightly to approximately 90 to 95 per cent of current levels. Most disturbing is the strong decreasing trend of future MAPs along the already water-stressed west coast of South Africa, where reductions in MAP in the order of 15 to 25 per cent are predicted. Fortunately, there could be some relief in an area from the North-West Province to the Drakensberg, including Lesotho and parts of the Southern Free State, as well as the north of the Eastern Cape where MAPs of up to 10 per cent in excess of those of the present could occur.

The recent IPCC projections (2007) confirmed these findings. Out of the 21 climate projection models, about half expected rainfall to increase in the eastern part of the country. For the west, only a few models expected an increase in precipitation (see Plate 25, centre pages). However, on average, a decrease in rainfall over the entire country is to be expected ranging from 5 to 20 per cent (see Plate 26, centre pages).

Possibly the only current indication of the localized impact of climate change on water resources in the Mgeni catchment is the regional hydrological modelling by Schulze and Perks (2000) as documented in Turpie et al (2001) and Naidu et al (2006). These reports suggest that:

- Maximum temperatures in Durban will rise by up to 3°C by 2070.
- There is likely to be a significant increase in demand for water for irrigation for the agricultural sector.

- Runoff into the main rivers is likely to be reduced over much of the country and will become less predictable.
- The Mgeni River is projected to have a 20 per cent reduction in outflow at the mouth by 2050.
- Water availability in the Mgeni River catchment is predicted to decrease by 157.8 million cubic metres for the period of 2070 to 2100.

Assessing local water resource impacts

While predictions of potential climate change impacts at the global and national level are being improved, little has been done at the regional and local levels. It is important that water utilities such as Umgeni Water are able to improve their predictions of timing and magnitude on water resources in key catchments such as the Mgeni. These predictions need to include the impacts of climate change upon water resources, water yield and infrastructure (augmentation and safety) requirements. Thus, an internal process (see Figure 12.2) was developed to determine the information needed to undertake the analyses.

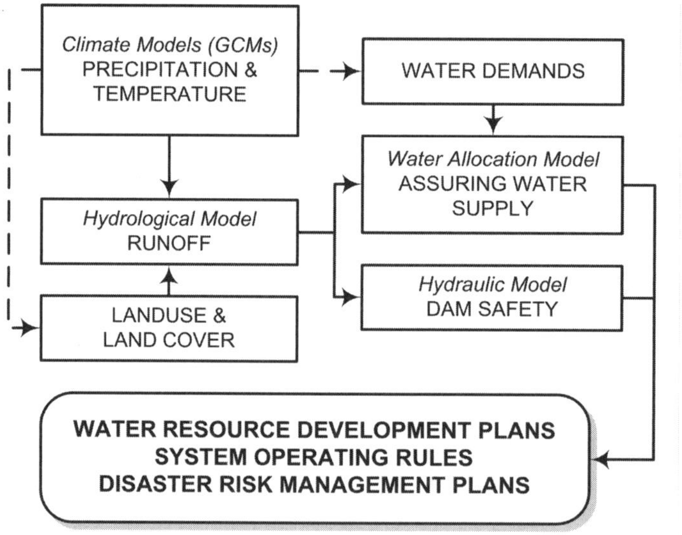

Figure 12.2 *Hydrological modelling process to determine the impacts of climate change upon local water resources*

Source: Summerton and Gillham, 2007

At the heart of the tools being used for these analyses is the ACRU hydrological model (Schulze, 1995), which is underpinned by a long-standing relationship with the model developers at the University of KwaZulu-Natal, based in Pietermaritzburg, South

Africa. ACRU is being configured at a finer resolution than in previous studies, including more than 100 sub-catchments for the 4440km^2 catchment and land cover at a 1:5000 scale. Updating the catchment characteristics and features to the present conditions provides a base against which modelled changes can be assessed.

In order to model the impacts of climate change on hydrology, key variables, including temperature and precipitation, will be adjusted based on global climate models (GCMs) from the CSAG. Currently, researchers at the University of KwaZulu-Natal are enhancing techniques to downscale the results from the GCMs to a regional scale that would be relevant to local catchments. They will be providing the adjusted ACRU temperature and precipitation files needed for the analyses. Resulting runoff sequences will then be compared to base conditions. By using these runoff sequences to represent the hydrology, together with water demands, in specialized water resources planning and yield models, it will be possible to determine the potential impact that climate change will have on the utility's current and future ability to supply bulk potable water at the required level of assurance. These results will then be incorporated within the review of the utility's water resource development plans and system operating rules.

Climate change runoff sequences will also be routed through hydraulic models to determine revised flood line, backwater and tail-water levels, and the adequacy of infrastructure such as dam spillways, bridges and culverts. These results will then be used to update the utility's disaster risk management plans and, in accordance with the utilities' disaster risk management framework, will be brought to the attention of the relevant local authorities to ensure integration with their plans for sustainable future development (Umgeni Water, 2006c).

As a next step, and enhancement to the process, the direct impacts of climate change (temperature and precipitation) on land use and land cover at a local level need to be established. It will then be possible to adjust these input files into the ACRU model and generate a further set of scenarios for analyses. These results would most likely be more representative of future conditions since the potential impacts within a catchment will have multiple components. Similarly, by accounting for the direct impact that climate change (temperature and precipitation) will have on water demand patterns, it will be possible to add another layer of complexity to the various analyses and to produce another set of (more realistic) scenario results. These enhancements to the process will be included as soon as a better understanding of the relationships has been obtained.

Climate change in context

Although the implications of climate change on water resources could well be one of the biggest risks requiring adaptation to ensure uninterrupted supply water (Umgeni Water, 2006c), it is only one of several stressors that have the potential to affect the utility's business. The results from the scenario analyses described in the preceding section, once completed, will assist in establishing whether the potential climate change impacts will be more significant than other influencing factors, or whether they will be negligible in comparison, or whether they will jointly contribute towards a significant impact. It is suspected that the last option will hold true.

By way of illustration of the significance of other factors on the utility's water resource development plans, the variability in water demand projection results is highlighted. Over the past eight years, the water demand projection for the Mgeni system has varied significantly in an upward trend each year when revised. This variation was as a result of changes to the external demand drivers of local water demand – in particular, the implementation of water demand measures by the city of Durban – which resulted in initial successes and was then followed by successive years of an inability to meet targets. Consequently, in 2000 it was predicted that the next water resource development (required to ensure levels of assurance were maintained) for the Mgeni system only needed to be commissioned in 2024 (providing more than sufficient lead time for the planning, design and construction of the scheme). However, by 2006, the water demand projection had moved dramatically higher, indicating that this next scheme would be required in 2010 – suddenly the lead time available was only four years! This is clearly insufficient to meet the deadline and the scheme development process has had to be accelerated in an attempt to minimize the overrun. These water demand projections have a significant impact upon the utility's ability to conduct its business, and the accuracy of the projections is paramount.

Thus, the dilemma is complex. In evaluating the potential impacts that the utility is faced with, there are further questions that need to be addressed, including:

- How significant are the uncertainties of one influencing factor over the others, and, thus, where should the utility's focus lie?
- What can be accepted as correct, and how are the decision-makers to be convinced of this?

It is therefore important that potential climate change impacts are compared, by means of an impact, risk and vulnerability assessment, to other stressors (such as increasing water demands, HIV/AIDS and other disasters) in order to determine the utility's priorities, after which appropriate adaptation measures can be implemented.

The way forward

As described in an earlier section, Umgeni Water has developed a process to assess the hydrological impacts of climate change, and the water utility is currently at the early stages of implementing this process. Completing the process and tabling the results is not considered to be the end since the process is dynamic and further analyses will be required as the driving information changes.

As the accuracy of climate change prediction models improves over time (and as their results start converging better) – particularly as the ability to downscale results to a regional and local level is refined – so will the accuracy of the information used in the assessment process improve.

The water resource planners within Umgeni Water are continually striving to broaden their knowledge and understanding of climate change. Through this learning and scenario-planning experience, new ideas will emerge, as well as the means to improve the input information, analysis process and interpretation of results.

An integrated water resources management (IWRM) approach to assessing the potential impacts will be fundamental to obtaining realistic results. Changes to other catchment factors, such as land use, land cover, water quality and consumption patterns, need to be accounted for in the process.

Climate change also has impacts beyond Umgeni Water Utilities. It is important that its water service authorities, customers and end users are also educated about climate change and its potential impacts. Thus, there is a need to increase regional awareness, possibly through workshops such as the one conducted internally. These will need to be arranged at an appropriate time when the results on the initial analyses are complete. Following on from this is the need to coordinate future coping strategies and plans with the objectives of other relevant stakeholders in the region.

Since the results of the analyses will be based on uncertainty, as is the case with other influencing factors, it is important for the utility to develop flexible/adaptable strategies to cope with these potential impacts. It is undesirable for the utility to be faced with issues such as unachievable deadlines for water resource development or loss of supply potential; therefore, the scenario planning results should be used to adequately prepare for sustainable solutions.

Regardless of the outcome of these local assessments, the systematic inclusion of climate change risk, on both the supply and demand side, will be a necessity for all water utilities in the future.

References

Gillham, S. (2003) 'Managing for droughts: Umgeni Water's *modus operandi*', in R. E. Schulze (ed) *The Thukela Dialogue: Managing Water Related Issues on Climate Variability and Climate Change in South Africa*, Report to International Dialogue on Water and Climate, University of Natal, Pietermaritzburg, RSA, School of Bioresources Engineering and Environmental Hydrology, ACRUcons Report, South Africa, vol 44, pp95–98

IPCC (Intergovernmental Panel on Climate Change) (2001) *Third Assessment Report: Climate Change*, Cambridge University Press, Cambridge

IPCC (2007) *Fourth Assessment Report: Climate Change 2007: Synthesis Report – Summary for Policymakers*, IPCC, Cambridge University Press, Cambridge

MDG Monitor (2008) *Millennium Development Goals Monitor*, www.mdgmonitor.org/

Naidu, S., R. Hounsome and K. Iyer (2006) *Climatic Future for Durban*, CSIR NRE, Pretoria, South Africa

Schulze, R. E. (1995) *Hydrology and Agrohydrology: A Text to Accompany the ACRU 3.00 Agrohydrological Modelling System*, Water Research Commission, Pretoria, South Africa, Report TT 69/9/95

Schulze, R. E. and L. A. Perks (2000) *Assessment of the Impact of Climate Change on Hydrology and Water Resources in South Africa*, University of Natal, Pietermaritzburg, School of

BEEH, Report to South African Country Studies for Climate Change Programme, ACRUcons Report, vol 33, p118

Schulze, R. E., T. G. Lumsden, M. J. C. Horan, M. Warburton and M. Maharaj (2005) 'An assessment of the impacts of climate change on agrohydrological responses over Southern Africa', in R. E. Schulze (ed) *Climate Change and Water Resources in Southern Africa: Studies on Scenarios, Impacts, Vulnerabilities and Adaptation*, Water Research Commission, Pretoria, South Africa, WRC Report 1430/1/05, Chapter 9, pp141–189

Summerton, M. J. and S. G. Gillham (2007) *A Water Utility's Approach to Addressing the Potential Impacts of Climate Change.* Proceedings, 13th SA National Hydrology Conference, Cape Town, South Africa

Turpie, J., H. Winkler, R. Spalding-Fecher and G. Midgley (2001) *Economic Impacts of Climate Change in South Africa: A Preliminary Analysis of Unmitigating Damage Costs*, Research paper sponsored by USAID and administered by the Joint Centre for Political and Economic Studies Inc, under a subcontract agreement from Nathan Associates Inc. Southern Waters Ecological Research & Consulting, and Energy Development Research Centre, University of Cape Town, Cape Town, South Africa

Umgeni Water (2006a) *Recommendations for Addressing Climate Change at Umgeni Water*, Planning Services, Engineering and Scientific Services, 9 January 2006

Umgeni Water (2006b) *Zero-based Risk Management Report*, Version 3, July 2006

Umgeni Water (2006c) *A Framework for Disaster Risk Management*, Prepared for the Umgeni Water Risk Committee by Planning Services, Engineering and Scientific Services, Version 1, November 2006

13

Adaptation Measures for Metropolitan Water Supply for Perth, Western Australia

Bryson C. Bates and Graeme Hughes

Introduction

Climate change and the impact upon already scarce water resources are important issues in the public debate in Australia. During 2006, Prime Minister John Howard refused to meet Al Gore when he visited Australia to promote his documentary *An Inconvenient Truth*. Severe droughts combined with changes in public mood softened the prime minister's rhetoric and later he declared that he broadly accepts the science behind climate change. Restrictions on urban water use have frequently been imposed in the past and Water Corporation Perth is starting to incorporate climate change within its planning processes by a combination of increasing the supply and simultaneously trying to decrease the demand. This case study will demonstrate what options are available for drinking water planners in terms of adaptive responses to climate change using the Perth case as an example.

Western Australia

Physical setting

Perth is the capital city of Western Australia, Australia's largest state (see Figure 13.1). The city has a population of about 1.5 million, with current and projected population growth rates of 1.8 per cent per year (Power et al, 2005). Globally, it is the most isolated city, with a population greater than 1 million. Perth is located in the south-west region of the state, situated on a narrow coastal plain (the Swan Coastal Plain) between the Indian Ocean and a low coastal escarpment known as the Darling Range to the east. Average heights of the escarpment range from 250m to 300m: the highest point (Mount Cooke) has a height of 582m. The range runs parallel to the south-west coast east of Perth for 200 miles (320km) from the Moore River (north) to Bridgetown (south). It is dissected by ravines, cut by rivers flowing to the sea. Some of these rivers are dammed to provide domestic and industrial water supplies. Water supplies are

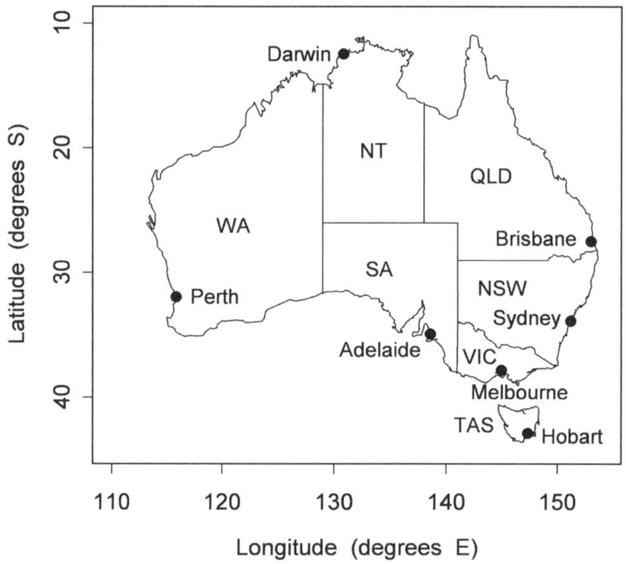

Figure 13.1 *Location map of Australia*

Note: QLD = Queensland; NSW = New South Wales; NT = Northern Territory; SA = South Australia; TAS = Tasmania; VIC = Victoria; WA = Western Australia.

Source: Bryson C. Bates and Graeme Hughes

also sourced from shallow and deep aquifers below the coastal plain and, since November 2006, a seawater desalination plant.

Climate of the south-west

Generally, the western edges of landmasses at the same latitudes as the south-west (about 30° to 35°) receive relatively low rainfall, partly because most oceans have a cool Equator-ward directed current along their eastern boundary, which reduces available atmospheric moisture. The south-west corner of Australia is unique in that the total rainfall it receives is comparatively high. Here, a warm ocean current (the Leeuwin Current) flows strongly southwards along the coast of Western Australian. It turns eastwards at Cape Leeuwin and continues into the Great Australian Bight, where its influence can extend as far as Tasmania (see Figure 13.1).

Most rain falls within the cooler winter months (June and July), with over 80 per cent falling between the months of April and October (see Figure 13.2). Rainfalls over the Swan Coastal Plain and the catchments of metropolitan dams are enhanced by the uplift of onshore moisture-laden winds caused by the Darling Range. Typically, these winds are generated by cold fronts associated with low pressure systems to the south. There is a marked increase in rainfall from north to south, with a marked rainfall gradient from the south-west to the north-east. In summer, the subtropical belt of high

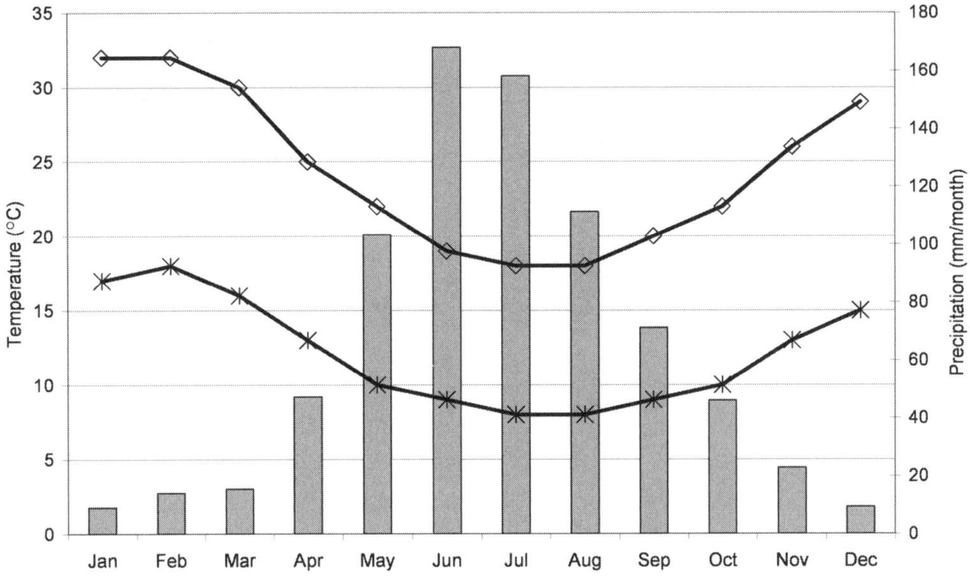

Figure 13.2 *Mean minimum and maximum temperature and precipitation for Perth*

Source: www.wunderground.com

pressure extends across the region, reaching its southernmost extension in January or February. Consequently, summers are dry and hot. During autumn, the high pressure belt gradually moves towards the north and lies almost wholly outside the south-west during the winter months. Over the last 14 years, Perth's average rainfall and maximum daily temperatures for December to February were 30mm and 30°C, respectively. The average rainfall and daily minimum temperature for June to August were 420mm and 8°C, respectively (Australian Bureau of Meteorology, 2007).

The Integrated Water Supply Scheme (IWSS)

Water for the city of Perth is supplied by the Integrated Water Supply Scheme (IWSS) (see Figure 13.3), which services the area from Quinn Rocks to Mandurah, includes several small south-west towns from Binningup to Waroona, and incorporates the Goldfields and Agricultural Water Supply Scheme to the east. Thus, the IWSS extends more than 600km east–west and 200km north–south, allowing sources to be substituted if they fail for any reason. Like most Australian urban water supply schemes, the IWSS was originally built as a one-pass system in which water flowed from storage to treatment to consumers, and then to disposal as treated wastewater (Kaspura, 2006).

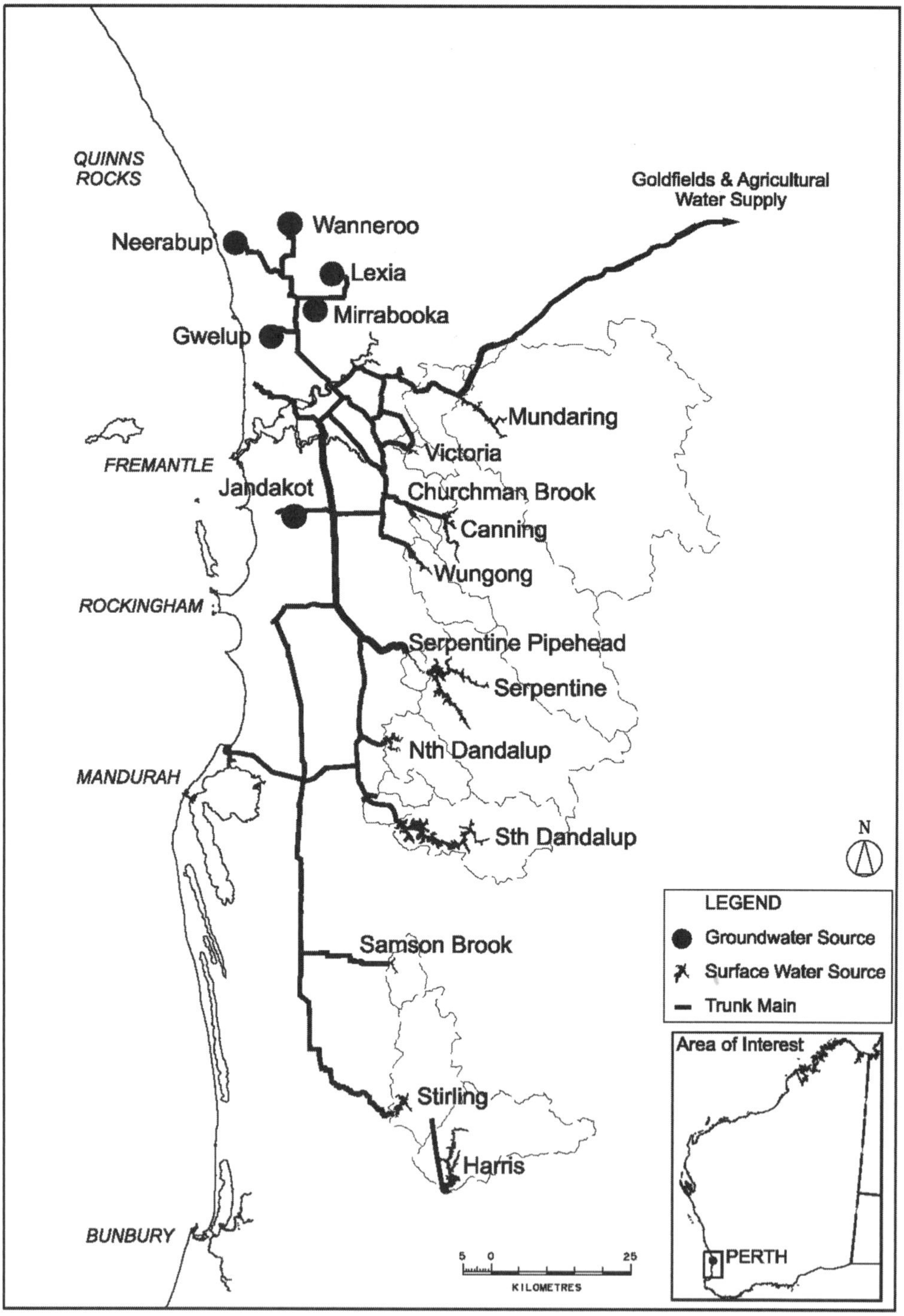

Figure 13.3 *Schematic diagram showing Perth's Integrated Water Supply Scheme (IWSS)*

Source: GWA (2003)

The IWSS supplies water to 1.6 million of the 2 million people living in Western Australia. Approximately 70 per cent of the water supplied is used domestically and 50 per cent of that is used in household gardens (Power et al, 2005). 18 per cent of the state's water use is tied to the IWSS: the remainder is for private and industry supply (McFarlane, 2005). Currently, there are four main sources of water: shallow unconfined aquifers, confined aquifers, surface reservoirs and seawater desalination. Public groundwater supplies are largely sourced from the northern metropolitan area.

The amount of water supplied into the IWSS in 2005/2006 was 265 billion litres, with more than half obtained from groundwater, with the balance from surface water sources. In November 2006, commissioning began for the 45 billion litres per year seawater desalination plant located at Kwinana, about 40km south of the centre of Perth (see Figure 13.4). At the start of 2007, the water supply to the IWSS was derived from 11 main dams, a total of 136 groundwater bores located in the confined Yarragadee (18 bores) and Leederville (18 bores) aquifers, the unconfined (100 bores) aquifers in the Perth region and the new seawater desalination plant.

Figure 13.4 *Perth seawater desalination plant, Kwinana*

Source: Water Corporation, Western Australia

The Department of Water (established 26 October 2005) is responsible for managing water in Western Australia and grants a licence to the Water Corporation to operate the IWSS. The corporation is the largest water service provider in Western Australia.

It is a state-owned commercial entity that returns its profits to the state government (Power et al, 2005).

In the context of the above settings, this case study examines the adaptive response of urban water planners to climate variability and change for the Perth region. This response has taken place in the context of urban water reform processes, driven by the paradigm of doing more with what already exists while accepting the importance of environmental health and the sustainable development of water supplies (Kaspura, 2006).

Water planning in Australia

Water planners have two options for restoring and maintaining the balance between water demand and supply: encourage water conservation (which may be the cheapest and most environmentally sensitive option), or increase supply. Past water planning strategies for the cities of Sydney and Melbourne (see Figure 13.1), in particular, and, to a lesser extent, Perth had a heavy reliance on reducing per capita consumption. However, concern about current drought conditions and projected climate change is leading to the reassessment of climatic risk, the development of publicized contingency plans (MJA, 2006) and growing interest in community attitudes towards water restrictions and reuse.

National urban water reform

Since the mid 1980s, per capita consumption in Australia's cities has decreased markedly due to pricing reforms, growing use of water-efficient appliances and fixtures, and water conservation programmes, such as public education campaigns and the provision of subsidies for the purchase of water-efficient appliances (WSAA, 2005). Over the past decade, in particular, it has been increasingly recognized that Australia's water resources are limited and that traditional approaches to meeting water demand by expanding supply (e.g. through the construction of new dams) are not environmentally or financially sustainable (ACIL Tasman, 2005). In 1994, the Council of Australian Governments (COAG) consisting of the prime minister, state premiers, chief ministers of Australia's territories, and the president of the Australian Local Government Association agreed to a Water Reform Framework with the aim of achieving efficient and sustainable urban and rural water industries. The framework initiated substantial policy and institutional change: it included provisions for water entitlements and trading, environmental requirements, institutional reform, public consultation and education, water pricing and research. Achievements to date include:

- the separation of responsibility for policy- and standard-setting, regulatory enforcement and service provision in urban water management;
- institutional reforms that encourage water utilities to become more accountable and transparent to their customers;

- the introduction of two-part tariffs in which users pay a delivery charge and a charge for the volume of water used;
- with few exceptions, the stopping of new water allocations from overexploited rivers and aquifers and the construction of dams that are ecologically unsustainable; and
- water management plans that provide for environmental flows in both surface and groundwater and aim to preserve ecologically significant environments.

The Water Reform Framework was extended by the National Water Initiative, which was agreed to by most state governments in June 2004 (Tasmania agreed in June 2005 and Western Australia in April 2006). The outcomes sought from the framework include:

- *The provision of healthy, safe and reliable water supplies.*
- *Increased water-use efficiency in domestic and commercial settings.* Measures that improve efficiency include the introduction of water efficiency standards for household fixtures and appliances, and mandatory product labelling; a review of the effectiveness of temporary low-level water restrictions and associated public education campaigns, and the scope for their extension as standard practice; improved management of leakage and other water losses; and the use of water-sensitive urban design (i.e. the integration of urban planning and development), with the management, protection and conservation of water considered within the context of the water cycle (ATSE, 2004).
- *Pricing policies that stimulate the efficient use of recycled water and storm water in cost-effective settings.*
- *Water trading between and within urban and rural sectors.* With the exception of Sydney, all capital cities can access water from rural areas without the construction of substantial infrastructure (WSAA, 2005).
- *Provision of more detailed information on water accounts.*

Urban water reform in Perth

As early as 1978, the Metropolitan Water Supply Sewerage and Drainage Board, a predecessor of the Water Corporation, introduced user-pays pricing with a fixed service charge and a pay-for-use tariff for household consumption above 150 kilolitres (kl) per year. This reform, coupled with a complete ban on sprinkler use between July 1977 and September 1978 in response to low dam inflows since 1975 (see Figure 13.5), had an immediate and noticeable effect on water demand. By the mid 1980s, about 30 per cent of domestic customers had installed private bores and per capita consumption had decreased from a peak of 233kl per year in 1975 to about 170kl per year. Regulations requiring the use of dual-flush toilets were introduced in 1993, and in 1994 a ban was placed on the use of lawn and garden sprinklers supplied by the IWSS between the hours of 9.00 am and 6.00 pm. Since the mid 1990s, there has been an ongoing education campaign to promote water conservation awareness (Hughes, 2003).

In 1995, the then Water Authority of Western Australia released a strategic review of water supply issues relevant to Perth (Stokes et al, 1995). The review incorporated significant public participation and proposed activities to manage water demand (e.g. leak reduction, reduction of the amount of unmeasured water in the supply system, use of efficient showers and water appliances, use of water-sensitive landscaping, and efficient irrigation of lawns and gardens) and to prepare for source development. It recognized the need for the provision of water to meet environmental demands, and the threats posed by climate change and changing customer expectations.

The separation of responsibilities for policy- and standard-setting, regulatory enforcement and service provision in urban water management occurred in Western Australia on 1 January 1996 with the establishment of the Water Corporation, the Water and Rivers Commission and the Office of Water Regulation. Commercial objectives and environmental targets and accountabilities for the corporation were established through a statement of corporate intent and a system of licences through several regulatory agencies. Legislation to enable water trading in Western Australia was introduced in January 2001.

Managing climatic risk and uncertainty

Climatic risk is reduced when a city has diversified sources of water and/or sources that are climatically and hydrologically distinct. Western Australia and the Northern Territory are the only state and territory in Australia in which the main urban water source is not surface water. Brisbane (see Figure 13.1) and Sydney are essentially reliant on one-dam systems: the Wivenhoe and Warragamba dams, respectively.

Traditional practice relies on the assumption that the historical record provides the best basis for water-supply planning rather than a scenario approach. However, current events – such as the multi-decadal hydrological drought in the south-west; the driest ten-year period on record for the city of Melbourne; the record-breaking multi-year sequence of low flows in Australia's largest river (the Murray); and ongoing water restrictions in the urban and rural sectors – are leading some planners to question this basic assumption. Moreover, it is becoming clear that water conservation programmes and water restrictions alone are no longer sufficient to cope with current and projected water supply shortfalls. The COAG has agreed that it is time to act collaboratively on adaptation to unavoidable climate change.

Across Australia, there appears to be major differences in the way that water planning and decision-making has dealt with climate risk and uncertainty. Distinguishing characteristics include the (MJA, 2006):

- extent to which past changes in stream-flow regime are recognized;
- degree of reliance on the complete historical record for planning purposes;
- extent to which climate change projections are considered;

- level of service requirements (intensity, frequency and duration of water restrictions);
- willingness to consider non-traditional sources (e.g. water recycling for agricultural, industrial or potable use; water trading and desalination);
- extent to which contingency plans and their triggers have been identified and articulated; and
- degree of reliance on per capita demand reductions through behavioural change or improved urban design.

Through a combination of foresight and necessity, the Water Corporation has revised its planning baselines and contingency plans for the IWSS over the last 20 years because of a multi-decadal rainfall decline over the south-west. This has led to changes in the level of service requirements and deeper exploration of issues surrounding the use alternative sources of water.

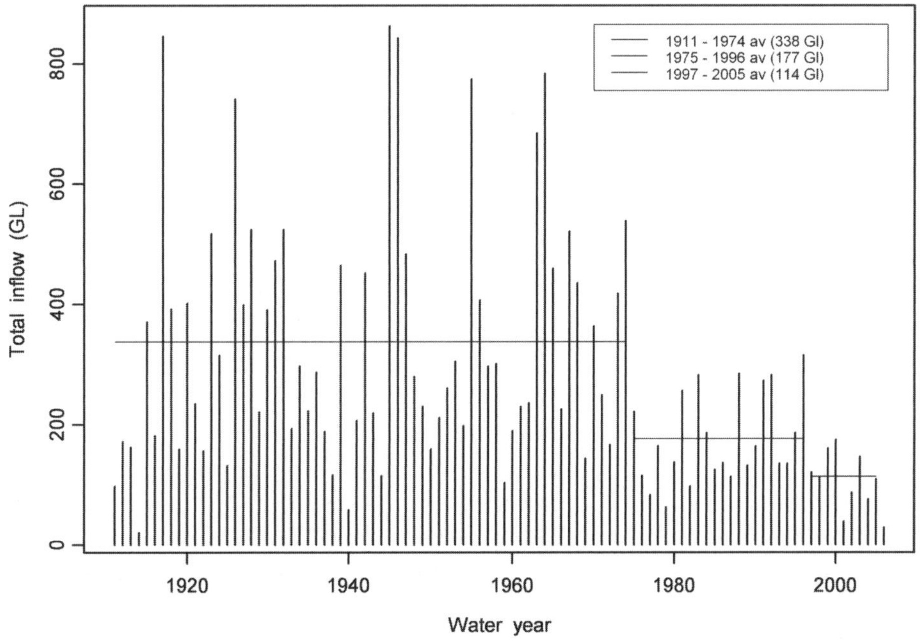

Figure 13.5 *Dam inflow series for the Integrated Water Supply Scheme (1911–2006)*

Source: Bryson C. Bates and Graeme Hughes based on data from Water Corporation, Western Australia

Community attitudes towards alternative sources of supply

Australian communities have lived through droughts, water restrictions and water conservation campaigns for a number of years. As affluence grows, urban communities demand higher levels of service and develop limited tolerance to water restrictions that impact upon their standard of living (WSAA, 2005). A recent survey of 3500 residents in Perth, Adelaide, Darwin, Melbourne and Sydney revealed that (Roseth, 2006):

- While one third of the community is worried about the above situation, there is no sense of urgency.
- 93 per cent of respondents agree or strongly agree that water is a scarce resource that should be carefully conserved.
- 88 per cent believe that individuals can make a difference to the amount of water saved – 79 per cent believe that they can do a little more to conserve water, and 22 per cent a lot more.
- 70 per cent regard having a healthy green garden as important.
- One third would be annoyed if existing restrictions were tightened.

Community attitudes to water restrictions have been studied extensively in the south-west. There, it has been found that in the context of greater experience of restrictions with increasing severity, there was greater support for regular restrictions every year to conserve water, but diminishing support for outright sprinkler bans (Nancarrow et al, 2002, 2003). The last complete ban on sprinkler use in the south-west occurred during 1977 to 1978. Complete sprinkler bans generate considerable debate and fuel perceptions of mismanagement and poor planning by water managers. It has also been estimated that a complete ban would lead to losses of Aus$300 million and 4000 jobs (Power et al, 2005). A recent study has found that Perth households are willing to pay 20 per cent more on their water usage bill to be able to use their sprinkler up to three days a week. Moreover, households would rather pay higher water bills of up to 40 per cent more to finance a new source of supply instead of enduring severe water restrictions.

Until recently, storm water and treated wastewater were viewed as a nuisance or a threat requiring disposal, rather than as a potential resource. A concept that is gaining popularity in Australia is integrated water cycle management (IWCM) in which urban water supply, roof runoff, the recycling of sewage, and the capture and use of storm water are managed simultaneously (ATSE, 2004). Typical goals of IWCM programmes include the efficient and sustainable use of natural resources through demand management; the use of non-traditional water sources and loss control; the achievement of water quality targets; protection from flooding due to surcharge from hydraulic structures; and minimization of the volumes and adverse impacts of discharges to the environment and other water users. A review of 15 demonstration sites in Eastern and Southern Australia has shown that the level of benefits ranged from modest to significant and that the degree of integration could be improved upon (Mitchell, 2006).

However, while Australian communities consistently support the concept of reclaiming wastewater, they are frequently unwilling to use it personally, particularly for potable purposes. In fact, the closer recycled water comes to human contact, the more the level of support diminishes. Emotion ('the yuck factor'), the source(s) and specific use(s) of recycled water, the opinions and influence of others, the cost to the consumer, the issue of choice, and the level of trust in assurances from authorities on health and environmental risks play a significant role in acceptance (Po et al, 2005). Industry concerns include access entitlements to reclaimed wastewater; lack of financial incentives and the undermining of traditional revenue bases; uncertain demand; the cost of supply to the retailer; and the cost of regulatory compliance (ACIL Tasman, 2005).

Contemporary and projected climate change in the south-west

Winter rainfall over the south-west was once considered the most consistent and reliable in Australia in that it exhibited lower inter-annual rainfall variability, relative to its total rainfall, than any other part of the continent (Nicholls et al, 1997). However, the region has experienced a substantial decline in the May to July rainfall since the mid 20th century. Average rainfalls over the region for the periods from 1925 to 1975 and 1976 to 2003 were 323mm and 276mm, respectively. The 170mm and 300mm isohyets have moved 70km to 100km closer to the south-west corner of the state, and the 500mm isohyets up to 200km. The form of the decline is a step rather than a gradual change (IOCI, 2004). The effects of the decline on natural runoff have been severe, as evidenced by the significant reduction in annual inflows to dams in the IWSS (see Figure 13.5). Although rainfall is likely to be the dominant factor reducing runoff, catchment management (e.g. forest and fire management and mining) and the spread of dieback disease could account for some changes as well (McFarlane, 2005). Similar pressures have been imposed upon groundwater resources. The rainfall decline was accompanied by a 20 per cent increase in domestic usage in 20 years (IOCI, 2002). With hindsight, it can be concluded that IWSS water demand has been close to – or at times exceeded – supply capacity over the last 30 years (Power et al, 2005).

For the A1B and A1FI emissions scenarios, the 10th and 90th percentiles of annual average temperatures for Perth are projected to increase from 1990 levels by 0.6°C to 1.2°C by 2030 and 1°C to 3.8°C by 2070. Similarly, the number of days per year with maximum temperatures exceeding 35°C is projected to increase from 28 in 1990 to 33 to 39 by 2030, and 36 to 67 by 2070. Winter (June to August) rainfall is projected to decrease by –14 to –1 per cent by 2030, and by –39 to –2 per cent by 2070 (Plate 27, centre pages).

Adaptive management of the Integrated Water Supply Scheme

The 1980s to 1999

By the mid to late 1980s water planners were concerned about the emerging low dam inflow sequence and 'de-rated' the supply capacity of the IWSS by 13 per cent (Power et al, 2005). In this context a de-rating means a reduction in the estimated long-term mean annual inflow to a water supply system. The construction of the Aus$59 million North Dandalup Dam (with an estimated yield of 25 billion litres per year) was approved and water efficiency measures were promoted to delay further source development. Additional groundwater production capacity was also constructed to service increased water demand due to population growth in the north-west corridor of the city. It was envisaged that no additional sources would be required prior to 2002 (Hughes, 2003).

In 1996, a further 'de-rating' of 54 billion litres was adopted on the basis that estimates of long-term mean annual inflow could no longer be safely based on the entire historical record. At that time it was estimated that the supply capacity was 40 billion litres per year below the expected demand (275 billion litres per year), and water restrictions could be expected in 40 per cent of years rather than the target of 1 in 10. The de-rating resulted in the corporation investing US$430 million in a programme of accelerated source development to provide for the loss in supply capacity and to meet demand growth. This programme included continued development of groundwater resources north of Perth and the construction of a new dam at Harvey to provide an alternative supply to irrigators and to facilitate the connection of Stirling Dam to the IWSS in 2002 (Hughes, 2003).

2000 to 2007

The winter of 2001 saw the worst inflow to the metropolitan dams since 1914 (40 billion versus 21 billion litres) and the 2001 to 2002 winters witnessed the worst two-year inflow sequence on record. From the 1998–1999 to 2001–2002 summers, some 28 to 40 superficial aquifer production bores were taken off line to reduce abstraction from environmentally sensitive areas (Hughes, 2003). A drought response plan was instigated which:

- reduced water demand by restricting the use of lawn and garden sprinklers to two days per week from 8 September 2001; this led to a reduction in residential consumption from 333 litres per person per day to 282 litres per person per day in 2005 to 2006 (15 per cent) – over the same period, non-residential water use from the IWSS fell by 18 per cent per capita;
- temporarily increased groundwater production mainly from the confined aquifers beneath the Gnangara Mound; and
- further augmented the supply capacity of the IWSS through the construction of 12 new groundwater bores and two new small dams at a cost of Aus$142 million.

Despite the additional investment in new supply capacity, the water situation was not secure. Community interest and concern led to a greater political involvement in water issues and resulted in the development of a *State Water Strategy* in 2003 (GWA, 2003). A wide range of programmes was initiated in response to the strategy. A key activity within the Water Corporation was to continue work on developing contingency plans for extra source capacity.

Prior to the mid 2000s, the IWSS had a sophisticated drought management capability with interconnected sources ranging from run-of-the-river large surface reservoirs and unconfined and confined aquifers (ATSE, 2005). Regular assessments of the supply situation highlighted increased concern over a further decline in inflows since 1996 (see Plate 28, centre pages). In July 2004, the government announced its decision to proceed with the Perth Seawater Desalination Plant as the next major source for the IWSS. Planning for the long-term supply needs of the IWSS was again reviewed. The plan established the imperative for immediate action by outlining the growth drivers and the uncertainty of climate risk and the associated impact upon the IWSS (Water Corporation, 2005).

The even-lower inflow sequence after 1996 adopted by the corporation to assess supply capacity was combined with a demand scenario for the IWSS of 155,000 litres per person per year, which assumes a continuing level of community support for ongoing water-use efficiency measures. The plan also aimed to increase the supply reliability by reducing the acceptable frequency of total sprinkler bans from 1 in 33 years to 1 in 200 years in response to guidance set in the *State Water Strategy*. The plan identified a range of planning responses that show a 'security through diversity' approach to meet the future water needs of the IWSS (Water Corporation, 2005; MJA, 2006).

The 2006 winter was worse than that of 2001, with inflows of only 29 billion litres recorded for the major metropolitan dams. This event necessitated another review of current risks and the summary below highlights current initiatives that are key elements of the continuing efforts to establish a safe and reliable water supply for the IWSS:

- *Seawater desalination.* Western Australia is the first Australian state to use a desalination plant as a major public water source: the Perth Seawater Desalination Plant (see Figure 13.4), located at Kwinana on the shores of Cockburn Sound. It was commissioned in November 2006, uses the reverse-osmosis technique to produce 45 billion litres per year and is Perth's largest water source. Water from the plant is mixed with groundwater and surface water in the IWSS. Hyper-saline return water from the plant (180 million litres per day) is discharged back into the sound. The annual power requirement for the plant is 24MW. This requirement is offset by the (Aus$180 million) Emu Downs Wind Farm located some 260 km north of Perth. The facility consists of 48 wind turbines (13 more than the number required to run the desalination plant) that are connected to the state's electricity grid. In March 2007, the government announced its decision to proceed with a second desalination plant south of Perth. The Southern Seawater Desalination Plant is to be built at Binningup, 154km south of Perth by late 2011. It will be very similar in operation to the first plant at Kwinana, but is being planned to allow expansion from an initial capacity of 50 billion to 100 billion litres per year. Renewable energy contracts are planned to provide power for the new plant.
- *Recycling of treated wastewater.* The proportion of wastewater reused in Perth (3.5 per cent) is very low when compared with other capital cities. The *State Water*

Strategy (GWA, 2003) set a target of 20 per cent reuse of treated wastewater by 2012. The strategy encourages 'fit for purpose' water consumption. Because of environmental, economic and public health considerations, it commits the state to large-scale scheme-based reuse options rather than reuse at the household level. Thus, water-consuming industries in Kwinana will be supplied with treated wastewater from the Woodman Point Wastewater Treatment Plant, either directly or after further treatment. The 6 billion litres per year Kwinana Water Reclamation Plant, using the reverse osmosis technique, already supplies key industrial customers with high-quality product water. Opportunities to expand the capacity of this recycling scheme are being pursued.

- *Managed aquifer replenishment.* A 1.5 billion litres per year trial using reverse-osmosis treated wastewater injected into a confined aquifer beneath the plant is being investigated at the Beenyup Wastewater Treatment Plant north of Perth. The trial seeks to determine the suitability of a larger-scale proposal for indirect potable reuse of highly treated wastewater for the IWSS.
- *Thinning of selected trees in crowded native re-growth forests and forests that have been rehabilitated after bauxite mining.* Initially, thinning operations will be limited to the Wungong catchment for a period of 12 years. The expected increase in water yield is 4 billion to 6 billion litres per year, which is 25 per cent of average inflow to the Wungong Dam. If the trial proves successful, with community and regulator support for an economic operation, forest thinnings will be undertaken in other metropolitan catchments.
- *Water trading with irrigation co-operatives.* Negotiations between the corporation and Harvey Water are proceeding for the purchase of 17 billion litres per year by the end of 2007. A further 17 billion litres per year could be obtainable from the Collie Irrigation Area.

At the start of 2007, per capita expenditure on water supply infrastructure for Perth over the previous five years was, for example, at least twice that in Sydney, Melbourne, Brisbane and Adelaide. Prior to the imposition of water restrictions in 2001, the corporation had a full-cost recovery rate of 85 per cent: this fell to 74 per cent in 2004/2005 due to increasing costs and lower water sales. However, the Water Corporation is in a strong financial position and could effectively borrow up to Aus$5.3 billion before reaching the international benchmark of 60 per cent debt to total assets (MJA, 2006).

2008 and beyond

Water consumption is about 45 billion litres per year below previous demand estimates due largely to the introduction of sprinkler-use restrictions in September 2001. With strong economic growth, population growth is forecast to continue at 1.7 per cent per year, increasing annual water demand on the IWSS by 90 billion litres over the next 20 years.

Water demand is expected to double in the Perth region in less than 50 years and there is little potential to augment the number of reservoirs as most feasible surface water sources are already dammed (ATSE, 2005).

Partnerships between the climate and water communities

While the sequence of de-rating decisions made by the Water Corporation and its predecessors was largely driven by the observed dam inflow sequence, their timing and magnitude were also informed and underpinned by advances in climate science. The impact of these advances on de-rating decisions during the period from the late 1980s to 2002 is discussed in detail elsewhere (Power et al, 2005). Only a brief outline for this period will be given here, with additional material for 2003 to 2007.

During the late 1980s, climate change scenarios included a 20 per cent rainfall decline by 2040 for the south-west which would result in a 40 per cent reduction in dam inflows. In recognition of the uncertainties involved, it was considered appropriate to adopt an adaptive response involving a gradual de-rating of the expected supply from the IWSS. It was also recognized that the imposed de-rating of 13 per cent would not have been as large had it not been for the climate change scenarios available at that time (Sadler et al, 1988; Power et al, 2005).

With the continuation of the low inflow sequence into the mid 1990s, a national Climate Variability and Water Resources Workshop was held in Perth in 1996. Participants included invited water managers, representatives of state government agencies, and climate and water scientists from Australia and overseas. The major outcomes of the workshop were that (Ruprecht et al, 1996; Hughes, 2003; Power et al, 2005):

- There had been a marked decrease in rainfall over much of the south-west, and that a sustained decline would have serious implications for the reliability of water supplies.
- Little else was known about the fundamental nature of climate variability within the south-west region.
- Although the observed rainfall decline could be regarded as a manifestation of natural climate variability, the role of global warming in the decline needed to be considered.
- The Water Corporation's approach to urban water planning was sound under the prevailing circumstances.
- A comprehensive adaptive response to climate change would require a sustained and integrated programme of research.

The 1996 workshop led to the formation of the Indian Ocean Climate Initiative (IOCI) in January 1998, a partnership between several state agencies, the Commonwealth Scientific and Industrial Research Organization and the Australian

Bureau of Meteorology, which ran from January 1998 to July 2006. The IOCI was led by a panel comprised of representatives from the state and research agencies. The role of the panel was to set the strategic programme of research; facilitate effective communication between decision-makers and scientists; and provide clarification of science-related issues (e.g. what can and cannot be delivered; relevance of proposed research to resource management needs) (Power et al, 2005).

The major research findings from the IOCI include the following:

- Most of the reduction in rainfall has occurred in the first half of the winter half-year (May to July, June and July being the wettest months). There is also an absence of very wet years, which were relatively common prior to the mid 1970s.
- There has been a 20 per cent reduction in the strength of the subtropical jet over Australia and an associated reduction in the likelihood of synoptic disturbances developing over the region since the early 1970s.
- There has also been a concurrent and ongoing increase in the frequency of dry weather patterns, and this frequency will increase with increasing atmospheric concentrations of greenhouse gases.
- Climate simulations indicate that at least some of the observed drying is due to the enhanced greenhouse effect.
- The 20th-century warming in the south-west is largely the result of the enhanced greenhouse effect.
- Even with the most optimistic greenhouse gas emission scenarios, the region is projected to be drier and warmer later this century.

Overall, the research outputs from the initiative have provided strong and acknowledged guidance for the *State Water Strategy* (GWA, 2003) and the *Western Australian Greenhouse Strategy* (WAGTF, 2004). While accusations of poor management are levelled by some sections of the community, water managers have been able to reassure the majority that increased investment in source development and water restrictions are necessary because the rainfall and dam inflow declines have been unusually large, abrupt and sustained. This has encouraged water managers to become more familiar with climate issues, and to seek explanations and clarification of the sources and levels of uncertainty (Power et al, 2005). There has also been very high public demand for IOCI products and information because of growing concern about global as well as regional climate change.

Conclusions

Urban water planning involves consideration of a portfolio of risks. In this chapter we have outlined the impact that recent climatic changes, growing concern about projected climate change, water reform, and community attitudes towards water restrictions and reuse are having upon urban water planning in Perth, Australia.

Australia's urban water industry is responding to these stressors by changing its operating environment, developing or at least considering additional and alternative sources of water (e.g. water reuse and desalination) and being sensitive to the views and issues of concern to its customers. With the exceptions of the cities of Perth and (to a lesser extent) Melbourne, the underpinning of decision-making with information obtained from the latest developments in climate science is still in its infancy.

The paradigm of an adaptive response that involves a gradual de-rating of the expected water supply has enabled the Water Corporation to:

- rapidly develop a source development programme in response to observed and projected climatic risk and population growth;
- review and amend the programme in response to short-term, very low-inflow events, and to climatic information that is complex, evolving and uncertain; and
- sensitize its customers to water supply issues, particularly the ongoing need for water conservation.

While the adaptive response of water planners in Perth to a multi-decadal hydrological drought may set a pragmatic precedent for water planners elsewhere, the nature of their adaptive response will be shaped by the physical, hydrological, socio-economic and political settings that they confront and the financial resources available. Perth has also benefitted from a coastal location (making seawater desalination a feasible option), ready access to shallow as well as deep groundwater supplies, and an extensive array of dams and pipelines facilitating inter-basin transfer of water supplies.

References

ACIL Tasman (2005) *Research into Access to Recycled Water and Impediments to Recycled Water Investment*, ACIL Tasman Pty Ltd, Melbourne, Australia

ATSE (Australian Academy of Technological Sciences and Engineering) (2004) *Water Recycling in Australia*, ATSE, Parkville, Australia

ATSE (2005) *Western Australia: Water Policy Issues in Climate Uncertainty*, ATSE, Parkville, Australia

Australian Bureau of Meteorology (2007) *Summary Statistics*, Perth Metro, 26 October 2007, www.bom.gov.au/climate/averages/tables/cw_009225.shtml

GWA (Government of Western Australia) (2003) *Securing our Water Future: A State Water Strategy for Western Australia*, GWA, Perth, Australia

Hughes, G. J. (2003) 'Meeting the challenge of climate variability in a major water supply system', *Water Science and Technology: Water Supply*, vol 3, no 3, pp201–207

IOCI (Indian Ocean Climate Initiative) (2002) *Climate Variability and Change in South-West Western Australia*, IOCI, Perth

IOCI (2004) *How Our Rainfall Has Changed – The South West*, Climate Note 5/05, Indian Ocean Climate Initiative, Perth, www.ioci.org.au/publications/pdf/IOCIclimatenotes_5.pdf

Kaspura, A. (2006) *Water and Australian Cities: Review of Urban Water Reform*, Institution of Engineers Australia, Canberra, Australia

McFarlane, D. (2005) *Context Report on South West Water Resources*, Prepared for expert panel examining Kimberley Water Supply Options, Client report to W. A. Government,

CSIRO, Water for a Healthy Country National Research Flagship, Canberra, Australia

Mitchell, V. G. (2006) 'Applying integrated urban water management concepts: A review of Australian experience', *Environmental Management*, vol 37, no 5, pp589–605

MJA (Marsden Jacob Associates) (2006) *Securing Australia's Urban Water Supplies: Opportunities and Impediments*, MJA, Camberwell, Australia

Nancarrow, B. E., J. D. Kaercher and M. Po (2002) *Community Attitudes to Water Restrictions Policies and Alternative Sources: A Longitudinal Analysis, 1988–2002*, CSIRO Land and Water Consultancy Report, CSIRO, Perth, Australia.

Nancarrow, B. E., J. D. Kaercher, M. Po and G. J. Syme (2003) *Social Values and Impact Study South West Yarragadee Blackwood Groundwater Area*, Australian Research Centre for Water in Society, CSIRO, Perth, Australia

Nicholls, N., W. Drosdowsky and B. Lavery (1997) 'Australian rainfall variability and change', *Weather*, vol 52, pp66–72

Po, M., B. E. Nancarrow, Z. Leviston, N. B. Porter, G. J. Syme and J. D. Kaercher (2005) *Predicting Community Behaviour in Relation to Wastewater Reuse: What Drives Decisions to Accept or Reject?*, Water for a Healthy Country Flagship, CSIRO Land and Water, Perth, Australia

Power, S., B. Sadler and N. Nicholls (2005) 'The influence of climate science on water management in Western Australia', *Bulletin of the American Meteorological Society*, vol 86, no 6, pp839–844

Preston, B. L. and R. N. Jones (2006) *Climate Change Impacts on Australia and the Benefits of Early Action to Reduce Global Greenhouse Gas Emissions*, Consultancy report for the Australian Business Roundtable on Climate Change, CSIRO Marine and Atmospheric Research, Melbourne, Australia

Roseth, N. (2006) *Community Views on Water Shortages and Conservation*, Research Report no 28, Cooperative Research Centre for Water Quality and Treatment, Salisbury

Ruprecht, J. K., B. C. Bates and R. A. Stokes (eds) (1996) *Climate Variability and Water Resources Workshop*, Water Resources Technical Report Series WRT5, Water and Rivers Commission, Perth, Australia

Sadler, B. S., G. W. Mauger and R. A. Stokes (1988) 'The water resource implications of a drying climate in south-west Western Australia', in G. I. Pearman (ed) *Greenhouse: Planning for Climate Change*, CSIRO, Australia, pp296–311

Stokes, R. A., J. A. Beckwith, I. R. Pound, R. R. Stone, P. C. Coghlan and R. Ng (1995) *Perth's Water Future*, Water Authority Publication no WP214, Water Authority of Western Australia, Perth, Australia

WAGTF (Western Australian Greenhouse Taskforce) (2004) *Western Australian Greenhouse Strategy*, WAGTF, Government of Western Australia, Perth, Australia

Water Corporation (2005) *Integrated Water Supply Scheme Source Development Plan 2005–2050: An Overview*, Water Corporation of Western Australia, Perth, Australia

WSAA (Water Services Association of Australia) (2005) *Testing the Water: Urban Water in Our Growing Cities – the Risks, Challenges, Innovation and Planning*, WSAA Position Paper no 1, WSAA, Melbourne, Australia

14

Benefits and Costs of Measures for Coping with Water and Climate Change: Berg River Basin, South Africa

John M. Callaway, Daniël B. Louw and Molly Hellmuth

Introduction

The Berg River Basin is located in the Western Cape Region of South Africa (see Figure 14.1). The upper Berg River Basin is an economically important water supply system in the Western Cape that provides the bulk of the water for household, commercial and industrial use in the Cape Town metropolitan region. It also provides irrigation water to the lower part of the basin to cultivate roughly 15,000ha of high-value crops, primarily deciduous fruits, table and wine grapes, and vegetables both for domestic and export use with strong multiplier effects in the domestic and national economy. Since the early 1970s, water consumption in municipal Cape Town has grown by around 300 per cent, fuelled largely by in-migration. As the population of the Metropolitan Cape Town region grows, the competition for water in the basin has become even more intense and farmers have responded by dramatically improving their irrigation efficiencies and shifting even more land into the production of high-value export crops.

The region has also recently experienced a number of unusually dry years, the most recent during the summer of 1994 to 1995, when peak storage in the upper basin was only about one third of average. At the same time, concerns about the effects of global warming on basin runoff have been growing, along with suggestions that recent climatic anomalies may be associated with regional climate change.

The Berg River Dam, with its 130 million cubic metre storage capacity, is expected to be operational sometime during the period of 2008 to 2010. It is, however, unclear to what extent reservoir operation is consistent with expected climate change.

Based on a combined water–climate–economy policy-planning model, a set of scenarios is analysed, resulting in alternative uses of the Berg River Dam. This case can serve as a typical example of how such a policy-planning tool could be used in other instances where water allocation issues should be assessed in the context of climate change.

Figure 14.1 *Map of South Africa*

Note: The Berg River is located just north of Cape Town in Western Cape province. It is approximately 294km long.

Source: based on a United Nations map

Case study objectives

In a recent interview, a leading climatologist in the region, Bruce Hewitson (2004), warned that the government should take a long-term view of changing climate conditions or face potential consequences that could 'seriously compound' the existing challenges facing South Africa. According to Hewitson (2004): 'We are still building society around what is considered to be normal climate in, for example, water usage and infrastructure. But we increasingly need to take the changing characteristics of climate into consideration.'

Triggered by a number of unusually dry years, the most recent in the summer of 1994 to 1995, when peak storage in the upper basin was only about one third of average, a number of national and regional commissions have been set up to investigate options for coping with the long-term water supply problems in the basin.

One outcome of these efforts was the authorization of the Berg River (originally Skuifraam) Dam. In June 2004, after almost 20 years of debate about its economic feasibility and environmental impacts, a final agreement was reached on the construction of the dam. It will consist of a 130.1 million cubic metre storage reservoir and a pumping site to pump water from below the dam back to it. The dam is expected to be operational some time during the period of 2008 to 2010.

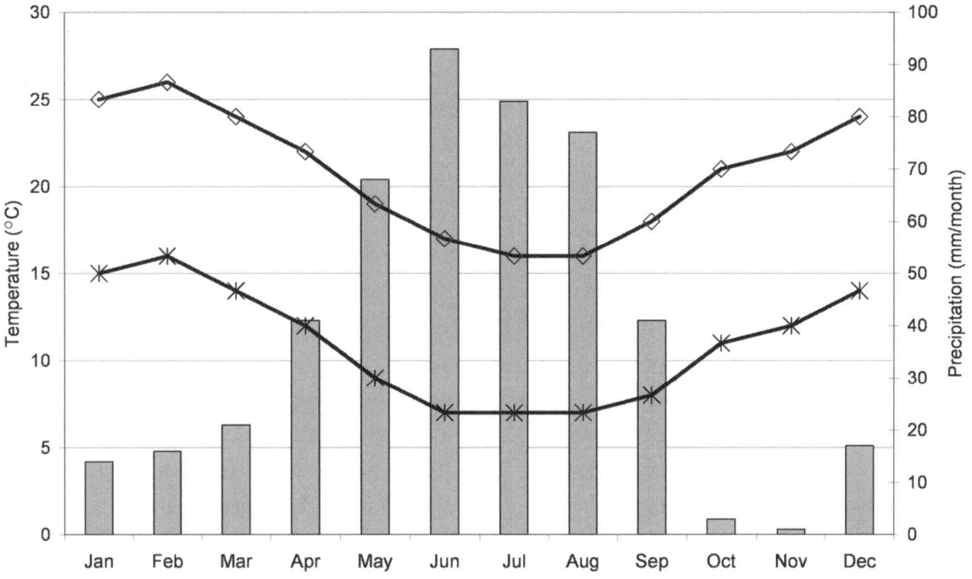

Figure 14.2 *Mean minimum and maximum temperature and precipitation for Cape Town, South Africa*

Source: www.wunderground.com

However, planning for the Berg River Dam and other water supply and demand options in the basin has, up until this point, failed to take into account the possibility that the build-up of greenhouse gases in the global atmosphere is already affecting, and will continue to affect, the regional climate by reducing runoff in the basin.

The context for this case study consists of three main elements in the Berg River Basin. The first is the increasing competition for water between urban and agricultural water users due to growing urban water demands; the second is the threat of unusual climate variability and/or climate change to exacerbate that competition; and the third is the planning and policy responses to these issues. Putting the three together, we came up with the need to develop a policy-planning model that can be used to evaluate a wide range of structural, non-structural and technological measures for coping with basin water shortages. Ideally, such model should be able to:

- characterize the effects of climate variability and climate change on spatially distributed monthly runoff, evaporation and crop water use;
- characterize important spatially differentiated features of basin hydrology, including runoff, storage and on-farm reservoirs, points of water diversion and use, and the flow connections between them;
- simulate all aspects of storage and on-farm reservoir operations;

- simulate the economic decision-making objectives of water planners, water managers and different types of water users, balanced by the ability to assess the opportunity cost of equity-based policies;
- assess the physical effects of alternative options over time and space, such as additional reservoir storage, alternative water allocation systems, reductions in conveyance losses, in-stream flow requirements and a variety of demand-side options; and
- estimate both the physical impacts and the economic benefits and costs of these options, including the benefits and costs of avoiding economic losses due to unusual climate variability and climate change.

To do this, we have developed a policy-planning tool called the Berg River Dynamic Spatial Equilibrium Model (BRDSEM). The objectives of this case study are, first, to describe the structure of this model; second, to illustrate how the model can be used to compare the net benefits of avoiding climate change damages by increasing maximum storage capacity in the Berg River Dam and/or implementing a system of efficient water markets; third, to present the results and major conclusions of this analysis for three deterministic climate scenarios; and, fourth, to describe the limitations of the current version of the model and analysis methods and to outline future plans for improving the model and analytical methods.

In summary the overall objective of this case study is to provide a detailed example of quantifying the economic effects of avoiding climate change damages from an economic research perspective.

The model

BRDSEM is a dynamic multi-regional, non-linear programming (DNLP) model patterned after the 'hydro-economic' surface water allocation models developed by Vaux and Howitt (1984) for California; by Booker (1990) and Booker and Young (1991, 1994) for the Colorado River Basin; and by Hurd et al (1999, 2004) for the Missouri, Delaware and Apalachicola–Flint–Chattahoochee River Basins in the US. This type of model is a more specific application of spatial and temporal price and allocation models due originally to Samuelson (1951) and Takayma and Judge (1971) that have been widely applied in many natural resource sectors (McCarl and Spreen, 1980). Hydro-economic models have been used by Hurd et al (2004) to estimate the economic value of the climate change damages in the four large US river basins; however, no effort was made to assess the benefits of various measures in avoiding the effects of climate change or their effectiveness in reducing damage. BRDSEM was designed specifically to do this (among other things), and this chapter represents the first attempt that we know of to quantify the benefits of avoiding climate change damages in economic terms in a water basin context using such a model.

BRDSEM is an extension of a static spatial equilibrium model developed by Louw (2001, 2002) for the Berg River Basin to examine the potential of water markets in the region (Louw and Van Schalkwyk, 2001). Much of the data from that model is used in BRDSEM, but is being updated on a continuous basis. Significant modifications were made to the original model to add the spatial relationships between runoff, water storage, water conveyance, transfers, return flows and water use in the natural and man-made hydrologic system. The original static model was also modified to account for the inter-temporal aspects of reservoir operation both in the upper part of the basin and for the farm reservoirs in the lower part of the basin. In addition, the model was extended to run on a monthly basis and to solve simultaneously for all of the endogenous variables over a 30-year time horizon (or longer, as needed). Finally, the regional farm models developed by Louw for the original model were recast in a dynamic framework and the necessary hydrologic connections were added to 'mate' these models to the hydrologic structure of the spatial equilibrium model.

One of the important features added for BRDSEM is that it can determine, endogenously, the 'optimal' (i.e. economically efficient) capacity of planned reservoirs and other structural works, and capacity can be fixed exogenously. The model does this by finding the capacity level that is equal to the sum of the discounted current values of storage in all periods where future storage levels are at this maximum, based on the relevant Kuhn-Tucker conditions. The maximum capacity is determined in year one and remains fixed thereafter.

Model overview

Figure 14.3 is a schematic diagram of BRDSEM and the models that feed information to it. The core of BRDSEM, shown in the box labelled 'Dynamic programming model' consists of three linked modules. The three 'modules' are interconnected in the framework of a dynamic non-linear programming model, which was constructed using the General Algebraic Modelling System (GAMS). Each module can be developed and modified separately, with only minor adjustments to other modules and elements in the non-linear programming model. These modules are:

1. *An inter-temporal spatial equilibrium module.* This module consists of a series of linear equations that characterize the water balance over time in specific reservoirs, and the spatial flow of water in the basin, linking runoff, reservoir inflows, inter-reservoir transfers and reservoir releases to urban and irrigated agricultural demands for water.
2. *An urban demand module.* This module simulates the demand for urban water for six urban water uses (lower-income households, higher-income households, garden and lawn water use, industrial consumers, commercial water users and public-sector water use).
3. *A regional farm module.* This module consists of seven regional dynamic linear farm models (one for each farm region) that simulate the demand for agricultural water in the upper section of the Berg River.

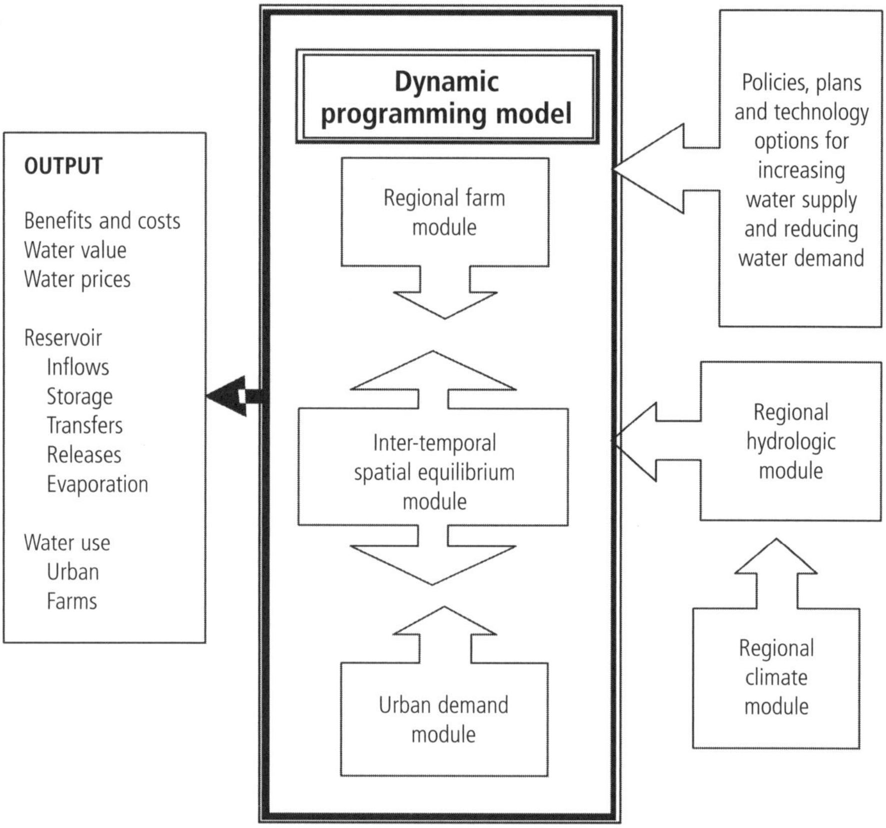

Figure 14.3 *Berg River Spatial Equilibrium Model (BRDSEM) schematic diagram*

Source: John M. Callaway

These three modules are linked, dynamically and spatially, at different points of use in the basin and are solved together as a dynamic non-linear programming model, using the inputs from the hydrology module. The model then simulates investment in reservoir capacity, all aspects of monthly reservoir operation, and water allocation to urban and irrigated agricultural demands based on the objective of economic efficiency through regulation or through a mixture of the two. The output of the model consists of:

- measures of the economic value of water for water users, broken down by urban sector and farm regions;
- various shadow prices[1] for water transfers between reservoirs, water transfers from the upper to the lower section of the basin, for urban and agricultural uses, and for in-stream flows from which water 'prices' can be constructed;
- monthly reservoir storage releases and transfers, and reservoir evaporation for main storage and farm reservoirs;

- monthly water diversions and consumptive use by the urban sector, farm regions and irrigated crops in each region;
- crop production and area as well as other resources used by farm regions; and
- monthly return flows by farm regions, low flows by farm regions, and various system and conveyance losses.

Figure 14.3 also shows three external sources of information inputs to BRDSEM:

1. *A regional climate model.* This model supplies the hydrologic model with information about monthly temperature and precipitation at specific points in the basin for climate variability/change scenarios.
2. *A regional hydrologic model.* This model – WATBAL (Yates, 1996) – converts the monthly temperature and precipitation data from the regional climate model into monthly runoff at different runoff gauges for each climate scenario and all estimates: reservoir evaporation coefficients for each storage and farm dam, and crop water use adjustment factors based on variations in potential evapotranspiration.
3. *Inputs about policies, plans and technologies*: This represents the source of information that can be used to alter various parameters in the programming model in order to reflect alternative demand- and supply-side policies, plans and technologies.

Spatial equilibrium structure

BRDSEM is a spatially differentiated trade model in which runoff nodes, reservoirs and points of water use or diversion are physically (and algebraically) connected by flows of water in a way that is physically faithful to the natural and man-made hydrologic system. A schematic overview of the inter-temporal spatial equilibrium module, showing the physical connections between runoff points, major storage reservoirs and water users in BRDSEM, is presented in Figures 14.4 and 14.5 (upper and lower sections). A more detailed hydrologic representation of both the upper and lower sections of the Berg River Basin can be found in Hellmuth and Sparks (2005). There are six 'sites' in the upper basin of BRDSM. Three of these sites constitute the major dams in the model, each associated with a storage reservoir: Theewaterskloof (TWAT), Wemmershoek (WMRS) and the Berg Dam (BERG). The final site is the Berg Supplemental Site (BERGSUP), which is a pumping station below the Berg River Dam that collects runoff from below the dam and pumps it back to the Berg Reservoir.

Figure 14.5 depicts the arrangement of water deliveries from the lower part of the basin, the runoff sources in the lower part and the allocation to, and use of, these supplies by the seven regional farms. The seven regional farms are located sequentially downstream of one another – Berg1, Suid-Agter Paarl (SAP), Berg2, Noord-Agter Paarl (NAP), Berg3, Perdeberg (PB) and Riebeek-Kasteel (RK). The available supply for each downstream farm (after Berg1) is equal to the sum of:

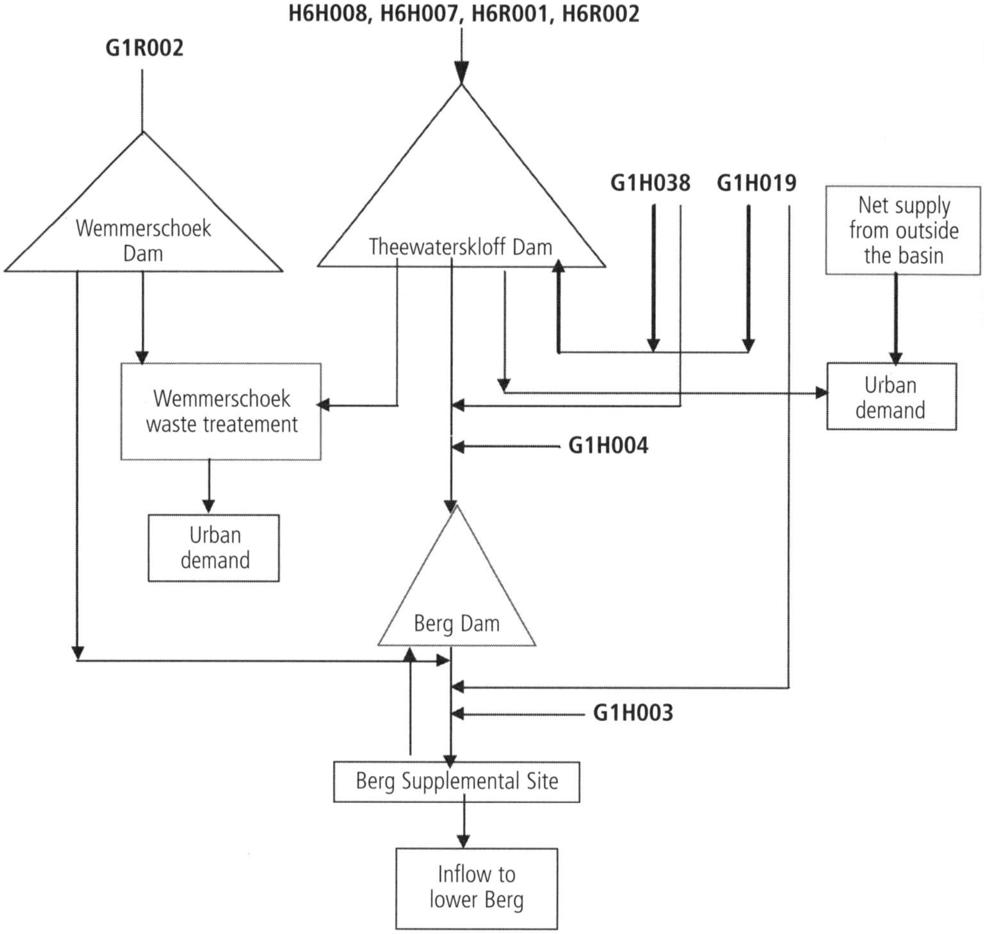

Figure 14.4 *Schematic diagram of the Berg River Basin: Upper section*

Source: John M. Callaway

- the undiverted portion of the supply available to the previous user (or in-stream flow) as designated by F1 to F7 in Figure 14.5;
- the return flows from the previous user; and
- runoff from sources between the two users.

The on-farm use of water is depicted in Figure 14.6. Each farm has the following mix of options for using water. It can:

- divert and pump water from the river to irrigate crops; and
- divert and transfer water to a farm reservoir for irrigation use later in the season.

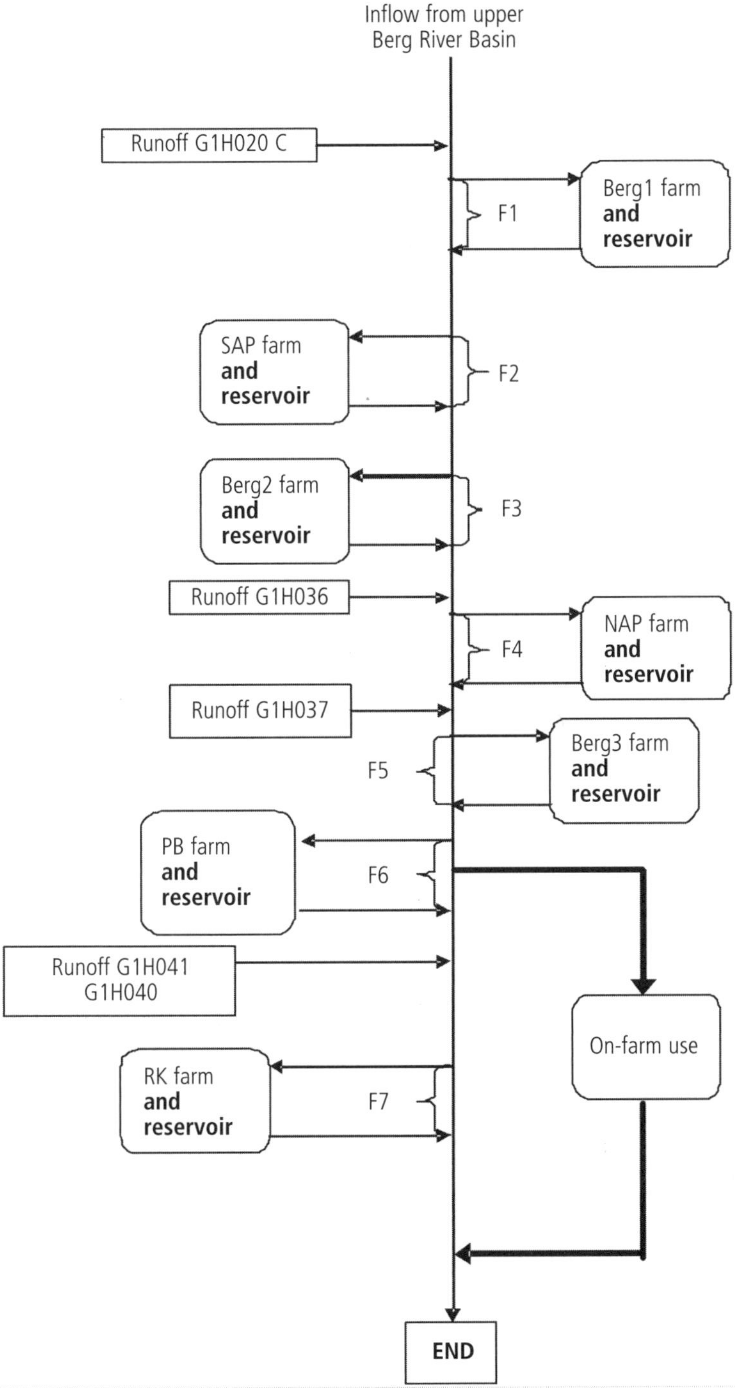

Figure 14.5 *Schematic diagram of Berg River Basin: Lower section as depicted in BRDSEM*

Source: John M. Callaway

Part of the water used to irrigate crops, whether it comes directly from diversions or farm dam storage, is used consumptively by crops (as determined endogenously by BRDSEM) and the remainder returns to the river as return flow.

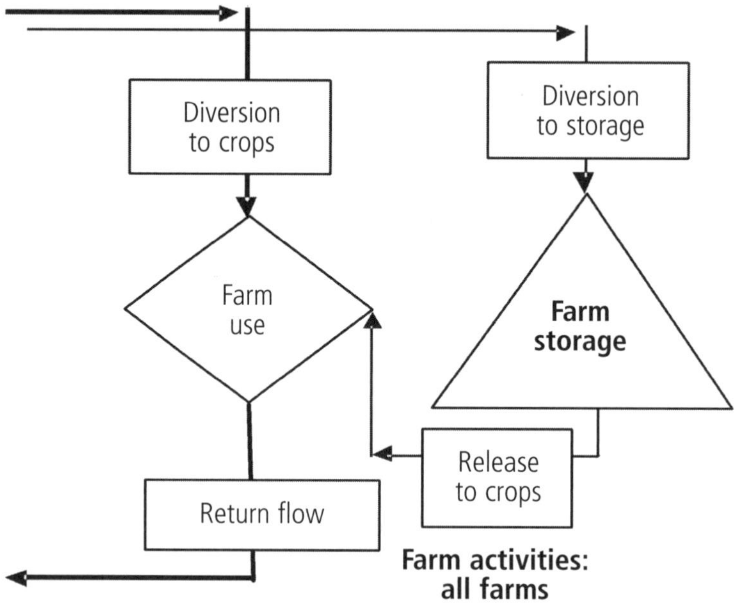

Figure 14.6 *On-farm use of water as represented in the model*

Source: John M. Callaway

As previously discussed, BRDSEM consists of three modules, linked together in a mathematical programming framework. For presentation purposes, the structure of the programming model must be broken down a little differently into four linked components as follows:

1 a non-linear (quadratic) objective function that characterizes the normative objectives of the agents in the model;
2 an inter-temporal spatial equilibrium module/matrix that characterizes the spatially distributed flow of water and water storage in the basin;
3 an urban water demand module/model that is linked directly to the objective function and the inter-temporal spatial equilibrium matrix; and
4 a regional farm/irrigation module/demand model that is linked directly to the objective function and the inter-temporal spatial equilibrium matrix.

These four linked components will be discussed in the following sub-sections.

The objective function

The objective function of the model is to maximize the net present value of the net returns to water in the basin over 30 twelve-month periods. In this form, the objective function serves two purposes. First, it is consistent with welfare maximization by water consumers, farmers and water managers and, thus, simulates the competition for water in efficient markets. The optimization purpose of the objective function can be partially overridden by constraining, among other things, runoff to reservoirs, reservoir transfers, reservoir releases, water diversions, water use and in-stream flows depending upon how 'tight' the constraints are. Second, the objective function is an accounting convention that measures the economic value of water, no matter what constraints are applied.

The net returns to water in BRDSEM are defined as the discounted sum of the following monthly stream of benefits minus costs:

- Benefits:
  - willingness to pay for water by the six urban consuming sectors in Cape Town and the municipalities in the basin; and
  - long-term farm income for the seven regional farms.
- Costs:
  - the costs of operating the reservoirs and delivering water to both municipal consumers and the seven regional farms, as well as pumping and transactions costs;
  - long-term (investment) and short-run (variable) costs for the seven regional farms, including water delivery and on-farm pumping costs; and
  - the capital cost of the Berg River Dam and Berg Supplemental Site (when the capacities are determined endogenously by the model).

Urban water demand

The quadratic willingness-to-pay function for water in the objective function for each of the six urban demand sectors is represented as the integral over monthly, linear Marshallian inverse water-demand functions. Urban water-demand functions do not exist for Cape Town or the municipalities in the region, nor are there any empirical estimates for the price elasticity of demand. Gathering and assembling billing data and estimating new demand functions for BRDSEM is planned for the future.

Inter-temporal spatial equilibrium module (matrix)

This module is actually a matrix of linear equations and constraints in the mathematical programming model that characterizes the water balances in the basin reservoirs – both storage and regional farm dams – and the spatially distributed physical linkages between runoff, water storage and points of water use. This matrix is depicted schematically in Figures 14.4 and 14.5. The model consists of 14 distinguished calculation blocks with a total of over 25,000 equations.

The regional farm module

The regional farm module is based on the linear programming regional farm formulation in Louw (2001, 2002). There are seven dynamic farm linear programming models in the agricultural sector of BRDSEM, one for each of the seven farm regions as previously identified. Each regional farm model is linked to the spatial equilibrium model through the objective function, the on-farm consumptive use balances, and dynamic balances for the on-farm reservoirs.

Each regional farm model contains production possibilities for seven dryland and nine irrigated crops. Crops are further broken down on a short-term (annual) and long-term (perennial) basis. Each crop has a crop budget associated with it that specifies crop yield per unit area, input requirements per unit area, variable input costs per area unit and crop price. Long-term crops (perennial crops) include the same information by growth stage from the establishment of the crop to re-establishment once the trees and vines have reached the maximum age at which they can be cultivated, plus investment costs for newly established trees and vines and carrying costs for the initial inventory already established. There are three irrigation technologies in the model (regular, supplemental and deficit). Monthly irrigation intensities (consumptive water use) used in the farm models varied by crop, month, irrigation type and growth stage (for perennial crops, except pasture) were taken directly from Louw (2001, 2002). A monthly annual adjustment was made for climate change, using crop factors (Hellmuth and Sparks, 2005) based on the potential evapotranspiration of each crop under higher/lower temperatures for each climate scenario.

Each farm model also includes flexibility constraints required to set the upper and lower bounds from observed crop production areas. These restrictions are also used to provide for risk since it is impossible to capture individual farmers' risk behaviour in such an aggregated model. However, these restrictions – particularly the lower bounds – can also have the effect of preventing reallocation of water from farm to urban areas as water becomes scarcer. Therefore, the lower bounds on both long-term and short-term crop area were reduced in the climate scenario simulations to 10 per cent of the land available for cultivation. Since almost all of the possible land that can be cultivated in the basin is currently under cultivation, there was no need to change the upper bounds on crop area.

Model application: Methods and scenarios

Economic methods

An economic framework for evaluating the costs and benefits of measures to avoid climate change damages is presented in Callaway et al (1998). It was extended to link adjustments (adaptation) to climate variability and climate change and to situations in which 'regrets' occur when the climate that occurs is not the same as the climate that planners and policy-makers anticipated in formulating and implementing their plans and policies (Callaway, 2004a and 2004b). This is sometimes referred to as *ex-ante* and *ex-post* planning and impact situations.

Table 14.1 *Framework for estimating benefits and costs associated with climate change adaptation*

Coping options	**Climate change**	
	No climate change (C_0)	**Climate change (C_1)**
No coping ($K[C_0]$)	Business as usual (W(BU))	Caution error (W(PE))
Coping ($K[C_1]$)	Precaution error (W(CE))	Correctly adapted (W(CA))

Note: W refers to those benefits and costs referred to as welfare.

Table 14.1 presents the basic framework used in this chapter for just two climate states, the existing climate (C_0) and climate change (C_1), and a single long-run measure that avoids climate change damages: K[C]. This could represent investment in reservoir capacity or institutional arrangements for allocating water. The long-run measure, shown in the table, is sensitive to climate change. This is true for many water resource investments, such as investment in reservoir capacity, but is not always the case for institutional measures, such as water allocation policies.

Each of the four cells measures net welfare, represented as W(XX). From a planning perspective, the upper left cell, W(BU), characterizes welfare for the existing climate, with current long-run measures adapted to the existing climate and in place. The cell in the upper right, W(PE), represents the short-run 'partial adjustments' that can be expected if the climate changes, but no new long-run measures are adopted. For example, even though reservoir storage capacity is fixed, operating policies can be changed. The cell in the lower right, W(CA), depicts the long-run welfare consequences that take place when long-run measures in relation to expected climate change are taken.

From a planning perspective, some of the cell entries labelled in Table 14.1 can also be used to characterize the welfare 'regrets' associated with making planning errors of 'caution' and 'precaution'. In this planning context, the cell entry in the upper right, W(PE), also depicts the welfare consequences of planning for climate C_0 when the climate turns out to be C_1. This represents an error of 'caution'. The cell entry in the lower left, W(CE), represents the welfare consequences of planning for climate C_1 by adjusting long-run measures for this expectation when the climate is actually not changing. This represents an error of 'precaution'. So-called 'no regrets' (Smith and Lenhart, 1996) measures do not suffer from errors of caution and their implementation improves welfare relative to the reference case. Whether a long-run measure is a 'no regrets' measure depends upon two factors:

1 whether the measure is climate sensitive or not; and
2 whether one uses a counterfactual 'optimal' base case to measure welfare improvements or the observed reference case.

The implementation of many climate-sensitive development options – such as increasing reservoir storage capacity – can only be considered 'no regrets' when they are measured from an observed suboptimal reference case, while non-climate-sensitive measures, such as efficient water markets, usually improve welfare in relation to both the observed and counterfactually optimal reference case.

Using Table 14.1, one can construct the following series of definitions for various benefits and costs (referred to as welfare) associated with planning (or not) for climate change:

- *Climate change damages*: the welfare losses caused by climate change without coping compared to the reference case: W(PE) – W(BU).
- *Net benefits of adaptation*: the welfare gains associated with reducing climate change damages by optimally adjustment: W(CA) – W(PE).
- *Imposed climate change damages*: the welfare losses (climate change damages) that cannot be avoided by optimally coping: W(CA) – W(BU).
- *The cost of precaution*: the welfare losses that will occur if the reservoir capacity and or allocation policies are adjusted in expectation of climate change but climate does not change: W(CE) – W(BU).
- *The cost of caution*: the welfare losses that occur without coping, but the climate does change. This is equal to the net benefits of adaptation, with the sign reversed: W(PE) – W(CA).

Policy scenarios and options

We used BRDSEM to apply this framework to the following four policy options for (coping with) the basin:

1. *Option A (no water markets, free water policy, no dam).* Lower bounds are placed on summer and winter diversions by the seven regional farms as in Louw (2000, 2001) and on household urban water consumption, consistent with the government's current 'free water' policy. The capacity of the Berg River Dam storage reservoir was set at zero.
2. *Option B (water markets, no free water policy, no dam).* We removed the allocation and free water constraints in option A, but not the zero capacity constraint on the Berg River Dam in order to simulate the economically efficient allocation of water to both urban and agricultural users without the Berg River Dam.
3. *Option C (no water markets, free water policy plus dam).* Using the same allocation and free water constraints as in option A, we allowed BRDSEM to find the economically efficient storage capacity of the Berg River Dam.
4. *Option D (water markets, no free water policy plus dam).* We removed the allocation and free water constraints in step D1 and the reservoir constraint in step D2 to estimate the partial welfare contributions of water markets, with no Berg Dam in D1, and optimal storage capacity to the Berg River storage reservoir in addition to water markets in D2.

Climate scenarios

We simulated the welfare consequences of the different policy options under both full (optimal) and partial adjustment for three deterministic, transient climate scenarios. A detailed explanation of the climate scenarios and how they were used to develop inputs for the economic model is discussed in detail in Hellmuth and Sparks (2005). For this assessment, we used information provided by WATBAL for the following climate-hydrology scenarios:

- CSIRO SRES B2 REF Case (REF): 1961–1990 (applied to 2010–2039);
- CSIRO SRES B2 Near Future Case (NF): 2010–2039;
- CSIRO SRES B2 Distant Future Case (DF): 2070–2099 (applied to 2010–2039).

The selection of climate scenarios was limited entirely by the availability of downscaled climate scenarios available for the region. BRDSEM was designed as a planning tool to be used with stochastic downscaled climate scenarios, but none exist. Each of these deterministic scenarios is time dependant (or transient), as indicated above, applying to specific years. The inconsistency in the temporal applicability of the climate scenarios made it necessary to apply them to a common period (2010 to 2039), thus avoiding the need to take into account long-term structural changes in the region's 'water economy' over the different time periods. As a result, we decided to retain the transient character of the scenarios in that they depict the hydrologic effects of climate change over time; however, we simulated all of the scenarios for the same time period, 2010 – 2039. Thus, REF is a counterfactual reference case, assuming the same underlying runoff as in the period 2010 to 2039 as 1961 to 1990, while DF (distant future), instead of being a longer-term continuation of CSIRO B2, can be viewed as a more adverse climate scenario, producing lower runoff and higher evaporation, compared to NF (near future) for the same time period.

The water balance model was used to convert the downscaled climate model into the following climate-sensitive information that was then passed to BRDSEM in the form of exogenous parameters:

- monthly runoff for 30 years at upper- and lower-section basin runoff gauges;
- monthly reservoir evaporation coefficients for 30 years for the three major storage reservoirs and seven regional farm dams; and
- monthly consumptive water use adjustment factors for 30 years for each of the seven farm regions.

Water-demand growth scenario

Agricultural area in the basin has been relatively stable for the last half decade and is not expected to grow much more due to limited land availability (Louw, 2001 and 2002). However, urban water demand in Cape Town has been growing rapidly. An annual water demand growth rate in base level water consumption was considered assuming 200 per cent increases over 30 years.

Other assumptions

Economic values in this case study are calculated in accordance with constant South African rand in the year 2000 (Louw, 2001 and 2002). This assumes that all input and output prices in the model are inflating at the same constant rate. A constant real discount rate of 6 per cent was used to convert future value flows into constant present values. In sensitivity trials, reducing/increasing the discount rate had predictable effects on water use, increasing/reducing future consumption and, thus, increasing/reducing the endogenously determined maximum optimal storage capacity of the Berg River Dam.

Results and main conclusions

The main results of this study are presented in Tables 14.2 through to 14.6. Table 14.2 shows the net returns to water and the four different policy scenarios for three different climate scenarios (REF, NF and DF). The simulated optimal water storage for the Berg River Dam is also shown for policy scenarios C and D. All of the net returns to water depicted in this table represent the optimum values that can be achieved for each climate and urban-demand scenario. The most important general conclusions that can be drawn from this table are that:

- Climate change will reduce total water availability by 8058m^3 (or –11 per cent) in the near future (NF) case and 16,609m^3 (or –17 per cent) in the distant future (DF) case (see Table 14.2).
- Climate change reduces basin-wide welfare for all four of the policy scenarios, between 6.3 and 8.4 per cent for the NF climate scenario and between 11.5 and 15.6 per cent for the DF climate scenario.

The pattern of the welfare changes in response to climate change under the different policy scenarios is quite complex. More specifically, if we look at the welfare comparisons at the bottom of Table 14.2, we can see that if we hold our assumptions about additional storage capacity constant (dam or no dam) and vary the allocation and pricing policies (comparisons B–A and D–C in the first two rows), we get much smaller welfare changes for each climate scenario than if we hold the allocation and pricing policies constant and vary our assumptions about additional storage capacity in the basin (comparisons D–B and C–A in the last two rows). What this means is that adding storage capacity in the basin to cope with any climate produces larger economic benefits than varying the allocation and pricing policies in the basin to cope with climate. This is not consistent with what we had expected – namely, that switching to efficient markets and marginal cost pricing as a means of coping with climate

Table 14.2 *Welfare results (net returns to water) for four planning options under three alternative climate scenarios*

		Climate scenarios	
	REF	**Near future (NF)**	**Distant future (DF)**
Average annual runoff (million m³)	75.501	67.443	58.892
Coping options		*Present value of net returns to water* (South African rand millions)*	
A Fixed farm allocations and free water policy to households	76,869	70,450	65,193
No Berg Dam		–8.4%	–15.2%
B Efficient water markets and no free water policy	77,437	71,235	66,113
No Berg Dam		–8.0%	–14.6%
C Fixed farm allocations and free water policy to households	79,928	74,886	70,625
Optimal storage for Berg Dam (thousand m³)		–6.3%	–11.6%
	151.0	124.6	164.4
D Efficient water markets, no free water policy	79,994	74,948	70,833
Optimal storage for Berg Dam (thousand m³)		–6.3%	–11.5%
	130.1	115.8	153.9
*Welfare comparisons: Differences in present value of net returns to water**			
Option B–option A	568	785	920
Option D–option C	66	62	208
Option D–option B	3125	4498	5640
Option C–option A	3059	4436	5432

Note: * All monetary estimates are expressed in present values for constant South African rand for the year 2000, discounting over 30 years at a real discount rate of 6 per cent. Exchange rate: 1 rand is approximately US$0.13 (as per April 2008).

change could be about as economically efficient as increasing water storage capacity. However, to properly look at this issue we need to look more carefully at the interaction between the effects of these options to cope with existing climate variability, the development pressure created by growing water demand and climate change (Callaway et al, 2008). Still, we can see the importance of markets and marginal cost pricing as a coping mechanism by noting that the optimal additional water storage capacity required to cope with climate change in policy scenario D is smaller (markets) than in C (no markets), while welfare is actually higher in D than C. This means that markets and marginal cost pricing have effectively substituted for water storage capacity due to the policy changes.

Finally, we can also see from Table 14.2 that the optimal additional water storage capacity required to cope with climate change for policy scenarios C and D is non-linear with respect to climate change. Less storage capacity is needed to cope with the NF than the REF climate scenario, but more storage capacity is required to cope with the DF than the REF climate scenario. This is because the pattern of runoff changes

Table 14.3 *Current value estimates for climate change damages, net benefits of adaptation/cost of caution, imposed climate change damages and cost of precaution for option B compared to option A (South African rand millions)*

Benefit and cost measures	*Climate scenarios*	
	REF to NF	***REF to DF***
Climate change damages	–6419	–11,676
Absolute percentage of reference case welfare (A)	8.3%	15.2%
Net benefits of adaptation	785	920
Absolute percentage of climate change damages	12.2%	7.9%
(Cost of caution – reverse sign)	–785	–920
Imposed climate change damages	–5634	–10,756
Cost of precaution	568	568

Table 14.4 *Current value estimates for climate change damages, net benefits of adaptation/cost of caution, imposed climate change damages and cost of precaution for option C compared to option A (South African rand millions)*

Benefit and cost measures	*Climate scenarios*	
	REF to NF	***REF to DF***
Climate change damages	–6419	–11,676
Absolute percentage of reference case welfare (C)	8.3 %	15.2%
Net benefits of adaptation	4436	5432
Absolute percentage of climate change damages	69.1%	46.5%
(Cost of caution – reverse sign)	–4436	–5432
Imposed climate change damages	–1983	–6244
Cost of precaution	3038	3048

Table 14.5 *Current value estimates for partial climate change damages, net benefits of adaptation and imposed climate change damages and cost of precaution for option D compared to option A (South African rand millions)*

Benefit and cost measures	*Climate scenarios*	
	REF to NF	***REF to DF***
Climate change damages	–6419	–11,676
Absolute percentage of reference case welfare (A)	8.3%	15.2%
Partial net benefits of adaptation:		
D1 Adding efficient water markets	785	920
Percentage of contribution to total	17.5%	16.3%
D2 Adding optimal storage capacity	3713	4720
Percentage of contribution to total	82.5%	83.7%
Total partial net adaptation benefits	4498	5640
Absolute percentage of climate change damages	70.1%	48.3%
Total cost of caution – reverse sign	–4498	–5640
Imposed climate change damages	–921	–6036
Total cost of precaution	3124	3105

in the lower basin vary in such a way relative to the upper basin in the DF scenario that it is less costly to meet water demands in both parts by moving water around via transfers between dams and the two parts of the basin than by building additional storage capacity.

Estimates for climate change damages, the net benefits of adaptation, imposed climate change damages and the costs of caution and precaution are displayed for policy scenario B (Table 14.3), C (Table 14.4) and D (Table 14.5) for the climate changes REF to NF, and REF to DF. For these calculations, we assumed that the no coping strategy was consistent with the REF climate in policy scenario A (see Table 14.2). An alternative is to use the REF climate in the policy scenario that one is examining. The latter is unrealistic (but perhaps theoretically correct) because the true current situation is depicted by the REF climate for policy scenario A. The assumption we use means that the climate change damages will be the same in all three comparisons. Overall, the results in the three tables reinforce the finding in Table 14.2 that adding storage capacity is a better strategy for coping with climate change (at this level of urban water demand) than using water markets and marginal cost pricing to allocate water. This is because the net benefits of adaptation are higher for policy scenario C than B (Tables 14.3 and 14.4) and higher in step D2 than step D1 (Table 14.5). Another important finding in these three tables is that for all of the coping options, the cost of caution is negative and reasonably large, while the cost of precaution is actually positive. Taken together, these two findings mean that all of these coping policies will improve basin-wide welfare, even if the climate does not change (or is not changing as we speak). In that case, they fit one definition for being 'no regrets' policies.

As mentioned earlier, simulating the 'free water' policy of the South African government and then relaxing the free water constraints in the model had only moderate effects on total welfare in the basin. But, as might be expected, the simulated distributional consequences were substantial in terms of changes in revenue from simulated water sales to households. To examine this, we calculated four pieces of information:

1 average annual market value of the free water sold to households in option A;
2 average annual market value of the water sold to households in option A;
3 average annual market value of the water sold to households in option B; and
4 average annual value of simulated actual household water sales in option B less the average annual value of actual sales in option A.

There are two ways to look at the revenue losses that occur as a result of the free water policy:

1 Measure the hypothetical revenue losses in option A.
2 Measure the difference in simulated actual revenue sales between options A and B, which is probably a better measure for assessing the cash flow consequences of the free water policy since it takes in account market price adjustments when the free water policy is relaxed.

These results are presented in Table 14.6 in terms of future values to give a better idea of the impacts to policy-makers in terms of current revenues. The simulated future value of the hypothetical lost revenues from free water sales in option A (row 1) is around 1 billion South African rand under all of the climate scenarios. These hypothetical revenue losses decrease slightly as the simulated climate worsens, as do the simulated actual revenues received in option A (row 2). This is due to simulated increases in the market price of water for both revenue sets and reductions in the quantity of non-free water as the simulated market prices rise in response to climate change when consumption is partially constrained. The actual revenues received in option B (row 3) increase in response to climate change and higher urban demands because in the unconstrained case (option B) the effects of price increases in demand as climate worsens the effects of simulating urban water-demand growth outweigh the effects of simulating urban water-demand growth. Finally, in row 4, we can see that the free water policy causes substantial losses in actual revenues in option A compared to option B, except under the REF climate scenario.

Table 14.6 *Revenue implications of free water policy, comparing the hypothetical revenues from free water sales to households and actual revenues in option A with the simulated actual revenues in option B*

Revenue calculation and option	*Climate scenarios*		
	REF	***NF***	***DF***
	Future value of average annual revenues (South African rand millions)		
Hypothetical revenues: Sale of free water to households in option A	981	980	978
Actual revenues from sale of water to households in option A	791	706	626
Actual revenues from sale of water to households in option B	651	742	816
Actual revenue losses in option A: Option B – option A	–141	36	189

The main limitation of the presented case, as in many other policy model studies, is that the 'work is in progress'. Typically, models like this are never 'final' and undergo numerous revisions as new data becomes available and new questions are asked. However, BRDSEM has reached the point in its development where it can be used to illustrate some of its policy uses, as done in this chapter. The limitations of the current version of BRDSEM and its application in this chapter can be summarized as follows:

- The parameters of the urban water demand functions are assumed to be elastic and are estimates of base-level consumption.
- The model lacks a water works supply function.

- The set of climate-coping options and policies needs to be expanded.
- The characterization of the existing water allocation rules in the basin needs to be improved upon.
- We need to combine coping with climate change, with coping with urban demand growth.
- The analysis was deterministic with regard to climate scenarios.

The first four limitations can be overcome by expanding the model by including more realism and obtaining more empirical data to update our equations. Combining coping with climate change, with coping with development pressure is a topic that we have begun to study in Callaway et al (2008) and work is currently in progress. This work sheds more light on the importance of water markets as a coping mechanism. The final limitation – the deterministic nature of the illustrative analysis in this chapter – requires a bit more discussion. We used deterministic climate change scenarios because downscaled stochastic climate scenarios do not currently exist for the region. When such information becomes available, it will be possible to propagate the runoff, evaporation and crop water-use distributions through BRDSEM by maximizing the expected value of net returns to water for a single or for mixed climate distributions using the methods illustrated in Callaway (2004b). This will also allow us to explore the economic and physical consequences of runoff sequences that depart from mean values – that is, drier and wetter than average periods than reflected in mean runoff. Finally, it will allow us to explore more thoroughly the stochastic nature of regrets and the possibility of minimizing these regrets by policies and plans that are flexible over a wide range of mixed runoff distributions.

Note

1 Shadow price is the maximum price that consumers are willing to pay for an extra unit of a given limited resource.

References

Booker, J. F. (1990) *Economic Allocation of Colorado River Water: Integrating Quantity, Quality, and Instream Use Values*, PhD thesis, Department of Agricultural and Resource Economics, Colorado State University, Fort Collins, CO

Booker, J. F. and R. A. Young (1991) *Economic Impacts of Alternative Water Allocations in the Colorado River Basin*, Colorado Water Resources Institute, Report 161, Colorado WRI, Fort Collins, CO

Booker, J. F. and R. A. Young (1994) 'Modelling intrastate and interstate markets for Colorado River water resources', *Journal of Environmental and Economic Management*, vol 26, pp66–87

Callaway, J. M. (2004a) 'Adaptation benefits and costs: Are they important in the global policy picture and how can we estimate them?', *Global Environmental Change*, vol 14, pp273–282

Callaway, J. M. (2004b) 'The benefits and costs of adapting to climate variability and change', in *The Benefits and Costs of Climate Change Policies: Analytical and Framework Issues*, OECD Press, Paris, Chapter 4, pp111–158

Callaway, J. M., L. Ringius and L. Ness (1998) 'Adaptation costs: A framework and methods', in J. Christensen and J. Sathaye (eds) *Mitigation and Adaptation Cost Assessment Concepts, Methods and Appropriate Use*, UNEP Collaborating Centre on Energy and Environment, Risø National Laboratory Press, Roskilde, Denmark, pp197–120

Callaway, J. M., D. B. Louw, J. C. Nkomo, M. E. Hellmuth and D. A. Sparks (2008) 'Benefits and costs of adapting water planning and management to climate change and water demand growth in the Western Cape of South Africa', in N. Leary, C. Conde, J. Kulkarni, A. Nyong and J. Pulhin (eds) *Climate Change and Vulnerability and Adaptation*, Earthscan Publications, London, UK, Chapter 3, pp53–70

Hellmuth, M. E and D. Sparks (2005) *Modeling the Berg River Basin: An Explorative Study of Impacts of Climate Change on Runoff*, AIACC Project Completion Report, Project no 47, UNEP Collaborating Centre on Energy and Environment, Risø National Laboratory, Roskilde, Denmark

Hewitson, B. (2004) 'Scientist warns of climate change', www.iafrica.com, 6 May, Cape Town, South Africa

Hurd, B. J., J. M. Callaway, P. P. Kirshen and J. Smith (1999) 'Economic effects of climate change on US water resources', in R. Mendelsohn and J. Neumann (eds) *The Impacts of Climate Change on the US Economy*, Cambridge University Press, London, pp133–137

Hurd, B. J., J. M. Callaway, P. P. Kirshen and J. Smith (2004) 'Climatic change and US water resources: From modeled watershed impacts to national estimates', *Journal of the American Water Resources Association*, vol 2, pp130–148

Louw, D. B. (2001) *Modelling the Potential Impact of a Water Market in the Berg River Basin*, PhD thesis, University of the Orange Free State, Bloemfontein, South Africa, January

Louw, D. B. (2002) *The Development of a Methodology To Determine the True Value of Water and the Impact of a Potential Water Market on the Efficient Utilisation of Water in the Berg River Basin*, Water Research Commission Report (WRC) No 943/1/02, WRC, Pretoria, South Africa

Louw, D. B. and H. D. van Schalkwyk (2001) 'Water markets an alternative for central water allocation', *Agrekon*, vol 39, no 4, pp484–494

McCarl, B. A. and T. H. Spreen (1980) 'Price endogenous mathematical programming as a tool for sector analysis', *American Journal of Agriculture and Economics*, vol 62, pp88–102

Samuelson, P. P. A. (1951) 'Spatial price equilibrium and linear programming', *American Economic Review*, vol 42, pp283–303

Smith, J. B. and S. S. Lenhart (1996) 'Climate change adaptation policy options', *Climate Research*, vol 6, pp193–201

Takayama, T. and G. G. Judge (1971) *Spatial and Temporal Price and Allocation Models*, North Holland, London

Vaux, H. J. and R. E. Howitt (1984) 'Managing water scarcity: An evaluation of interregional transfers', *Water Resources Research*, vol 20, pp785–792

Yates, D. N. (1996) 'WatBal: An integrated water balance model for climate impact assessment of river basin runoff', *Water Resources Development*, vol 2, pp121–139

15

Institutional Adaptation to Climate Change: Current Status and Future Strategies in the Elbe Basin, Germany

Sabine Möllenkamp and Britta Kastens

Introduction

Current water management systems are characterized by complexity and increasing uncertainty: water has to be allocated to competing uses and new legal or managerial requirements pose challenges, such as the implementation of European directives. Water management needs to adapt to fundamental changes in the physical and human environment. Climate change is one of the most important challenges of current and future water management and requires adaptive management strategies in order to cope with its impacts.

Although uncertainty prevails about the extent of future climate change, and especially on the consequences for precipitation, some lines seem to evolve: for Germany, a future trend towards increasing winter rainfall and decreasing spring and summer rainfall could be possible, while spatial distribution can differ a lot within Germany (Jacobs et al, 2008). Current observations show that in Central Europe the number of flood events has increased to the same extent as long dry periods in summer (Leipprand et al, 2006).

Management systems for water resources must accommodate these challenges and must be adaptable to changes in climatic systems (Gunderson and Holling, 2001). Adaptive management can be considered as a 'systematic process for continually improving management policies and practices by learning from the outcomes of implemented management strategies' (Pahl-Wostl, 2007). In this context, the concept of integrated social–ecological systems (Berkes and Folke, 1998) is of major importance. It stresses that social and ecological systems are linked and the delineation between them is artificial and arbitrary (Berkes et al, 2003).

Adaptive management explicitly acknowledges both uncertainties and complexity of a social–ecological system. River basins are typical examples of such a system which requires adaptive management for successful governance. Adaptation strategies thus also refer to institutional structures that are capable of generating long-term sustainable policy solutions to complex and dynamic natural resource problems through collaboration among diverse resource users and governmental agencies (Scholz and Stiftel, 2005).

The aim of this case study is to investigate whether the current river basin management institutions in the Elbe Basin allow for adaptation to climate change impacts. Institutions are considered as a broad 'set of rules, decision-making procedures, and programmes that define social practices, assign roles to the participants in these practices, and guide interactions among the occupants of individual roles' (Young, 2002).

Based on a literature study, we identify basic elements of institutional adaptation, among them knowledge and information, polycentric governance, participation, sectoral integration, flexibility, openness for experimentation, and political willingness. Making use of these elements, we investigate how adaptive the current institutional arrangements in the Elbe Basin are, which aspects need improvement, and how better adaptation could be achieved. The study focuses on water management in the German part of the Elbe Basin. Here, flood management and strategies to deal with droughts have caught particular attention after the major flood and drought events in 2002 and 2003.

The chapter is organized as follows: in the next section we provide some background on the Elbe Basin and the impacts of climate change; we will then briefly describe our empirical methods used in this study and elaborate upon the mentioned elements of adaptive institutions. The fifth section is based on our empirical analysis and shows that some elements of adaptive institutions already exist in the Elbe Basin. The chapter closes with conclusions on the current status of institutional adaptation in the German Elbe Basin and some thoughts on the use of the developed framework for further studies.

The Elbe Basin: Impacts of climate change and general settings

The Elbe Basin lies entirely within European Union (EU) territory and has a catchment area of 148.268km^2. It is shared by four states: Germany, the Czech Republic, Austria and Poland, with the latter two covering less than 1 per cent of the catchment (see Plate 29, centre pages). About 25 million people live in the catchment area (FGG Elbe, 2004).

Climate change impacts upon the Elbe Basin

Climate change has been discussed for several years at global and national levels. We will not repeat this discussion here, but want to provide a short overview on the expected changes and impacts in the Elbe Basin.

While studies on climate change are quite advanced on a global level, the resolution of the general circulation models (GCMs) is currently too rough for a correct representation of the hydrological cycle variations within river basins. This problem can partly be solved by downscaling the GCM outputs onto the regional or river basin level. Krysanova et al (2005) have modelled the interactions between climatological, hydrological and ecological processes on different scales in the Elbe Basin by making use of the eco-hydrological Soil and Water Integrated Model (SWIM) (Krysanova et al, 2000). Based on the Intergovernmental Panel on Climate Change (IPCC) emission scenario A1, the climate change scenario used here is characterized by an increase in temperature of 1.4°C by 2050 and a moderate decrease in mean annual precipitation in the basin. According to this scenario, a 17 per cent decrease in average annual precipitation is expected for the total German part of the Elbe Basin in 2046 to 2055 compared to the reference period of 1991 to 2000. Lower precipitation will be most marked in the central and southern parts, with less marked increases in precipitation in the northern part of the basin (Krysanova et al, 2005).

Krysanova et al (2005) have evaluated the impacts of climate change upon water quantity as well as upon some aspects of water quality. Concerning water quantity, different components of the water balance were analysed. Evapotranspiration is expected to decrease, on average, by 4 per cent in the Elbe Basin, with significant sub-regional differences corresponding to the change in precipitation. Runoff and groundwater recharge show a decreasing trend, whereas groundwater recharge responds most sensitively to the anticipated climate change (–37 per cent, on average). A significant reduction in river flow is predicted.

Krysanova et al (2005) also studied the consequences of climate change on water quality in the Elbe Basin. They found a notable decrease in diffuse pollution in some parts of the basin, showing that climate change could also have positive impacts.

The overall result of the study is that the mean water discharge and the mean groundwater recharge in the Elbe Basin will most likely decrease, and diffuse pollution will be diminished. At the same time, it has to be acknowledged that the uncertainty in hydrological and water quality responses to changing climate is too high for conclusive predictions. The uncertainties also differ in three Elbe sub-regions: the mountainous area, the loess sub-region and the lowland area (Krysanova et al, 2005). However, the future impacts of climate change on water resources in the Elbe Basin are likely to increase (Krysanova et al, 2006).

It is important to mention that the model used by Krysanova et al builds upon the A1 scenario and is thus to be considered under this focus. A recent study by Jacob et al (2008) makes use of scenarios A2, A1B and B1 and the regional climate model REMO. Jacob et al conclude that temperatures in Germany could increase up to 4°C and summer rain could decrease, while winters could get wetter in the future. The analysis of extreme events is ongoing.

While differing in many ways, both studies detect considerable future change from the current climate regime in different parameters. Thus management needs to be able to apply new adaptation strategies to cope with these changes. For water management, changes could become especially visible in form of increasing frequency and unpredictability of extreme events involving floods or severe droughts (also see Becker and Grünewald, 2003). Recent events such as the floods during the summer of 2002 and the droughts in the summer of 2003 have already shifted attention to such extremes.

Organizational structures for managing the Elbe

Several organizations are dealing with the management of the Elbe at different levels. On a large basin scale, the International Commission for the Protection of the Elbe (ICPE) is a major role player. Germany and the Czech Republic are the main contracting parties with the ICPE, with Austria, Poland, non-governmental organizations (NGOs) and the International Commission for the Protection of the Rhine (ICPR) as observers. The ICPE issues recommendations for river basin management; its main goal is to improve the status of the Elbe and its main tributaries and to increase the ecological value of the Elbe Valley. Moreover, the implementation of the European Water Framework Directive (WFD, 2000) is being coordinated under the auspices of the ICPE.

On a national level in Germany, the Federal Ministry for the Environment, Nature Conservation and Nuclear Safety (Bundesministerium für Umwelt, Naturschutz und Reaktorsicherheit, or the BMU) is the highest authority for water policy and management. The Federal Environment Agency (Umweltbundesamt, or UBA), as the scientific authority for environmental issues, reports to the Federal Ministry for the Environment. The federal states (*Länder*) have major competences in water management and are in charge of implementing water law and water management issues. Ten out of the 16 German *Länder* are part of the Elbe Basin.

Within the *Länder* the ministries dealing with water management, as well as the respective state agencies, play a major role in managing the Elbe. The state agencies act as technical and scientific bodies for the ministries of environment and/or agriculture in specific states. In 2004, the ten *Länder* in the basin and the federal government of Germany established the River Basin Community Elbe (Flussgebietsgemeinschaft, or FGG Elbe) in order to coordinate the German part of the WFD implementation. The aim of the FGG is to establish a systematic and mutually accepted programme of management activities and measures for the German part of the Elbe Basin. Even though the FGG has no formal decision-making power, its technical guidance for cooperation between the federal states is widely acknowledged (Borowski et al, 2004; Raadgever, 2005, p7). Organizational structures such as the Working Group for the Protection of the Elbe (Arbeitsgemeinschaft zum Schutz der Elbe, or ARGE Elbe), which earlier had a similar standing in the German Elbe Basin, are becoming less influential.

Expert interviews

This chapter draws on empirical data gained from 11 in-depth interviews conducted in the spring of 2007 with representatives from different organizations at international, national and sub-national levels. The experts from the German Elbe Basin were chosen in order to elicit knowledge and obtain perceptions of different groups concerned with institutional adaptation to climate change impacts. The interviews were open, guided and explorative. The interview guideline was broadly structured along the different elements (criteria) of adaptation as identified from the literature.

Figure 15.1 shows the choice of experts, indicating their level of work and their affiliation in governmental or non-governmental organizations. Governmental experts were chosen from the German national level as well as from three German *Länder*. Governmental experts belong to either ministries or state agencies. Three non-governmental actors from the German Farmer's Association, German Society for Nature Protection (NABU) and the Munich Re Group (*Münchener Rück*) were also interviewed. The agricultural association constitutes the non-governmental representation of regional or local agriculture. NABU is an environmental NGO. The Munich Re Group is a German re-insurance company that has no direct link to the Elbe Basin, but which has wide experience on adaptation to climate change. It was also strongly involved in discussions on how to deal with the Elbe flood in 2002. All interviews were conducted during spring 2007, making use of predefined interview guidelines. Interviews were conducted in a semi-structured way, leaving room for closer examination of issues that were important to the interviewees. The interviews were then evaluated using a pre-established analytical framework (see the following section).

Elements of adaptive institutions: An analytical framework

Adaptation of water management institutions to climate change impacts comprises various aspects, from information management to institutional flexibility. In order to structure our study, we opted to group seven elements as criteria for adaptation to climate change. The choice was grounded on analysis undertaken by scholars of both

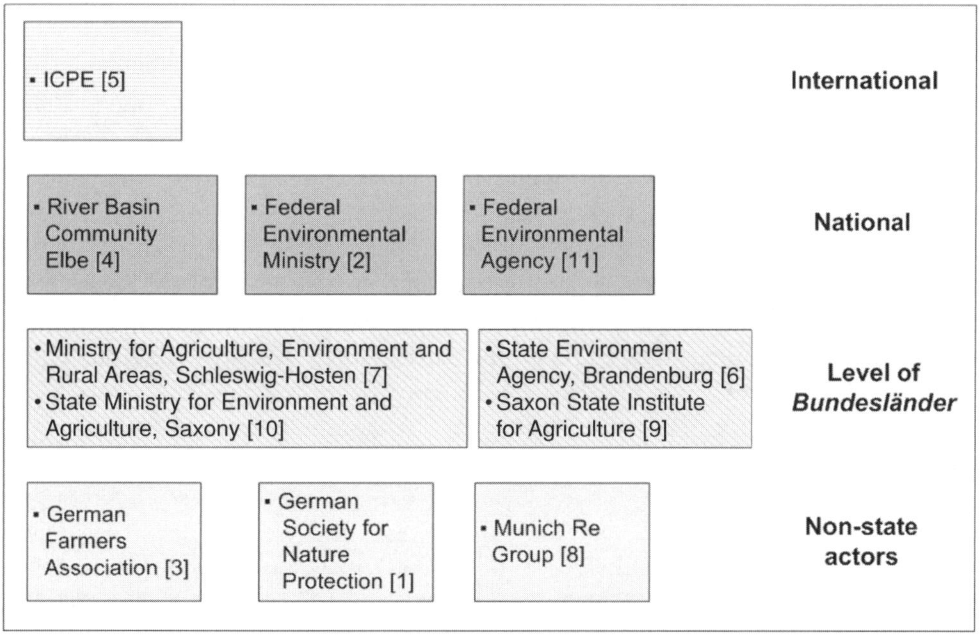

Figure 15.1 *Organizations from which experts were drawn for interviews for the study: Numbers in brackets refer to citations in the text*

Source: Sabine Möllenkamp and Britta Kastens

institutional adaptation, in general, and adaptation to climate change, in particular (Adger et al, 2005; Folke et al, 2005; and Berkhout et al, 2006; Huitema et al, in review). The list of elements is not exhaustive, but reflects the main issues of governance. The proposed elements are meant to facilitate the investigation of different cases and are based on various studies on adaptation. Yet, within these studies, the categories and the terminology are by no means consistent. We were aware of some overlap between the elements of adaptive institutions. We deliberately accepted these in order to avoid an artificial separation of closely linked institutional aspects. The elements were used as a heuristic to evaluate the status of the institutional setting in terms of adaptation in the Elbe Basin, and to identify potential needs for improvement.

Availability and communication of information

Adaptation often requires knowledge of unknown future developments and of uncertainties (Adger et al, 2005, p81). Stern (2007, pp430ff) emphasizes the specific role of governments in establishing policy frameworks to encourage adaptation by private individuals and firms. The need to address information uncertainties to ensure transparency of transactions, and to tackle constraints that will reduce the capacity for autonomous adaptation, are also their concern. Complex social–ecological systems are largely unpredictable due to their variable and non-linear behaviour. Knowledge about such systems is usually insufficient and requires constant updating (Brugnach et al, 2007, pp5ff). Rigorous up-to-date and relevant information is a vital prerequisite for keeping institutions adaptive (Dovers 2001, p217; 2003, p9). In other words, information on climate change and its impacts has to be sufficient and reliable.

Comprehensive knowledge acquisition, improved understanding, filling of information gaps, and dealing with uncertainties can be achieved 'by open, shared information sources that fill gaps and facilitate integration' (Pahl-Wostl, 2005, p6). Information management thus also requires adequate communication and exchange of information. Various tools, such as risk maps or technical risk information systems, are useful vehicles for communication and encourage exchange between the competent authorities, as well as between authorities and non-state actors (Pahl-Wostl, 2007, p55; Raadgever et al, 2006, p3). Local knowledge is necessary to acquire quality information. The latter is dependent upon public participation, which will be dealt with in the sub-section below.

Polycentric governance

In contrast to monocentric or hierarchical systems, polycentric governance consists of different centres of management and control (McGinnis, 1999; Ostrom, 2001). Ostrom (2001, p2) describes polycentric systems as being the 'organization of small-, medium- and large-scale democratic units that each may exercise considerable independence to make and enforce rules within a circumscribed scope of authority for a specific geographical area'. These units may be located at different geographical levels and can be either general purpose authorities or specialized authorities with specific tasks (Hooghe and

Marks, 2003). While polycentric governance offers a flexible system that promotes experiments within small-scale units (Ostrom, 2005) and has a good capacity to cope with external shocks (Ostrom, 2001), it risks being inefficient because of fragmentation or duplication of authority. Coordination and collaboration between the different spheres of authority are thus essential for polycentric systems to be effective. In this context, cross-boundary integrators (Roberts and King, 1996) – individuals or collectives who connect centres, levels and sectors – are pivotal for the coordination of the individual units.

Participation

Participation of non-state actors, either as stakeholders or as civil society in general, is formative in institutional adaptation. Despite some critique of the participation process (Cooke and Kothari, 2001), the involvement of non-state actors is generally seen as a key for policy development and implementation (deLeon and deLeon, 2002; Lee and Abbot, 2003; Olsson et al, 2004; Crabbé and Robin, 2006, p125). The argument in support of public participation is that it facilitates legitimacy of decisions and enhances goal achievement (Newig, 2007). Public participation also helps to widen the range of interests to be included in adaptive processes such as ecosystem services and risks (Lebel et al, 2006). Participatory approaches establish the basis for learning processes and creative adaptation solutions (Folke et al, 2005). There is an overwhelming literature pool on factors that determine the success of participatory processes (see Ridder et al, 2005, p5; Newig, 2007, p63).

Many authors stress that it is of utmost importance to involve stakeholders as well as the broad public from the very beginning. Early involvement helps to prevent actors from feeling left out and to create a sense of belonging to a group, which ultimately results in commitment and compliance by actors (Folke et al, 2005; Ridder et al, 2005, p5; Newig, 2007, p63). Non-state actors at lower levels might have stronger incentives to take adaptive action in response to climate change impacts since they are affected directly. Farmers, for example, will have lower crop production due to extended dry seasons (Crabbé and Robin, 2006, p105). We did not study the various success factors because participation processes concerning adaptation to climate change are embryonic in the Elbe Basin. Our investigations focused mainly on whether public participation, related to adaptation to climate change, is perceived as a problem by interviewees and, if so, why.

Sectoral integration

Effective climate adaptation policy cannot be made by environmental policy-makers in isolation because 'the effectiveness of specific institutions often depends not only on their own features, but also on their interactions with other institutions' (Young 1999, p49). Institutional response to climate change thus demands sectoral integration. Expected climate change impacts and macro-economic projections need to be incorporated within planning processes. They also need to be incorporated within other policies, such as forests, water resource and coastal zone management. All of this will require institutional arrangements of similar policy issues to be adjusted to each other (Burton et al, 2002; Stern, 2007, p432). Cross-sectoral activities are more likely to identify and address emergent problems effectively (Pahl-Wostl, 2005, p6). Such interplay is assumed to increase resilience and to strengthen adaptive capacity (Dryzek, 1987). Rosendal (2001) states that regime interactions are potentially synergetic by building on compatible norms and giving rise to mutually reinforcing or complementary regulations. Moreover, when parties of these regimes become aware of the interplay, they may seek to coordinate their activities in order to tap the potential for synergy (Kim, 2003, pp2f). However, where the effects of one regime contradict another regime's policy direction, the interplay might also have a conflictive component (Gehring and Oberthür, 2000; Kim, 2003, pp2f).

Openness for experimentation

As Folke et al (2005, pp462ff) point out: 'adaptive governance focuses on experimentation and learning'. Openness for experimentation is paramount for institutions to successfully adapt to climate change by learning from past experience. Openness for experimentation goes along with the social learning concept (originally by Bandura, 1977; further developed by various scholars, e.g. Pahl-Wostl et al, 2007; Mostert et al, 2007). The essence of this concept according to the European project HarmoniCOP is 'learning together to manage together' (Ridder et al, 2005). It stresses a common understanding of management processes within a multiparty collaboration. A feedback loop between outcomes and the context of such a collaborative process takes into account structural changes in a cyclic and iterative fashion and allows for new insights to enter the process.

Experimentation can be seen as a research methodology, but also as a management approach (see also Lee, 1999). Experimentation as a research methodology is the most commonly known form of experimentation. In water management, this can extend to experimentation in icon sites, such as restoration areas along river stretches or in pilot areas (EC/JRC, 2005). These experiments aim to achieve knowledge about the system in order to design policies that are better able to cope with changing situations.

Experimentation as a management approach (as followed in Huitema et al, in review) entails the use of policy itself as a set of experiments (Folke et al, 2002, p52). Action is admittedly taken without perfect knowledge, but on the basis of scenarios and likelihoods and with close monitoring of results. The important aspect here is that the management should subsequently be able to adapt to evaluation results and changing background conditions. Such policy experiments are often discussed in conjunction with the idea of policy or social learning (also Folke et al, 2002, p47), which is also of great interest from the perspective of adaptive management (Huitema et al, in review). The concept of experimentation as a management approach incorporates the idea that governance should allow for learning without foreclosing future development options (Folke et al, 2002, p9).

Flexibility

Another important element is institutional flexibility. Governance systems concerned with the development and preservation of resilience 'need to be flexible and open to learning' (Folke et al, 2002, p52). Institutions should be able to change and to adjust to changing external conditions. This means that management procedures and management structures might need adjustment to new (environmental) conditions or new (scientific) knowledge (Folke et al, 2002, p45), stressing the idea of learning from past experience:

> Flexible adaptation options reduce vulnerability to risks of climate change, and variability and function in light of a range of climate conditions, not simply a particular projected condition. (Dolan et al, 2001, p18)

Seeing that natural conditions in ecosystems change frequently, it is obvious that the social systems linked to them also need to be flexible and able to change. Flexible systems are characterized by the ability to incorporate the results of monitoring, evaluation or experimentation within the system and to change accordingly. Equally important is the ability to reverse decisions taken in such a system.

In the following sub-section we focus on the capacity of organizational structures to cope with new challenges and their ability to change. We distinguish between structural flexibility (the creation of new structures, such as working groups) and functional flexibility (the ability to include new functions and goals in existing structures).

Planning horizons, political support and economic resources

Many impacts of climate change will only be experienced a couple of decades from now. In order to stay beyond purely reactive emergency responses, climate change has to be seen as a long-term policy challenge, which usually cannot be tackled over short terms and within fixed periods, such as a legislative period or one implementation cycle of the WFD. Adaptation therefore needs to involve long-term planning of required resources to ensure a comprehensive response strategy to climate change. Such resources refer, first of all, to financial and labour resources (Burton et al, 1998; Homer-Dixon, 1999; Dovers, 2001, p217), but also to professional, technical and, particularly, political support (Allman et al, 2004).

If water managers perceive other policy problems as more urgent, owing to implementation deadlines or public pressure, adaptation to climate change impacts may be neglected at the daily operational level. Consequently, adaptation to climate change might not gain sufficient support. The water manager's working priorities are, however, also guided by the political atmosphere. Where the issue of adaptation gets clear support by political decision-makers, water managers will usually have wider options to implement adaptation strategies at the daily operational level. Strong political commitment is often combined with demonstrated governmental leadership, showing that adaptation actions are possible. Political commitment and willingness to achieve adaptation to climate change can also make roles and responsibilities for the implementation of adaptation strategies at lower levels more understandable (Smit and Pilvosova, 2001, p898).

Institutional adaptation in the Elbe Basin: Empirical results

For the investigation of current institutional adaptation in the German Elbe Basin, interview partners were asked to describe the actual status of adaptation in terms of the different elements explained above. The interviewees were also asked to suggest strategies for improving the current situation and to explain how they deal with uncertainties in some of the cases.

Availability and communication of information

The analysis focused on three aspects of information management in the Elbe Basin: first, on the question of sufficient information and current knowledge gaps; second, on the communication and exchange of knowledge; and, third, on how the interviewees would deal with uncertainties.

Basic information on climate change impacts and on flood management was generally perceived as sound and good. Nevertheless, some interviewees highlighted concrete information gaps: information on dealing with droughts, in particular, was noted as insufficient [4].[1] Interviewees expressed a need for more accurate modelling results. They also wanted information on the confidence that one can put to model predictions – particularly for lower levels such as the sub-basins and local catchments [2, 6, 8, 9]. Other interviewees emphasized the need for more information on the causes of climate change since it is virtually impossible to separate the impacts of climate change from other anthropogenic impacts [3, 11]. Current information on climate change in the Elbe Basin is mainly based on the results of GLOWA-Elbe research, a German-funded project conducted in the Elbe Basin.

As to the transfer of information on water management issues, the ICPE acts as a communication platform. Yet, to date, the ICPE has no strategy or tool to disseminate information on climate change adaptation among its members [5]. This is also attributed to the fact that some of the *Länder* have only just started to deal with climate change adaptation. The same is true for related research: many organizations of the *Länder* conduct or commission studies on water management [6, 9, 10, 11]; but only a very few of these studies are explicitly designed to deal with climate change impacts upon water resources [10].

For some organizations, however, the work on climate change impacts is already part of their daily operations [3, 6, 9]. This is particularly true for those organizations related to agriculture.

Interviewees supported the acceptance of the uncertainties related to climate change and its impacts in the Elbe. Instead of offering excuses, they recommended that the known, albeit meagre, information on climate change be used as a basis for further action. Strategies for improved quality of information are emanating from stronger cooperation between administration and science. Improved information exchange within existing networks of centralized data collection and the distribution of data sets add to the quality of information. Finally, many interviewees wanted to see more information on adaptation to climate change provided to the broad public. Currently, public awareness is confined largely to mitigation strategies, while adaptation issues seem to remain a discussion among water managers [6, 8, 10].

Polycentric governance

The study initially focused on the question of whether the current management system comprises different centres of management and control, and to what extent coordination among such centres and spheres of authority takes place.

The interviewees largely agreed that the current water management system in the Elbe Basin is polycentric in nature. It comprises different centres of management, such as the different national and regional ministries. These structures are embedded in the general administrative system of the German federal state. Since the implementation of the WFD, new coordination structures were added, such as those on the level of sub-catchment areas. The ICPE and the FGG – both with a hydrological orientation – are part of the management system that surpasses conventional administrative structures. They run coordinating activities in some thematic areas of water management. Current collaboration within these organizational structures is still imperfect but pragmatic [5]. Good relations between stakeholders and organizations are often attributed to the activities of the ICPE or the FGG Elbe. Interviewees suggested improvement of coordination within the existing structures rather than changes to systems *per se* [5]. This is recommended particularly for river basin and federal level [2, 10]. Better coordination is also crucial for basic areas such as funding and data exchange. The river basin commissions could act as cross-boundary integrators and connect the different centres of action even more than today. A need for better coordination within the federal structure is necessary to overcome the lack of action on the larger level [2].

Interestingly, some interviewees would like to see the EU take a leading role or even to increase pressure on lower levels to adapt [5, 8]. Such stronger leadership will encourage polycentric systems on lower levels. A survey published by the European Environment Agency (EEA) in 2007 also identified a need for EU-level action. It stated that EU activities could encompass a general framework for adaptation, monitoring and information exchange, as well as coordination between sectors and sectoral policies or educational measures. Contrary to the suggested increased EU involvement, countries included in the EEA report put emphasis on the subsidiarity principle and called for implementation of adaptation measures to remain the responsibility of the member states in order to ensure flexible response to the specific challenges in their countries (EEA, 2007, p47).

Participation

It was important for us to establish the presence of participatory processes in the Elbe Basin that addressed the impacts of climate change. As highlighted by Folke et al (2005, p447), it is social learning, in particular, that helps to develop adaptive expertise and processes. Social learning requires active participation (see EU, 2002, p56) and, thus, processes that clearly go beyond consultation and information-sharing.

Even though participatory approaches play a major role in European water management and its implementation, water managers at the European level are not adequately equipped to promote active involvement amidst high social and ecological uncertainty (Galaz, 2005, p6). There is, thus, no strategy to incorporate climate change within current participatory practices. At the time of our study in the Elbe Basin, there were no participatory processes that included the issue of climate change. Not even did the first Elbeforum (28–29 March 2007 in Usti nad Labem, Czech Republic), which had the aim to inform stakeholders about water management in the basin, include climate change adaptation on its agenda [4, 5].

There was also no consensus among the interviewees whether the involvement of stakeholders and the broader public is a necessary strategy for climate change adaptation at all. Critical voices cautioned against overstraining (potential) participants and stakeholders, and highlighted that a sufficient number of authorities and scientists were already available to deal with adaptation issues [3, 9, 10]. In the opinion of some interviewees, public interest in the discussion on climate change adaptation is currently only restricted to those users whose interests are negatively affected (e.g. concerning navigation, due to droughts or floods) [1, 11]. In contrast, proponents emphasized that a comprehensive adaptation strategy could not be applied without the backing of stakeholders and the broader public [1, 6, 8, 11], and that one should be open to include new stakeholders who emerge with the new issue of climate change adaptation [4]. The involvement of local stakeholders, in particular, would offer the chance to tackle climate change at a level where its impacts are prevalent and where local knowledge could help to close information gaps. The latter is particularly important since many interviewees stated that knowledge about the future effects of climate change at regional and local levels is, to date, not sufficient. Moreover, some stakeholders, particularly the environmental NGOs in the Elbe Basin, are proactive and have already developed their own adaptation actions [1, 8] (e.g. by disseminating information to the broader public).

Public participation with reference to adaptation to climate change has been initiated only at the level of information strategies. One important strategy for a better information flow from government to non-state actors was established by the Federal Environment Agency in Germany (UBA), which initiated an information and co-operation platform called KomPASS. This platform is supposed to act as a new competence centre on climate change adaptation and also to serve as an example for stakeholder involvement and information provision for the general public. One interviewee urged a stronger initiative by the media to inform the broader public on adaptation to climate change, similar to its initiatives to inform the public on mitigation strategies [8].

Sectoral integration

Current water resource regimes are often characterized by sectoral fragmentation and limited integration. This is viewed as the main reason for low adaptive capacity of these regimes (Pahl-Wostl, 2007). Recent work by scholars of adaptive water management expressed a need, in particular, to integrate climate change adaptation within the implementation processes of the WFD (EEA, 2007; Pahl-Wostl, 2007, p55). Explicit recommendations were incorporated within the WFD's implementation plans to address integration of flood management and to address low water levels. The six-year cycle of the WFD river basin management plans, in particular, offers opportunities for such sectoral integration. In this respect, it is important that different governmental actors, such as representatives of ministries, interact and coordinate their efforts.

During our interviews in the Elbe Basin we specifically asked about the degree of integration of climate change issues and especially about flood management during the WFD implementation in order to interlink adaptation and water management. The overall observation was that sectoral integration of climate change issues in the WFD is only beginning to develop. Strategies on sectoral integration within the Elbe Basin – in the view of some interviewees – were usually further developed for mitigation than for adaptation. Some interviewees urgently recommended a stronger integration of the WFD with climate change issues [1, 2, 6, 11] – for example, in terms of the integration of aspects of flood management in the river basin management plans and their iterations every six years. The possible change of criteria for the status of reference waters, the cross-cutting character of climate change, in general, and the connections between water stress and floods were identified as cross-sectoral issues. Concerning the interplay of the WFD and flood management, some interviewees expected the new European Directive on Floods to close this gap. This directive explicitly refers to a combined river basin management plan for the implementation of both the ecological status of water bodies and flood management [2, 6, 11]. Even though low water levels are already addressed as an issue in the framework of the WFD, the interviewees demanded a stronger focus on climate change impacts and future management strategies in discussions [1, 2, 6]. In some areas of the Elbe Basin there could be long-term climate change impacts on sectors such as navigation and agriculture. Research needs to further address prevailing open questions and future discussions on water management need to take the possible changes into account [2].

Finally, interviewees stressed that potential measures needed to integrate existing policies and initiatives (e.g. Common Agricultural Policy or the Flora, Fauna and Habitat Directive) [1, 2, 6, 11]. Yet, it was also emphasized that sectoral integration, and particularly the integration of climate change issues within the implementation process of the WFD, will only be a first step, but by no means a sufficient adaptation strategy [2].

Openness for experimentation

Interviewees alluded to a large variety of experiments in the classical scientific sense, such as pilot projects or research-related activities. For example, projects were initiated to gain insights into the use of drought-resistant plants and new irrigation technologies [8, 9]. In general, most interviewees perceived a need for more experimentation by allocating more human resources for cooperation with science. There also still seems to be a need for scientific consultancy on climate change adaptation. Many interviewees stated that science should provide relevant guidance for water management practices. This suggestion is by no means a new one. In a project conducted on the science–policy interface for climate change in The Netherlands at the beginning of the 1990s, policy-makers clearly demanded that scientists should get more actively involved in the public debate by disseminating their knowledge in the form of demonstration projects (Klabbers et al, 1996, pp81ff).

In general, some interviewees considered the cooperation between science and administration as being positive [6, 9]. The interviewee from the insurance company considered the insurance sector as one of the cross-boundary facilitators between science and political decision-makers [8]. Some perceived a need to push the current discussion from the scientific and administrative towards the political scene [1, 2, 6, 11]. Others urged scientists working on climate change adaptation to take over the role of service providers, developing tools and concrete recommendations for policy and administration.

Various pilot projects linked to the implementation of the WFD are also being conducted or have just been finalized in the Elbe Basin. One main institutional advantage is the potential for cooperation that these projects present. Being conducted within hydro-morphological units of river catchments and basin districts, the projects usually involve governmental as well as non-governmental actors from different federal states. Some of the projects focus strongly on policy issues. Even though these projects currently do not address climate change adaptation, the project researcher's experiences concerning other questions of water management can be drawn upon and partly transferred to future climate change adaptation actions.

Interviewees differed in their support for more pilot projects. Some were of the opinion that the results of these projects, which usually refer to the implementation of the WFD, can be transferred to other *Länder* [7] or topics. However, the funding of such projects is not yet secured [7]. There are currently no pilot projects dealing explicitly with climate change adaptation [2]. Other interviewees held the opinion that there are already many pilot projects and that new ones might be less effective [4].

While classical experiments are common, management experiments are not. Generally, management decisions are not considered as experiments by many decision-makers. They believe existing structures and personnel are not suitable for experimentation and a change in structures is only considered as an improvement [10]. A typical example is launching a committee for a specific task as long as that task is of pivotal interest [10]. Implementing results from experiments in the policy process can result in potential changes of institutional and organizational structures that will require additional flexibility.

Flexibility

The extent of changes in organizational structures as a result of climate change impacts, and the likelihood of such impacts, was discussed under this heading. We also investigated the perceived suitability of current structures in dealing with impacts of climate change without restructuring.

Most interviewees perceived the current organizational structures as being generally suitable to deal with climate change adaptation [e.g. 10]. Changes in current structures, should they be necessary, would currently be difficult, seeing that the actual impacts of climate change are not supported by adequate quantitative data. [4].

Instead, some interviewees expressed their wish to make better use of existing water management structures for climate change adaptation [2, 5] and to integrate the respective discussions on other aspects, such as flood management or agriculture [4]. Currently, there is no coordinated discussion on climate change adaptation in the ICPE [2]. Such discussions only take place on lower levels, especially within the *Länder*.

At the same time, the suggestion was made to bring about structural changes within the general system boundaries and not to change the basic system structures themselves. This could entail the creation of new additional structures in a flexible way – for instance, in the form of working groups. An example of this approach can be found in Saxony, where an integration of different water management aspects has taken place within one unit of the regional ministry. After an internal structural reform, the work on WFD implementation, floods and droughts, as well as climate change adaptation, was brought together in a single unit. This structural change was driven by an external auditor who aimed to minimize interfaces and to improve the workflow [10]. In this case, the general culture of the organization seems to be an important factor. While some organizations aim for constant quality improvement and, thus, are used to changes, others have a more stable tradition. Other interviewees reported having created new working groups dealing with water management and climate change adaptation [e.g. 1, 7, 9]. The already mentioned initiative KomPASS serves as a link between decision-makers in companies and administrations. In summary, we can say that structural flexibility exists within the general system boundaries, but a profound structural reform is not suggested by the interviewees.

Functional flexibility – for example, the inclusion of the discussion on climate change in existing structures – was also mentioned. Some interviewees suggested an expansion of tasks in their water unit in order to include climate change-related questions [6, 11] or linking it to existing foci, such as the discussion on biodiversity [1].

Some interviewees indicated that they have not (yet) included climate change adaptation in their work, either in existing or in new structural arrangements [3, 5].

Planning horizons, political support and economic resources

Galaz (2005, p6) discovered from his analysis on the adaptiveness of the WFD implementation in Sweden that some water directors still apply 'a "wait-and-see" strategy to climate change, and there are no concrete plans to adapt classification scales and river basin plans, taking into account the effects of climate change'. The same situation seems to be true for the Elbe Basin, where many interviewees highlighted the need for stronger political commitment to adapting to climate change. As one example, they expect the national and the *Länder* government to show more willingness for adaptation by scrutinizing current land- and water-use structures (e.g. creating retention areas).

The interviewees further characterized the current situation of climate change adaptation in the Elbe Basin as a short-term strategy reactive to external drivers, such as floods [1, 6, 11]. Currently, thinking in legislative periods and in WFD reporting timescales dominates political decisions and day-to-day work of the authorities in charge [5, 6, 7, 10, 11]. There is thus a need for long-term planning for climate change adaptation and action that is continuous and not just linked to strong external pressures [1, 6, 7, 9, 10]. Preliminary ideas for long-term planning were given by interviewees to integrate climate change issues within the WFD (e.g. by investigating implementation measures in the light of adaptation and by making use of the six-year cycle of the river basin management plans). The precautionary principle and no-regret measures are suggested as important elements for future developments [11].

Concerning the availability of resources to adapt to climate change, other studies on the Elbe Basin already emphasized that the implementation of measures regarding climate change adaptation is slow due to a lack of labour and finances (Kliot et al, 2001; Borowski et al, 2004). This impression was confirmed by our interviews. In the opinion of the interviewees, more flexibility is needed, particularly in funding mechanisms [6, 7, 9]. Moreover, actions towards tighter collaboration with both science and administrative actors on state and national levels demand more time and personnel resources [6, 7]. Only some organizations have managed to provide budgets for the less pressing environmental issues [11]. Currently, resources are concentrated on implementing the WFD. Climate change needs to receive higher priority in order to open up funding opportunities.

Conclusions

We aimed to study the current institutional adaptation to climate change impacts in the Elbe Basin from the perspective of interviewed experts. While the framework used for the study is rather all encompassing in its categories, we did not aim at a full picture of the situation in the basin, but rather at highlighting insights provided by a specific range of actors and focusing on the institutional situation. We only interviewed 11 experts and did not analyse the situation in the Czech part of the basin. Therefore, more intensive follow-up studies with a broader empirical base are needed for the whole basin.

Despite the restrictions, our study provides a general overview of the current situation of institutional adaptation in the German Elbe Basin. We were able to show that adaptation is still at an early stage, while relatively high issue awareness already exists. Remarkably, the information on, and discussion about, adaptation is not as prominent as the current discussions concerning climate change mitigation. However, at the same time, adaptation strategies already exist on lower organizational levels of water management. The current cooperation structures for water management in the Elbe Basin are polycentric; but there is a perceived need for better coordination of the different existing centres, requiring institutions such as the ICPE or the FGG, but also the national level, to strengthen their role as a cross-boundary integrator. Some experts expect stronger leadership from higher levels, especially from the EU, which could give more guidance on adaptation.

State actors consider adaptation mainly as an administrative or political task and see the state less often in the role of developing an enabling framework to encourage autonomous action by private actors. An initial requirement in this direction is the information provided by governments. Authorities and decision-makers in the Elbe Basin should, in our view, make further steps in that direction. This refers particularly to a broader discussion with non-state actors on the development of concrete adaptation strategies. The participatory process currently conducted within the implementation of the WFD may provide a suitable platform for these discussions. It is remarkable in this context that the interviewed private actors are already proactively tackling the issue and aiming at contributing to climate change adaptation on their own. It also became obvious during the interviews that many stakeholders see sectoral integration as a central prerequisite for adaptation to climate change and perceive the need to diversify the topics of discussion in the Elbe Basin. This was especially mentioned for the WFD implementation within which the impacts of climate change and, particularly, flood management should be integrated as soon as possible.

Finally, there is a range of background conditions that have to be fulfilled to make adaptation successful. It appears obvious that institutional adaptation has to be sufficiently resourced in term of finances, time and labour. Only then will water managers in the Elbe Basin (as well as elsewhere) be able to establish comprehensive long-term strategies that leave room for anticipated actions that do not purely react on specific weather events. Moreover, climate change aspects can be integrated if political will is sufficient. In other words, adaptation requires leadership and support by political decision-makers. Currently, in the Elbe Basin, thinking in legislative periods and, thus, short-term oriented planning – also according to stringent timescales of the Water Framework Directive – inhibit more action on institutional adaptation and can restrict options for an early integration of climate change adaptation within policies. The WFD plays a major role in daily water management practice and, in many cases, priority is given to the next implementation steps. The stringent deadlines set by the WFD are certainly one of the reasons for this. At the same time, the WFD implementation itself offers good opportunities to include measures for climate change adaptation – for example, by making use of the river basin management plans and by understanding that the implementation of the WFD will not stop with the end of its first policy cycle in 2015.

In conclusion, adaptation to climate change impacts in the Elbe Basin has a large potential for more action in the future. Prospective future activities by the European

Commission will certainly foster a desire for strategies and measures of adaptation at lower levels. The Green Paper (2007) on adaptation to climate change in Europe and the new Common Implementation Strategy (CIS) working group on the integration of climate change issues in the implementation of the WFD can develop as leading institutions in this respect and foster interest in intensified adaptation in the Elbe Basin.

Note

1 Bracketed reference numbers refer to the interviews and reflect the organization as represented in Figure 15.1.

References

Adger, W. N., N. W. Arnella and E. L. Tompkins (2005) 'Successful adaptation to climate change across scales', in *Global Environmental Change*, vol 15, pp77–86

Allman, L., P. Fleming and A. Wallace (2004) 'The progress of English and Welsh local authorities in addressing climate change', *Local Environment*, vol 9, no 3, pp271–283

Bandura, A. (1977) *Social Learning Theory*, Prentice-Hall, Englewood Cliffs, New Jersey, US

Becker, A. and U. Grünewald (2003) 'Flood risk in Central Europe', *Science*, vol 300, no 5623, p1099

Berkes, F. and C. Folke (eds) (1998) *Linking Social and Ecological Systems: Management Practices and Social Mechanisms for Building Resilience*, Cambridge University Press, New York, NY

Berkes, F., J. Colding and C. Folke (2003) 'Introduction', in F. Berkes, J. Colding and C. Folke (eds) *Navigating Social–Ecological Systems: Building Resilience for Complexity and Change*, Cambridge University Press, Cambridge, pp1–30

Berkhout, F., J. Hertin and D. M. Gann (2006) 'Learning to adapt: Organisational adaptation to climate change impacts', in *Climatic Change*, vol 9, no 1, pp135–156

Borowski, I., N. Kranz, E. Kampa and A. Vorwerk (2004) *Public Participation in the Elbe Basin*, Case study report produced under Work Package 5, Osnabrück

Brugnach, M., A. Dewulf, C. Pahl-Wostl and T. Taillieu (2007) 'Towards a relational concept of uncertainty: About knowing too little, knowing too differently and accepting not to know', Paper contributed to the 14th International Conference on Multi-organisational Partnerships, Alliances and Networks (MOPAN), 28–29 June 2007, Leuven

Burton, I., J. B. Smith and S. Lenhart (1998) 'Adaptation to climate change: Theory and assessment', in J. F. Feenstra, I. Burton, J. B. Smith and R. S. J. Tol (eds) *Handbook on Methods for Climate Change Impact Assessment and Adaptation Strategies*, United Nations Environment Programme and Institute for Environmental Studies, Vrije Universiteit, Amsterdam, pp5.1–5.20

Burton, I., S. Huq, B. Lim, O. Pilifosova and E. L. Schipper (2002) 'From impacts assessment to adaptation priorities: The shaping of adaptation policy', *Climate Policy*, vol 2, pp145–159

Cooke, B. and U. Kothari (eds) (2001) *Participation: The New Tyranny?*, Zed Books, London and New York

Crabbé, P. and M. Robin (2006) 'Institutional adaptation of water resource infrastructures to climate change in Eastern Ontario', *Climatic Change*, vol 78, pp103–133

deLeon, P. and L. deLeon (2002) 'What ever happened to policy implementation? An alternative approach', *Journal of Pubic Administration Research and Theory*, vol 12, no 4, pp467–492

Dolan, A. H., B. Smit, M. W. Skinner, B. Bradshaw and C. R. Bryant (2001) *Adaptation to*

Climate Change in Agriculture: Evaluation of Options, Occasional Papers in Geography no 26, Guelph, Ontario, Canada

Dovers, S. (2001) 'Institutional barriers and opportunities: Processes and arrangements for natural resource management in Australia', *Water Science and Technology*, vol 43, no 9, pp215–226

Dovers, S. (2003) *Scaling Governance and Institutions for Sustainability*, Academic Forum, Network of Regional Government for Sustainable Development: Regional Governance for Sustainability, Fremantle

Dryzek, J. S. (1987) *Rational Ecology: Environment and Political Economy*, Basil Blackwood, Oxford.

EC/JRC (eds) (2005) *Pilot River Basin Outcome Report*, Testing of the WFD Guidance Documents, Luxemburg

EEA (European Environment Agency) (2007) *Climate Change and Water Adaptation Issues*, EEA Technical Report no 2/2007, Copenhagen

EU (European Union) (2002) *Common Implementation Strategy for the Water Framework Directive (2000/60/EC)*, Guidance Document no 8, Public Participation in Relation to the Water Framework Directive, Luxemburg

European Flood Directive 2007/60/EC of the European Parliament and of the Council of 23 October 2007 on the Assessment and Management of Flood Risks, OJ L 288

FGG Elbe (Flussgebietsgemeinschaft Elbe) (ed) (2004) *Bericht an die EU-Kommission nach Art. 3 Wasserrahmenrichtlinie für die Flussgebietseinheit Elbe*, Magdeburg

Folke, C., S. R. Carpenter, T. Elmqvist, L.H. Gunderson, C. S. Holling, B. H. Walker, J. Bengtsson, F. Berkes, J. Colding, K. Danell, M. Falkenmark, L. Gordon, R. Kaspersson, N. Kautsky, A. Kinzig, S. A. Levin, K.-G. Mäler, F. Moberg, L. Ohlsson, P. Olsson, E. Ostrom, W. Reid, J. Rockström, S. Savenije and U. Svedin (2002) *Resilience and Sustainable Development: Building Adaptive Capacity in a World of Transformations*, ICSU Series on Science for Sustainable Development, no 3, International Council for Science, Paris

Folke, C., T. Hahn, P. Olsson and J. Norberg (2005) 'Adaptive governance of social–ecological systems', *Annual Review of Environment and Resources*, vol 30, pp441–473

Galaz, R. V. (2005) *Does the EC Water Framework Directive Build Resilience? Harnessing Socio–Ecological Complexity in European Water Management*, Policy Paper 1 by the Resilience and Freshwater Initiative, Swedish Water House, Stockholm, Sweden

Gehring, T. and S. Oberthür (2000) *Exploring Regime Interaction: A Framework for Analysis*, Final conference of the Concerted Action Programme on the Effectiveness of the International Environmental Agreements and EU Legislation, Barcelona, Spain

Green Paper (2007) Green Paper from the Commission to the Council, the European Parliament, the European Economic and Social Committee and the Committee of the Regions – Adapting to Climate Change in Europe: Options for EU Action, SEC (2007) 849

Gunderson, L. and C. S. Holling (2001) *Panarchy: Understanding Transformations in Systems of Humans and Nature*, Island Press, Washington, DC

Homer-Dixon, T. (1999) *Environment, Scarcity, and Violence*, Princeton University Press, Princeton, NJ

Hooghe, L. and G. Marks (2003) 'Unravelling the central state, but how? Types of multi-level governance', *Reihe Politikwissenschaft*, vol 87, Wien

Huitema, D., W. Egas, S. Möllenkamp, E. Mostert, C. Pahl-Wostl and R. Yalcin (in review) 'Adaptive water governance: Assessing adaptive management from a governance perspective', *Ecology and Society*

Jacob, D., H. Göttel, S. Kotlarski, P. Lorenz, K. Sieck (2008) 'Klimaauswirkungen und Anpassung in Deutschland – Phase 1: Erstellung regionaler Klimaszenarien für Deutschland', UBA Forschungsbericht 000969

Kim, J. A. (2003) *Institutional Interplay between Biodiversity and Climate Change: Toward Synergy Creation*, UNU/IAS Working Paper no 100, www.ias.unu.edu/sub_page.aspx?catID=7&ddlID=196

Klabbers, J. H. G., R. J. Swart, R. Janssen, F. Vellinga and A. P. van Ulden (1996) 'Climate change policy development: Enhancing the science/policy dialogue', *Mitigation and Adaptation Strategies for Global Change*, vol 1, pp73–93

Kliot, N., D. Shmueli and U. Shamir (2001) 'Development of institutional frameworks for the management of transboundary water resources', *International Journal of Global Environmental Issues (IJGENVI)*, vol 1, no 3/4, pp306–328

Krysanova, V., F. Wechsung, J. Arnold, R. Srinivasan and J. Williams (2000) *SWIM (Soil and Water Integrated Model), User Manual*, edited by PIK, Report no 69, Potsdam

Krysanova, V., F. Hattermann and A. Habeck (2005) 'Expected changes in water resources availability and water quality with respect to climate change in the Elbe River basin (Germany)', *Nordic Hydrology*, vol 36, no 4–5, pp321–333

Krysanova, V., Z. W. Kundzewicz, I. Pinskwar, A. Habeck and F. Hattermann (2006) 'Regional socio-economic and environmental changes and their impacts on water resources: An example of Odra and Elbe basins', *Water Resources Management*, vol 20, no 4, pp607–641

Lebel, L., J. M. Anderies, B. Campbell, C. Folke, S. Hatfield-Dodds, T. P. Hughes and J. Wilson (2006) 'Governance and the capacity to manage resilience in regional social–ecological systems', *Ecology and Society*, vol 11, no 1, www.ecologyandsociety.org/vol11/iss1/art19/

Lee, K. N. (1999) 'Appraising adaptive management', *Ecology and Society*, vol 3, no 2

Lee, M. and Abbot, C. (2003) 'Legislation: The usual suspects? Public participation under the Aarhus Convention', *The Modern Law Review*, vol 66, no 1, pp80–108

Leipprand, A., T. Dworak, F. Hattermann, V. Krysanova, J. Post and S. Kadner (2006) *Impacts of Climate Change on Water Resources – Adaptation Strategies for Europe*, Ecologic, Berlin

McGinnis, M. (ed) (1999) *Polycentric Governance and Development: Readings from the Workshop in Political Theory and Policy Analysis*, University of Michigan Press, Ann Arbor, MI

Mostert, E., C. Pahl-Wostl, Y. Rees, B. Searle, D. Tàbara and J. Tippett (2007) Social learning in European river-basin management: Barriers and fostering mechanisms from 10 river basins', *Ecology and Society*, vol 12, no 1, p19

Newig, J. (2007) 'Does public participation in environmental decisions lead to improved environmental quality? Towards an analytical framework', *Communication, Co-operation, Participation*, vol 1, no 1, pp51–71

Olsson, P., C. Folke and F. Berkes (2004) 'Adaptive co-management for building resilience in social–ecological systems', *Environmental Management*, vol 34, no 1, pp75–90

Ostrom, E. (2001) 'Vulnerability and polycentric governance systems', *Newsletter on the International Human Dimensions Programme on Global Environmental Change*, no 3

Ostrom, E. (2005) *Understanding Institutional Diversity*, University Press, New Haven, Princeton University, Princeton, NJ

Pahl-Wostl, C. (2005) *Transition towards Adaptive Management of Water Facing Climate and Global Change*, NeWater Working Paper no 6

Pahl-Wostl, C. (2007) 'Transitions towards adaptive management of water facing climate and global change', *Water Resources Management*, vol 21, no 1, pp49–62

Pahl-Wostl, C., M. Craps, A. Dewulf, E. Mostert, D. Tabara and T. Taillieu (2007) 'Social learning and water resources management', *Ecology and Society*, vol 12, no 2, p5

Raadgever, G. T. (2005) *Analysis of Transboundary Regimes – Case Study: The Elbe Basin*, Appendix to Deliverable 1.3.1. of the NeWater project, Delft, The Netherlands

Raadgever, G. T., E. Mostert and N. van de Giesen (2006) 'Measuring adaptive river basin management', Paper presented at the Adaptive Management of Water Resources Summer Speciality Conference (AWRA), 26–28 June 2006, Missoula, Montana

Ridder, D., E. Mostert and H. A. Wolters (eds) (2005) *Learning Together to Manage Together: Improving Participation in Water Management*, Handbook of the HarmoniCOP project

Roberts, N. C. and P. J. King (1996) *Transforming Public Policy*, Jossey-Bass, San Francisco

Rosendal, G. K. (2001) 'Overlapping international regimes: The case of the Inter-

Governmental Forum on Forests (IFF) between climate change and biodiversity', *International Environmental Agreements: Politics, Law and Economics*, vol 1, no 4, pp447–468

Scholz, J. T. and B. Stiftel (eds) (2005) *Adaptive Governance and Water Conflict, New Institutions for Collaborative Planning*, RFF Press, Washington, DC

Smit, B. and O. Pilvosova (2001) 'Adaptation to climate change in the context of sustainable development and equity', in J. J. McCarthy, O. Canziani, N. A. Leary, D. J. Dokken and K. S. White (eds) *Climate Change 2001: Impacts, Adaptation and Vulnerability*, Cambridge University Press, Cambridge, pp877–912

Stern, N. (2007) *The Economics of Climate Change: The Stern Review*, Cambridge University Press, Cambridge, UK

WFD (Water Framework Directive) (2000) Directive 2000/60/EC of the European Parliament and of the Council of 23 October 2000 Establishing a Framework for Community Action in the Field of Water Policy, OJ L 327/1

Young, O. R. (1999) *Science Plan for the Project of the Institutional Dimensions of Global Change*, IHDP Report no 9, Bonn, Germany

Young, O. R. (2002) *The Institutional Dimensions of Environmental Change: Fit, Interplay, and Scale (Global Environmental Accords: Strategies for Sustainability)*, MIT Press, Cambridge, MA

16

The Use of Seasonal Climate Forecasts within a Shared Reservoir System: The Case of Angat Reservoir, the Philippines

Casey Brown, Esther Conrad, A. Sankarasubramanian, Shiv Someshwar and Dulce Elazegui

Introduction

While much attention focuses on climate change, the most pressing challenge for water managers throughout the world is managing the year-to-year, month-to-month and even daily changes in the availability of water. For large-volume water users, reservoirs have provided the storage to reduce much of the variability in the supply from water sources. Growing demand for water that accompanies population growth and economic development, and the rise of minimum flow requirements for the environment, have made managing climate variability increasingly difficult. In addition, the recognition of the non-stationary nature of climate and the influence of anthropogenic global change on water resources raises concerns regarding the assumptions that were used to design and operate water infrastructure. As a result, there is a need for adapting the current practices of reservoir management to the current challenges of a changing climate, competing demands and finite water resources.

A currently underexploited source of assistance to the challenges of water managers is the use of climate information and, specifically, climate forecasts on seasonal to interannual timescales. Advances in our understanding of climate variability, stemming largely from increased understanding and observations of the evolution of ocean temperatures, have made skilful forecasts of precipitation possible in many parts of the world. Due to the chaotic nature of the atmosphere and the slow evolution of ocean temperatures, the skill of these long-lead forecasts is highest when averaged over a period of several months. In some cases, the lead time of the forecasts can be as high as one year. Stream flow tends to be more easily predicted than rainfall due to the smoothing of the spatial and temporal variability that occurs in rainfall over a watershed.

The potential to apply seasonal climate forecasts to water management appears straightforward. Water users could plan their use patterns according to the expected availability of water and water managers could plan release schedules in the same way. However, the probabilistic nature of the forecasts, the uncertainty associated with any new innovation and the institutional context within which water is managed all complicate the potential use of forecasts.

In this chapter, we present a case study of the development of seasonal climate forecasts for Angat Reservoir, which provides the primary source of water for Metropolitan Manila, the Philippines. This is the case of a relatively small reservoir that provides only seasonal storage, serving three competing demands within an often contentious institutional environment. The location of the Philippines in the western Pacific causes an added complication of strong climate variability, due largely to the El Niño Southern Oscillation (ENSO). However, the influence of ENSO is also the source of predictability and seasonal climate forecasts of inflows to the Angat Reservoir are found to be skilful. The challenge remains in bringing the potential benefits of forecast use to fruition.

In the next section, the institutional setting is described, including a description of the water users and the rules that govern water allocation. Following that, the physical setting is depicted, including a description of the seasonal climate forecast development. The next section describes the application of the forecast to reservoir management. Opportunities and constraints are discussed next, followed by some conclusions.

Setting: The shared water resources of Angat Reservoir

Angat Reservoir, located in Bulacan Province in Central Luzon, collects water from the Angat River Basin, with a drainage system covering approximately 568km^2 (see Figure 16.1) and supplying 4000 million litres per day. The reservoir, completed in 1969, is critically important to three sets of users: agricultural producers in Bulacan Province, municipal water users in Metro Manila, and hydropower managers. Angat water currently provides 97 per cent of Metro Manila's water, a large proportion of the irrigation water for Bulacan, and a critical back-up power source during heavy demand periods for the Luzon power grid.

In normal years, this water has met the needs of all three users. However, increased growth in Metro Manila and privatization of the power market has placed additional demands upon the reservoir, which become especially acute during periods of low rainfall. As a result, contentions over the allocation of water across users have increased. The Philippine Water Code provides that during normal conditions, first priority for water use goes to those holding the original rights to the water (in this case, the farmers of Bulacan, where the reservoir is located). In times of scarcity, municipal water use takes priority over uses for agriculture or hydropower generation, and those users who normally would have received priority should be compensated accordingly (Tabios and David, 2004).

In practice, however, the situation looks rather different. Given growing demands from Metro Manila and its strong political significance, in practice the situation of

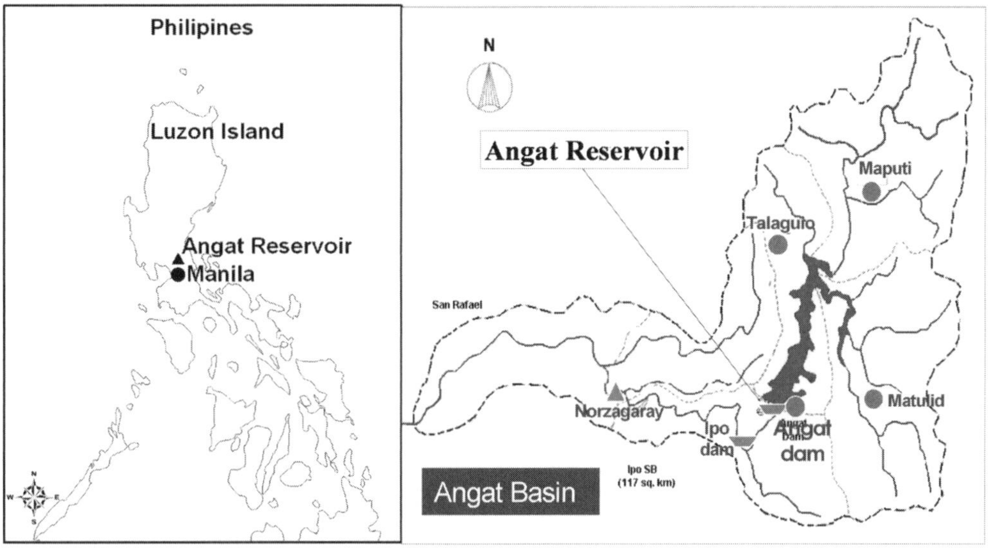

Figure 16.1 *Location of Angat Basin and Angat Reservoir in Luzon Island, the Philippines*

Note: The figure on the right shows the Angat Reservoir in the Angat Basin (shaded area). Circles indicate the location of four precipitation stations upstream of the Angat Reservoir.

Source: National Water Resources Board, Government of the Philippines

water 'scarcity' and priority for municipal water seems to apply more often than not. The rights to the 69m^3/s of available Angat water are formally divided as follows: 36m^3/s for agriculture in Bulacan and 31m^3/s for Metro Manila, with 2m^3/s reserved for environmental purposes, with hydropower generation assumed to be non-consumptive (Tabios and David, 2004). However, 15m^3/s of the allotted amount for agriculture is regularly given to Metro Manila, making the typical allocation 46m^3/s to Metro Manila and 21m^3/s for agriculture. Given political pressures, reducing Metro Manila's allocation during times of scarcity proves difficult, regardless of climate forecasts.

The responsibility for making regular water allocation decisions lies with the National Water Resources Board (NWRB), which coordinates regular meetings involving both stakeholders at PAGASA, the Philippines National Meteorological Service. Informal institutional behaviour can play an enormous role in determining how decisions are actually made, and these are shaped by the socio-economic, institutional and policy context that each user faces. In this section, we briefly discuss socio-economic context and institutional dynamics for each of the three uses for Angat water, and review the institutional dynamics of the water allocation decision process. These factors are critical to consider when assessing possibilities for the use of seasonal forecasts of reservoir inflow.

Agriculture in Bulacan

Bulacan Province, where the Angat Reservoir is located, lies in Central Luzon to the north of Metro Manila. Its proximity to the metropolitan area has stimulated growth in industry and service sectors over the past few decades. However, agriculture has remained a main source of livelihood for many households. Rice, as a staple food across the Philippines and the country's main indicator of food security, is the most important crop. Land-use data indicate that the total farmed area in the province is gradually decreasing; but almost all of this is rice (*palay*), and two-thirds is irrigated (see Table 16.1). Small amounts of vegetable and fruit crops, such as eggplant, tomato and mango, are grown. This could represent an important income opportunity for Bulacan farmers given their proximity to Metro Manila. However, interviews with the Bulacan Provincial Irrigation Management Office (BPIMO) suggest that farmers may continue to plant rice to ensure food availability for their own households, and simply because it is what they and their families have done for generations (BPIMO, 2006).

Table 16.1 *Agricultural land use, Bulacan Province, 1960–2002*

	1960	1971	1980	1991	2002
Number of farms	30,206	27,948	39,394	48,451	46,183
Area of farms (ha)	72,592	63,960	73,465	69,242	63,164
Average area per farm (ha)	2.40	2.29	1.86	1.43	1.37
Irrigated farms					
Number of farms reporting	10,586	13,087	27,681	24,858	30,386
Physical farm area irrigated (ha)	20,608	25,226	47,942	38,799	42,793
Average area of irrigated farm (ha)	1.95	1.93	1.73	1.56	1.41
Area planted/harvested (ha)					
Palay	66,161	68,796	105,006	77,002	58,221
Corn	1956	366	950	2263	869

Source: Bulacan Provincial Agriculture Office, adapted from Rola and Elazegui (2006)

Bulacan rice farmers plant two crops per year: a 'wet-season' crop in June, which is harvested in October, and a 'dry-season' crop in November, which is harvested in March or April. The 'dry' season is more important for farmer incomes because expenses are less (labour and transportation is cheaper) and crop harvests tend to be higher. However, this season is also the most risky because farmers must have irrigation water in order to plant. For this, they depend upon the Angat Reservoir (Rola and Elazegui, 2006).

Variability in both rainfall and the supply water from the Angat Reservoir has been associated with serious impacts upon crop harvests and, in turn, farmer livelihoods. The most dramatic example is the 1997 El Niño year, when a prolonged dry period led to a complete loss of the dry-season crop, when farmers received no water at all from the Angat Reservoir. In 2004, another El Niño year, dry-season production was 18 per cent below the average of 1990 to 2005, and 32 per cent below average in the wet season (Rola and Elazegui, 2006). This pattern of impacts is determined not only by variability in rainfall patterns, but also by institutional factors guiding water allocation decisions.

The National Irrigation Administration (NIA) is responsible for major irrigation infrastructure – called 'national' irrigation systems – in the Philippines. The Angat–Maasim River Irrigation System (AMRIS) is among these, covering Bulacan and some of the neighbouring Pampanga Province. Bulacan's provincial irrigation office assesses irrigation needs each year and submits 'irrigation diversion requirement' to the national office, usually about two months prior to each cropping season. The NIA's national office then represents Bulacan farmer interests in meetings with the National Water Resources Board to decide Angat water allocation. The NIA is funded substantially through irrigation service fees, paid by farmers according to the amount of water they receive.

The NWRB makes a tentative seasonal water allocation to the farmers, which the BPIMO then uses to create a water delivery schedule throughout the season. However, this water allocation could be delayed or reduced at any point in the season. In the dry season, the NWRB may delay or reduce initial water deliveries to Bulacan if uncertainty about the upcoming season is too great, forcing farmers to delay planting. If serious water shortage conditions develop after water deliveries have begun, allocation to farmers could be reduced later in the season. If reductions occur in February and March, this causes serious problems for farmers trying to sustain their crops until harvest at the end of March or early April (Rola and Elazegui, 2006).

When irrigation water is delayed or curtailed, Bulacan's Provincial Development Coordinating Office (PDCO) works with the agriculture and irrigation offices to coordinate response measures to help mitigate impacts, including programmes such as supply of additional agricultural inputs, adjustments in water delivery schedules, distribution of water pumps and planting alternative crops. Crop damage reports are used to target such assistance, and funding comes from several sources, including AMRIS and the provincial disaster relief fund (Rola and Elazegui, 2006). These funding levels are quite small, and Bulacan Province and NIA have for years been advocating for compensation when water deliveries are reduced. However, despite formal provision for this in the Water Code, mechanisms for compensation have never been developed.

Metro Manila water supply

Metro Manila, home to over 11 million people, is really a set of 17 cities clustered into a special administrative area called the National Capital Region (NCR). Given Metro Manila's political and economic importance for the Philippines, the president of the Philippines can exercise direct supervision over the NCR (Elazegui et al, 2007). This has important implications in decision-making about water allocation for Metro Manila; in several instances, the president has intervened to override decisions, usually to ensure sufficient water for Metro Manila (NWRB, 2006).

97 per cent of Metro Manila's water comes from the Angat Reservoir; 3 per cent is from groundwater. The number of water connections has steadily increased, doubling between 1986 and 2004 to over 1 million connections serving approximately 8 million people. Demand will continue to increase; by 2025, it is anticipated that over 12 million will be served, pushing demand up to 3570 million litres/day (see Table 16.2).

Table 16.2 *Projected water demand for the National Capital Region of Manila*

Water demand (million litres/day)	1981	1985	1993	1997	2001	2005	2010	2015	2020	2025
Domestic	1155	1232	1658	2003	2379	1584	1741	2169	2267	2336
Commercial	456	554	952	1272	1698	586	774	862	948	1031
Industrial	414	493	688	806	943	112	150	169	186	203
Total	2025	2279	3298	4081	5020	2282	2665	3181	3401	3570

Source: adapted from Elazegui et al (2007) – sources: *Study on Water Supply and Sewerage Master Plan of Metro Manila, 1996* cited by World Bank in *Philippines Environment Monitor 2000* (1981–2001 data); Sinclair Knight Merz (Philippines) Inc and DCCD Engineering (2005) *Water Supply, Sewerage and Sanitation Master Plan for Metro Manila: Volumes I – V* (2005–2010 data)

Water services in Metro Manila are managed by the Metropolitan Waterworks and Sewerage System (MWSS). While MWSS is a stakeholder in the decision-making process, as a government-owned and controlled corporation (GOCC), it has a certain degree of autonomy in its decision-making regarding Metro Manila's water supply. In addition, it is part owner of the Angat Reservoir, since it paid for one third of its construction costs.

The privatization of the Metro Manila water supply had an important impact upon the water allocation decision process. In 1997, MWSS was split into two entities: the corporate office, which owns and manages assets, and the regulatory office, with authority to regulate water and sewerage services, including setting tariff rates and monitoring private concession contracts for Metro Manila's water, without NWRB involvement. Two 25-year concession contracts were issued to Maynilad and Manila Water to handle the distribution of water in Metro Manila's west and east zones, respectively. These contracts guaranteed them a total of $46m^3$ of water from MWSS, in spite of the fact that the MWSS's formal water allocation from Angat is only $31m^3$. The additional $15m^3$ is water that is only 'conditionally' granted to the MWSS; the NIA is supposed to have first claim on it, and it can only be used by the MWSS if it is not needed by the NIA. However, according to the agreements, the MWSS must compensate the concessionaires if they deliver less than $46m^3$. Thus, in practice, the MWSS is typically unwilling to accept reductions of more than 10 per cent below this amount, even though there is no written rule to this effect (MWSS, 2006).

Part of the rationale for privatization was to reduce Metro Manila's substantial non-revenue water losses, which were over 60 per cent in 1997. The two concessionaires have performed quite differently in this respect; while Manila Water reduced non-revenue water in its zone to 24 per cent in 2007, rates in the Maynilad zone actually increased, reaching 66 per cent in 2007 (Cuevas-Miel, 2008; Manila Water Company, 2008). These continued water losses are regularly brought up by the NIA as they argue against cuts to water for agriculture. Concessionaires are vocal in the process as well; under the concession agreements, the MWSS has a responsibility to ensure that the concessionaires are represented in meetings convened by the NWRB, which may have implications for their interests.

Given the frequent contentions over water, accompanied by growing water demand, the MWSS has been seeking new sources of water for Metro Manila. Several large reservoir projects are under exploration; but these will probably take a decade or more to materialize. In January 2008, the MWSS announced that it was in the process of making an agreement with the governor of Bulacan to pay for the construction of a low-level dam near Calampit in Bulacan Province, which would supply to farmers the $15m^3$ that is now being regularly allocated to Metro Manila. It is unclear how long this project will take (NWRB, 2008).

Hydropower generation

The Angat Reservoir is also used for power generation, with a capacity of about 250MW. Although it contributes only about 5 per cent of the total power in the Luzon grid, hydropower from Angat and other reservoirs plays an important role in providing additional supply in high-demand periods and in restarting the grid following blackouts.

The Angat Hydroelectric Plant is operated by the National Power Corporation (NPC), which is also responsible for the physical operation and maintenance of the reservoir. Therefore, all decisions by the National Water Resources Board about Angat water allocations must be implemented by the NPC. Hydropower is formally considered a 'non-consumptive' use, to be generated from the flows being allocated to the NIA and MWSS. According to NWRB guidance, the NPC is not allowed to release water from the reservoir for the sole purpose of hydropower production without consent from the NWRB (NWRB, 2003).

However, the power privatization process in the Philippines has changed the institutional context in which the NPC operates. In June 2006, an electricity spot market was established, regulated by the Philippine Electricity Market Corporation (PEMC), which now controls production and dispatching of hydropower production. The NPC must submit bids in order to produce electricity during times determined by the PEMC. In order for hydropower production to be non-consumptive, the NPC must deliver water to the NIA and MWSS during times that correspond to the hours of dispatch, usually the hours of peak electricity demand (NPC, 2006). This presents a considerable challenge for the NPC. On a number of occasions, the NPC has released water to generate hydropower outside of NIA or MWSS uses, which has been a cause for considerable concern on the part of the NWRB, NIA and MWSS (NWRB, 2008).

Angat water allocation decision process

The National Water Resources Board was established in the Philippine Water Code to implement the nation's water policies, including water allocations for Angat Reservoir. Releases from the reservoir must be authorized by the board of NWRB, which is made up of a set of government agencies who are not direct stakeholders for water use. Angat Reservoir stakeholders are engaged in the process via committees coordinated by the NWRB that provide recommendations to the board. Typically, these recommendations are developed by the so-called technical working group, chaired by the NWRB with participation from the MWSS (and sometimes its concessionaires Maynilad and Manila Water), NIA, NPC and the meteorological service (PAGASA). The NWRB presents scenarios projecting water use over the next several months, using an Excel-based reservoir model comparing current conditions against historical averages. The NWRB receives a monthly forecast from PAGASA, which provides deterministic precipitation forecasts for the coming month. At the meeting itself, PAGASA offers further detail on climate conditions, while the MWSS, NIA and NPC each bring their particular interests to the table and advocate for their water needs. Meetings become more frequent – and more contentious – when water levels are low (e.g. NWRB, 2007).

Meeting notes and interviews with stakeholders suggest that PAGASA's forecasts do play a general role in release decisions. For example, forecasts of higher than average rainfall when reservoir levels are already high strengthen the NIA's argument for water releases, and when rainfall is anticipated to be below average, this would make the NWRB even more reluctant to allow releases for agriculture. PAGASA's classifications of particular years as 'El Niño' or 'La Niña' appear to be frequently used as a basis for judgement about expected rainfall behaviour, with El Niño years being associated with drought and La Niña years with excess rain (e.g. see NWRB, 2006–2007). However, the use of national-scale deterministic rainfall forecasts and general indications of ENSO conditions in order to assess possible future water levels in Angat does not allow for a clear consideration of the specific correlations between expected rainfall conditions and inflow; nor does it enable a risk management approach to decision-making.

Seasonal forecasts of Angat Reservoir inflows

Angat Reservoir is located on the island of Luzon in the northern Philippines. It collects water from a watershed area of 568km^2. The climate of the northern Philippines is tropical and the watershed receives an average annual rainfall of 292cm. The rainfall occurs in two seasons (see Figure 16.2). The summer rainy season is associated with the Asian south-west (summer) monsoon, lasting from June to September

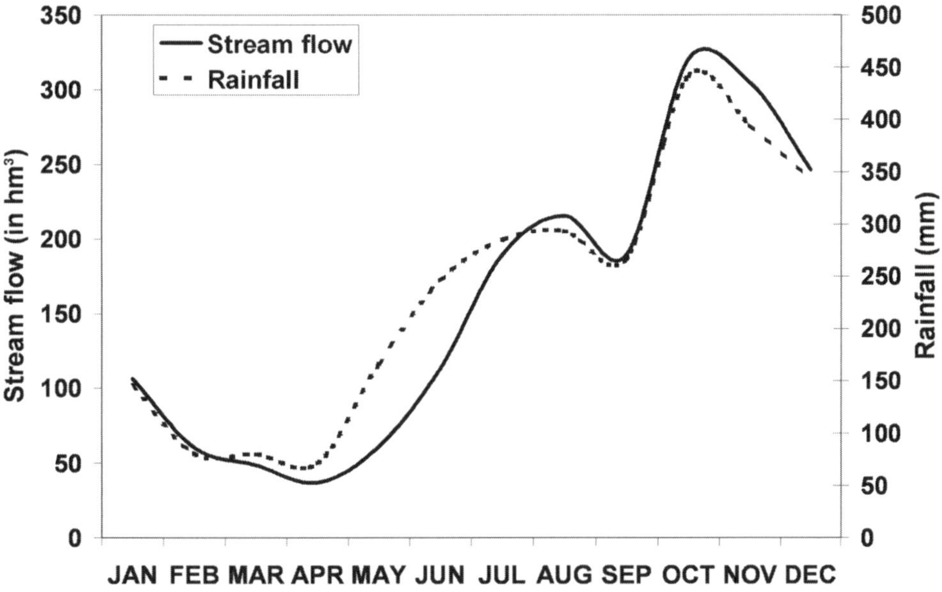

Figure 16.2 *Annual cycle of rainfall and inflows at Angat Reservoir*

Source: IRI/PAGASA

and accounting for 30 per cent of the annual inflow to Angat. The winter rainy season results from the north-east monsoon, beginning in October and continuing until February, contributing 55 per cent of the annual total inflow. Typhoons also occur during the winter rainy season and are an important source of water. The average annual inflow to the reservoir is 150 million cubic metres.

The amount of rainfall that occurs during the winter season is strongly influenced by conditions in the equatorial Pacific Ocean. An index representing those conditions, NINO3.4, shows a statistically significant correlation with the inflows to Angat ($r = -0.55$). This correlation indicates the magnitude and direction of influence that the El Niño Southern Oscillation (ENSO) has on rainfall and inflows to the reservoir. During El Niño events (warm phase), there is a heightened risk of drought in Luzon and inflows to Angat are below normal on average. During La Niña events (cold phase), there are typically above-normal inflows. The ENSO event of 1997 to 1999 exemplifies these effects. In the autumn of 1997, a strong El Niño developed, causing a major decrease in rainfall for the Angat area and a reduction of 60 per cent in inflows to the reservoir. The drought persisted throughout 1998. However, in the autumn of that year, the El Niño changed to a La Niña, and with it the rains returned. In fact, the inflows for the 1998 to 1999 winter season were 80 per cent above normal. The strong inter-annual variability of inflows to the reservoir, such as experienced during 1997 to 1999, is a pressing and continuing challenge to the stakeholders of Angat Reservoir.

While the influence of ENSO on reservoir inflows causes challenging variability, it also can be a source of predictability. ENSO is a more or less deterministic process that is relatively well understood, has been successfully represented in climate models and for which there exists demonstrated (although not perfect) predictive skill. ENSO manifests as a pattern of anomalous sea surface temperatures (SSTs) and winds in the equatorial Pacific, and through 'teleconnections' affects rainfall and temperatures throughout the world, most strongly in the tropics. Since SST patterns exhibit persistence, having a high correlation from one month to the next, and since the rainfall teleconnections tend to lag the SST pattern, they provide a source of predictability. ENSO patterns form in the autumn and persist throughout the winter until the following spring. Thus, if a particular pattern, such as El Niño, forms in the autumn, the rainfall teleconnection (e.g. reduced rainfall) can be expected to have influence throughout the winter.

Given an impetus for predictability, the actual prediction of reservoir inflows can be accomplished in a number of ways. There are two primary approaches: statistical forecasts and model-based forecasts. Statistical forecasts use empirical relationships between predictors, such as SSTs, and the predictand – in this case, reservoir inflow. Based on these relationships, statistical models are created that relate the predictors to the predictands. Statistical models are advantageous in that they are inexpensive and they allow explicit treatment of uncertainty. A disadvantage is the risk of over-fitting, where a model's skill is overestimated by matching the historical data with too few degrees of freedom, and of spurious correlations, which are exhibited in the data but do not have a physical basis. It is important when designing statistical models to explain the physical basis for the model and to reserve some of the historical data that was not used for model-fitting for validation of the model.

Model forecasts are produced by complex numerical models that simulate the physical processes in the atmosphere. The atmospheric model responds to the state of the underlying ocean (the 'boundary conditions') and the initial state of the atmosphere (the 'initial conditions'). The ocean conditions are often modelled separately, either with a separate ocean model or through simple persistence or damped persistence of the initial conditions. Some models, called 'coupled models', attempt to model both the ocean and atmosphere together.

Due to the inherent chaotic nature of the atmosphere, model predictions are typically made as ensembles: a series of model runs, each starting from slightly different initial conditions. In some cases, 'super ensembles' are created, combining ensembles from several models. This is done in an attempt to account for model error as all models make compromises in their representation of physical processes. Since the models often differ in their approaches, using several may improve the degree to which the climate signal is perceived and separated from the background noise that occurs due to randomness and model errors.

In the case of Angat Reservoir, the ECHAM 4.5 atmospheric general circulation model (GCM) was found to have skill in predicting monthly precipitation in the area of Angat for the winter season. The ocean boundary conditions were provided by assuming persistence of the initial ocean state. The monthly values of predicted precipitation fields for October to February were used as inputs to a statistical downscaling model. The downscaling model consisted of a simple regression between the first three empirical orthogonal functions (EOFs) from the predicted precipitation field with a spatial domain roughly covering Luzon Island (0–25°N; 115–130°E) and the inflows to Angat. The details of the forecast are provided in Sankarasubramanian et al (2008). The forecasts of reservoir inflows for October through February are available in September. The correlation between the mean of the forecast and observed reservoir inflows is on the order of 0.5 (see Plate 31).

Due to the limited skill of the seasonal reservoir inflow forecast, it is best represented in probabilistic terms. This is characteristic of seasonal climate forecasts, where the inherent uncertainty of the climate system makes a range of outcomes possible. If a forecast has skill, it reduces the uncertainty regarding the possible distribution of outcomes in comparison to the historical distribution of outcomes. In the current case, the forecasted quantity is the mean of the distribution of possible inflows to the reservoir. The distribution of inflows is modelled as a normal distribution and the scale parameter (variance) is invariant from year to year.

Reservoir decision-making based on inflow forecasts

Decisions to release water from the Angat Reservoir are made by the NWRB. These decisions are formally guided by a lower and an upper rule curve. This is the typical approach used in reservoir management throughout the world. The rule curve is designed such that the reservoir level will be able to meet the demand for water if a design drought were to occur. Rule curves are thus very conservative as they prepare the reservoir to withstand a very severe drought at all times. There is a cost associated with this conservative approach. Since water must be retained to meet demand in the drought, releases are curtailed that otherwise could be generating hydroelectricity and provided for irrigation.

In the case of Angat, the rule curves and the accompanying operations guidelines were formulated by the NWRB secretariat and approved by the board in December 1998, taking effect in March 1999. The latest amendments to the operations guidelines were approved by the board in February 2004. The rules provide that when water is above the upper rule curve, full domestic water supply, irrigation and hydropower generation requirements are granted. When the reservoir elevation is between the upper and lower rule curves, both domestic water supply and irrigation requirements are satisfied, but hydropower generation is limited to the releases made for domestic water supply and irrigation. When water is below the lower rule curve, domestic water supply requirement is granted first and releases for irrigation may be allowed only when the resulting water level will not fall below the minimum operating level of 180 metres (see Plate 30).[1]

The rule curves that guide Angat Reservoir, and those that guide reservoirs around the world, are static. They assume that the probability of the design drought is the same each year or that the probability is unknowable. They do not reflect the notion that deterministic elements of climate variability can change the probability of drought, and that this probability is, to a certain extent, knowable. With the advent of skilful seasonal climate forecasting in many parts of the world, it is possible to estimate the probability of the design drought for a given year. In a year in which the probability differs greatly from the long-term average, reservoir release decisions could be altered to reflect that information. For example, in years where the probability of drought was lower, the rule curve could be lowered and additional water could be released, generating additional benefits. Alternatively, in years when drought is more likely, the rule curve could be raised, retaining additional water to help meet demand if the drought did occur. The result would be a 'dynamic rule curve'.

This rationale was applied in a simulation of the operations of Angat Reservoir using the forecast of winter season inflows described above. Release decisions were made according to a dynamic rule curve calculated for each year based on the forecast. The rule curve was designed to retain enough water to meet demand in the case of the 5 per cent exceedance volume of inflows over the winter season as estimated according to the forecast (i.e. the volume of inflows that, on average, would be exceeded during 95 per cent of all years). Thus, the rule curve would shift with the year-to-year estimations of the 5 per cent exceedance volume. The 5 per cent exceedance volume is calculated from the estimated cumulative distribution function that is a function of the mean predicted by the inflow forecast.

The results of the simulation are presented in Figures 16.3 and 16.4. Figure 16.3 shows the additional hydroelectricity that would be produced by using the forecast of reservoir inflows. This is a result of forecasts of above-normal inflows that result in the dynamic rule curve being adjusted downward. More water can then be released while adhering to the adjusted rule curve. Figure 16.4 shows the additional water that could be delivered to the irrigation district. During certain years, such as 1987, 1989 and 1993, the forecast of above-normal flows again results in a downward-adjusted rule curve and more releases. The forecast of seasonal reservoir inflows allows the water manager to make better use of water in the above-normal years – water that is otherwise spilled over the top of the reservoir and wasted. The forecast essentially increases the size of the reservoir.

Constraints and opportunities in applying forecasts

The simulation analysis of seasonal reservoir inflow forecasts to Angat Reservoir indicated that there are significant potential benefits to their use. This does not mean, however, that the use of forecasts will necessarily follow this demonstration of potential benefits. As with any new technology, there is a degree of inertia that must be overcome before there can be changes to the status quo. In the case of water resources, this inertia is made stronger by the contentious institutional arrangement that often guides the allocation of water resources among competing uses. In addition, the risk

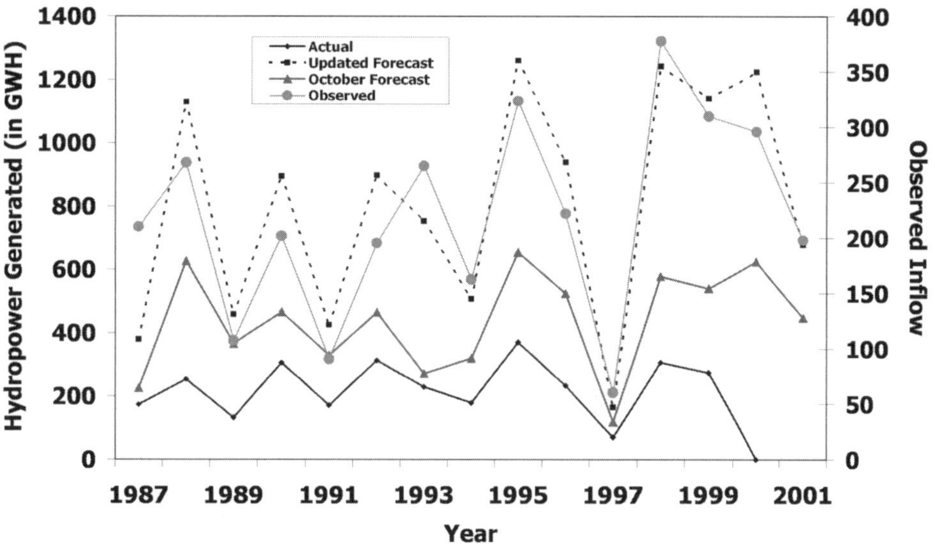

Figure 16.3 *Hydroelectricity production for the years 1987 to 2001 using a forecast delivered in October (triangles) and a forecast updated monthly from October to January*

Note: The black line indicates the actual hydroelectricity production and the circles indicate observed inflow to the reservoir (right axis).

Source: Elazegui et al, 2007

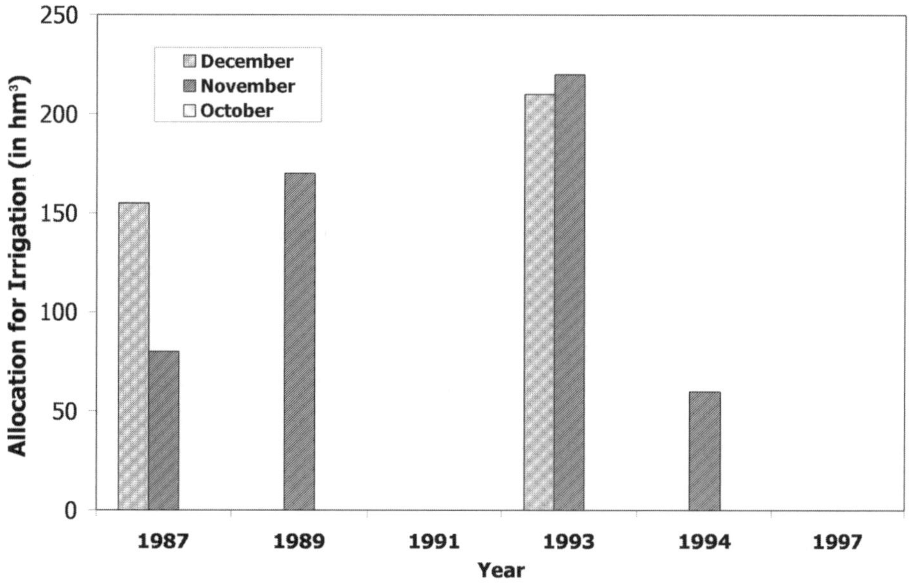

Figure 16.4 *Additional irrigation water that could potentially be delivered according to the forecasts available in October, November and December*

Source: Elazegui et al, 2007

aversion of water managers and the lack of incentives for providing additional benefits further impede adoption of promising innovations.

In the case of Angat Reservoir, the context is relatively conducive to innovation due to the severity of the effects of climate variability and the perceived need for improved water management. Past crises have been clearly linked with ENSO in the minds of the key stakeholders and there is acceptance that skilful forecasts are possible. There is also general acknowledgement that the manner in which previous droughts have been managed is suboptimal and has caused a differential distribution of costs that falls disproportionately on the farmers. Therefore, there is good potential for applying seasonal forecasts for improved water management.

The key constraint is the institutional arrangement that guides water allocation and disagreement over how the forecast is applied to that arrangement. At present, the seasonal forecast of rainfall is prepared by PAGASA and delivered to the technical working group. The seasonal reservoir inflow forecast is currently being introduced; however, the decision dynamic exhibited with regard to the rainfall forecast is informative. The incentive structure of the water allocation policy colours the perception of the forecast. Since the MWSS has first priority for the water and gains nothing from the water that is delivered to the other uses (agriculture and hydroelectricity), its incentive is to ensure that adequate water is kept in the reservoir to serve its needs. Of course, water that is kept in the reservoir cannot be delivered to the other stakeholders. Thus, the MWSS has a strong incentive to call for action on any forecast that indicates an enhanced risk of drought. This gives it a reason to argue for curtailment of deliveries to the other users. The curtailment comes at no cost to the MWSS, but at high cost to those stakeholders who have water deliveries reduced.

Alternatively, the MWSS has a strong disincentive to take action on forecasts of above-normal rainfall (or reservoir inflows). Above-normal rainfall would result in extra water released to the other stakeholders. The MWSS receives no benefit from releasing extra water. Furthermore, if the rainfall was not above normal, it could face risk of drought due to the extra releases (since the forecast is probabilistic, and uncertainty remains, there is always the chance that drought will occur, even if it is very unlikely). Thus, its tendency is to ignore forecasts of above-normal rainfall. Given its priority status and influence, this effectively eliminates the extra benefits that could be gained from using the forecasts.

The crux of this challenge is assuaging the concerns of the MWSS that releasing extra water will increase its risk of drought. The concern is strong and warranted because the consequences of water shortages are dire. As a result, its goal is to eliminate any risk of drought. The solution, then, is to create a drought management system that reduces the consequences of reduced reservoir inflows. One such system has been proposed (Brown and Carriquiry, 2007). The general idea is to use option contracts to allow the MWSS to purchase water from the NIA when drought is forecasted or occurs. In this way, the MWSS is ensured that it will receive all the water that is available without the political difficulties that occur in the current negotiated outcomes. The NIA would receive compensation in place of the water that was scaled according to the sunk investments that the farmers had made at the point of exercising the options. Index insurance based on the inflows to the reservoir could be used to smooth the cost to the MWSS of purchasing water during drought years.

Conclusions

Much of water resources engineering and management has been founded on the view of climate as a stationary process. The risk of climate extremes in any given year was assumed to be unchanging. During the last few decades, increasing understanding of climate variability and, in particular, ENSO, has undermined the rationale for the traditional view. Furthermore, the massive production of greenhouse gases by human activities is changing the climate system, probably leading to a warmer and more variable state of climate. The challenge of reconciling our new understanding of climate with our traditional means of management water resources remains a daunting one.

A case in point is the application of seasonal climate forecasts to reservoir management. Seasonal forecasts, based on our improved understanding of the year-to-year changes in probabilities of rainfall and extremes, offer a potential source of improvement to water management and the means to adapt to a changing climate. Slowly evolving ocean patterns influence rainfall in various parts of the world at lagged timeframes, providing the basis for predictability. This information can be used to enhance preparations for reduced water or to take advantage of the opportunities that arise from excess water. Reservoir management is a particularly auspicious application as the slow response times of reservoirs are well matched to the seasonal timeframe of these forecasts – the period over which they have most skill.

As demonstrated in the case of Angat Reservoir, the forecasts can be used to dynamically change the reservoir rule curve such that it reflects the probability of dry conditions in a given year instead of the long-term probability. Based on the conditions of the ocean and the atmosphere, it can be determined that the probability of dry conditions is greater than or less than the long-term average. In years when the probability of dry conditions is less than normal, more water could be released. A simulation of such a dynamic rule curve applied to Angat shows that additional hydroelectricity and irrigation water could be released during many years.

The implementation of seasonal climate forecasts depends not only upon the potential benefits, but also upon the institutional context. The goal of water managers is to avoid water shortages and the negative attention that accompanies them. Innovations that improve water management are not likely to be adopted unless they contribute to this ultimate goal. While seasonal forecasts can contribute to extra benefits in wet years, and can give advance warning of dry years, they cannot replace the need for a drought management system once the dry year occurs. Having such a system in place is a likely prerequisite for an innovation such as seasonal forecasting to be implemented. Seasonal climate forecasting will make any such system more effective.

Note

1 NWRB Resolution No 004-0204, Revision of the Operation Guidelines for the Angat Multi-Purpose Reservoir, 19 February 2004.

References

BPIMO (Bulacan Provincial Irrigation Management Office) (2006) Meeting with BPIMO officials, Bulacan, Philippines, August

Brown, C. and M. Carriquiry (2007) 'Managing hydroclimatic risk with option contracts and reservoir index insurance,' *Water Resources Research*, 2007WR006093, Bulacan Provincial Irrigation Management Office, Meeting with BPIMO officials, Bulacan, The Philippines, August 2006

Cuevas-Miel, L. C. (2008) 'Maynilad taps foreign firms for water-loss reduction', *The Manila Times*, internet edition, 12 May, www.manilatimes.net/national/2008/may/12/yehey/business/20080512bus9.html (accessed 21 May 2008)

Elazegui, D. D., M. J. M. Rabang, A. C. Rola, E. Ebrahimian and S. Someshwar (2007) *Managing Climate Risks in Metro Manila*, Working Paper no 07-02, Institute for Strategic Planning and Policy Studies, University of the Philippines Los Baños, the Philippines

Manila Water Company (2008) 'Investor guide', www.manilawater.com/investor-relations/investor-guide (accessed 21 May 2008)

MWSS (Metropolitan Waterworks and Sewerage System) (2006) Meeting with Leonor Cleofas, MWSS Corporate Office, August

NPC (National Power Corporation) (2006) Meeting with Virgilio Garcia, National Power Corporation, August

NWRB (National Water Resources Board) (2003) Minutes of the 6th Board Meeting of the Newly Reconstituted NWRB Board, 7 May

NWRB (2006) Meetings with National Water Resources Board, Metro Manila, the Philippines

NWRB (2006–2007) Technical Working Group Meeting Reports, February–December 2007, and NWRB Resolutions on Water Allocation, 2006–2007

NWRB (2007) Technical Working Group Meeting Reports, February–December, 2007, and NWRB Resolutions on Water Allocation, 2006–2007

NWRB (2008) Meetings with National Water Resources Board, Metro Manila, Philippines, February

Rola, A. C. and D. D. Elazegui (2006) *Climate Risk Management at the Local Level: Angat Reservoir Case Study*, Bulacan, Working Paper no 07-01, Institute for Strategic Planning and Policy Studies, University of the Philippines Los Baños, the Philippines

Sankarasubramanian, A., U. Lall and S. Espinueva (2008) 'Role of retrospective forecasts of GCM forced with persisted SST anomalies in operational streamflow forecasts development', *Journal of Hydrometeorology*, vol 9, pp212–227

Sinclair Knight Merz (Philippines) Inc and DCCD Engineering (2005) *Water Supply, Sewerage and Sanitation Master Plan for Metro Manila: Volumes I–V (2005–2010 data)*, the Philippines

Tabios, G. Q. and C. C. David (2004) 'Competing uses of water: Cases of Angat Reservoir, Laguna Lake and groundwater systems of Batangas City and Cebu City', in Rola, A. C. et al (eds) *Winning the Water War: Watersheds, Water Policies and Water Institutions*, Philippine Institute for Development Studies

Index

earthscan
publishing for a sustainable future

Sudden and Disruptive Climate Change

Exploring the Real Risks and How We Can Avoid Them

Edited by ***Michael C. MacCracken, Frances Moore*** *and* ***John C. Topping, Jr.***

'This book makes it clear to all of us that, while risks of abrupt climate change are increasing, opportunities to solve the climate crisis are abundant. It's an impressive accomplishment.' *Al Gore, co-recipient of the 2007 Nobel Peace Prize*

'The multiple lines of evidence ... show that climate change will have more serious effects on our civilization than could have been imagined even a couple years ago. Fortunately, it also points the way to accelerating our response to the growing threat posed by climate change.' *Lester R. Brown*

While changes in emissions and atmospheric concentrations of greenhouse gases are projected to be slow and smooth, the intensity and impacts of climate change could be abrupt and erratic. This book explores the possibilities of such changes, their significance for society and efforts to move more rapidly to limit climate change than current government measures.

Paperback £24.95 • 320 pages • 978-1-84407-478-5

Climate Change and Agriculture in Africa

Impact Assessment and Adaptation Strategies

Ariel Dinar, Rashid Hassan, Robert Mendelsohn *and* ***James Benhin*** *and others*

This book encompasses a comprehensive quantitative analysis of potential economic impacts of future climate change, and value of adaptation measures in Africa.

It examines how climate affects current agricultural systems in Africa, predicts how these systems may be affected in the future by climate change and suggests what role adaptation could play.

The study is the first to combine both country level and cross-country, spatially referenced household survey and climatic data for conducting this type of analysis.

Hardback £49.95 • 224 pages • 978-1-84407-547-8

Climate Policy

Editor-in-Chief ***Michael Grubb***

Climate Policy is a leading journal of analysis and debate on responses to climate change. It addresses both the mitigation of, and adaptation to, climate change. It also provides a forum for the communication of research, analysis, review and discussion of responses to climate change.

Climate Policy is of direct and vital relevance to academic, industrial and government researchers, consultants and negotiators, industrial and non-government lobby organizations, and to all those involved in making, developing and implementing climate change policy at the national, regional or global level.

NOW AVAILABLE ONLINE • For subscription information visit **www.earthscanjournals.com**

For more details and a full listing of Earthscan titles visit:

www.earthscan.co.uk

earthscan

publishing for a sustainable future

Climate Change and Adaptation

Edited by ***Neil Leary, James Adejuwon, Vicente Barros, Ian Burton, Jyoti Kulkarni*** *and* ***Rodel Lasco***

'Adaptation to climate change must become a top priority for policy makers and resource managers. This book provides valuable lessons that will improve public policy and the quality of decisions that will affect generations to come.' *Richard Moss, UN*

Climate Change and Adaptation covers current practices for managing climate risks, deficits between current practices and needs, the changing nature of the risks due to human caused climate change, strategies for adapting to changing risks, and the need to integrate these strategies into development planning and resource management.

Hardback £75.00 • 448 pages • 978-1-84407-470-9

Climate Change and Vulnerability

Edited by ***Neil Leary, Cecilia Conde, Jyoti Kulkarni, Anthony Nyong*** *and* ***Juan Pulhin***

'An excellent set of case studies and analysis on vulnerability to climate change in different sectors and countries in the developing world.' *Saleemul Huq, IIED*

Climate Change and Vulnerability discusses who is vulnerable and the nature and causes of their vulnerability for parts of the world that have been poorly researched until now. It also provides researchers with new examples of applications of vulnerability assessment methods, an approach that is of growing interest in the climate change area but for which there are relatively few applications in the literature.

Hardback £75.00 • 544 pages • 978-1-84407-469-3

BUY THE TWO-VOLUME SET AND SAVE £20.00

Climate Change and Vulnerability and Adaptation
Two Volume Set

'These two books are the result of a major research initiative proposed by the IPCC after its Third Assessment in 2001 ... The new knowledge acquired in this project is extensive and these books are important reading for students and practitioners alike.'
Martin Parry, Intergovernmental Panel on Climate Change (IPCC)

These authoritative volumes, resulting from the work of the Assessments of Impacts and Adaptations to Climate Change (AIACC) project launched by Nobel Prize-winners, the IPCC in 2002, are the first to provide a comprehensive investigation of the adaptation and vulnerability issues faced by developing countries.

Hardback £130.00 • 992 pages • 978-1-84407-480-8

For more details and a full listing of Earthscan titles visit:

www.earthscan.co.uk

earthscan

publishing for a sustainable future

The Earthscan Reader on Adaptation to Climate Change

Edited by ***Lisa Schipper*** *and* ***Ian Burton***

As the importance of adaptation becomes more apparent, there is need for wider appreciation and understanding of the concept and its potential as well as the obstacles to its effective deployment.

This reader, the first of its kind, pulls together and makes sense of the most significant writings on adaptation to climate change from the past two decades. An introduction maps out the field and traces the evolution of adaptation from a biological concept into a policy objective. The literature is divided into five sections.

Paperback £19.99 • 288 pages • 978-1-84407-531-7

The Atlas of Climate Change, Revised Edition

Mapping the World's Greatest Challenge

Kirstin Dow *and* ***Thomas E. Downing***

'This pioneering atlas will become an essential point of reference for anyone looking for a quick and accurate overview of this multi-disciplinary subject.' *John Ashton, Foreign and Commonwealth Office*

Rigorous in its science and insightful in its message, this atlas examines the drivers of climate change and its possible impacts.

With more than 50 full-colour maps and graphics, this is an essential resource on a wide range of topics, including warning signals, future scenarios, vulnerable populations, health impacts, renewable energy and emissions reduction.

Paperback £12.99 • 128 pages • 978-1-84407-522-5

Climate Change as a Security Risk

Hans Joachim Schellnhuber *and others,* ***German Advisory Council on Global Change (WBGU)***

'With *Climate Change as a Security Risk*, WBGU has compiled a flagship report on an issue that quite rightly is rising rapidly up the international political agenda.' *Achim Steiner, executive director, United Nations Environment Programme (UNEP)*

This book examines how climate change could overstretch many societies' adaptive capacities and result in internal destabilization processes and state failure. Classic security policy cannot respond adequately to these new threats and the authors examine how adaptation to climate change can become key elements of preventive security policy.

Hardback £75.00 • 256 pages • 978-1-84407-536-2

For more details and a full listing of Earthscan titles visit:

www.earthscan.co.uk

earthscan
publishing for a sustainable future

From Kyoto to the Town Hall

Making International and National Climate Policy Work at the Local Level

Edited by ***Lennart J. Lundqvist and Anders Biel***

'This book is essential reading for policy-makers everywhere and at all levels, and for citizens who elect them.'
Professor Robert Gifford, University of Victoria, Canada

International agreements offer some hope of addressing climate change, but often implementation is derailed at the local and individual level.

Drawing on original research from Sweden, a world leader in effective environmental solutions, this book examines the difficulties of aligning climate policy from international to national and sub-national levels.

Hardback £65.00 • 184 pages • 978-1-84407-423-5

Environmental Justice and the Rights of Indigenous Peoples

International and Domestic Legal Perspectives

Laura Westra

'Critical insights into international and comparative human rights jurisprudence.' ***Gaetano Pentassuglia, University of Liverpool***

This book reveals the lethal effects that ecological damage can have on communities. Using examples in national and international case law, Westra demonstrates how their lack of sufficient legal rights leaves indigenous peoples defenceless, time and again, in the face of governments and businesses who have little effective incentive to consult with them, let alone gain their consent.

Hardback £65.00 • 400 pages • 978-1-84407-485-3

Environmental Change and Human Security in the Arctic

Edited by Gunhild Hoogensen and Dawn Bazely

This book is the first comprehensive exploration of why human security is relevant to the Arctic and what achieving it can mean. It eschews the classic top-down (state-centric) view, instead taking a multi-disciplinary approach that allows non-dominant groups (e.g. women, indigenous peoples, Arctic researchers) to share their experiences, perceptions and capabilities.

The result is powerful document that not only reviews the potential threats to security but also situates them in an analytical framework and indicates a clear path for solutions.

Hardback £65.00 • 192 pages • 978-1-84407-549-2

For more details and a full listing of Earthscan titles visit:

www.earthscan.co.uk

earthscan
publishing for a sustainable future

Rationality and the Environment

Decision-making in Environmental Politics and Assessment

Bo Elling

'A much-needed, fundamental contribution.'
Wil A. H. Thissen, Delft University of Technology

This revolutionary work argues that ethical and aesthetic dimensions can only be brought into environmental politics and policies by citizens actively taking a stand on the specific matters in question. The author draws on Habermas' trisection of rationality as cognitive-instrumental, moral-practical and aesthetic-expressive, to suggest that truly effective environmental policy needs to activate all three approaches and not favour only the rational. To achieve this objective, the author argues that public participation in environmental policy and assessment is essential.

Hardback £65.00 • 296 pages • 978-1-84407-524-9

Public and Private in Natural Resource Governance

A False Dichotomy?

Edited by ***Thomas Sikor***

'Ten wonderful studies from around the world illuminate opportunities for advancing the theory, analysis and effective formation of sustainable systems of resource use.'
Jeff Romm, University of California at Berkeley

Bringing together a group of internationally respected researchers, this book provides a new perspective on prominent issues in resource governance, including the state, NGOs, civil society, communities, participation, devolution, privatization and hybrid institutions, highlighting the three-dimensional nature of relations between 'public' and 'private'.

Hardback £70.00 • 256 pages • 978-1-84407-525-6

Governing a Common Sea

Environmental Policies in the Baltic Sea Region

Edited by ***Marko Joas, Detlef Jahn*** *and* ***Kristine Kern***

How is a natural common resource such as a sea, which is shared by several countries, best governed? The potential for international conflict is immense, as each country may have different agendas with regard to issues such as exploitation and environmental protection.

This book uses a case study of the Baltic Sea region to examine this complex problem. Recently the bordering nations have establishing several new forums to manage the sea. The resulting political structures are reviewed and shown to provide a model that could be applied to other seas and natural resource systems elsewhere in the world.

Hardback £60.00 • 256 pages • 978-1-84407-537-9

For more details and a full listing of Earthscan titles visit:

www.earthscan.co.uk

Biochar for Environmental Management

SCIENCE AND TECHNOLOGY

Johannes Lehmann and Stephen Joseph

'An invaluable handbook for those interested in the science and economics of soil amendment using biochar. A vital tool for policy-makers and scientists working on climate change mitigation.'

- *Chris Goodall, author of the bestselling* ***How to Live a Low-Carbon Life***

'An essential reference in fostering pro-poor policies on energy efficiency and GHG mitigation based on land and soil improvements that could bring more justice into the climate change agenda.'

- *Luc Gnacadja, Executive Secretary, United Nations Convention to Combat Desertification (UNCCD)*

The positive agricultural and environmental impacts of biochar are many. Its stability in soil and superior nutrient-retention properties make it an ideal soil amendment to increase crop yields.

In addition to this, biochar sequestration can be carbon-negative and therefore used to actively remove carbon dioxide from the atmosphere, with major implications for mitigation of climate change.

Biochar production can also be combined with bioenergy production through the use of the gases that are given off in the pyrolysis process.

The book's interdisciplinary approach, which covers environmental sciences, agricultural sciences, engineering, economics and policy, is a vital tool at this stage of biochar technology development.

This comprehensive overview of current knowledge will be of great interest to advanced students, researchers and professionals in a wide range of disciplines.

Hardback £49.95 • 384 pages • 978-1-84407-658-1 • February 2009

For more details and a full listing of Earthscan titles visit

www.earthscan.co.uk